Communications
in Computer and Information Science   767

Commenced Publication in 2007
Founding and Former Series Editors:
Alfredo Cuzzocrea, Xiaoyong Du, Orhun Kara, Ting Liu, Dominik Ślęzak,
and Xiaokang Yang

More information about this series at http://www.springer.com/series/7899

Mārīte Kirikova · Kjetil Nørvåg
George A. Papadopoulos · Johann Gamper
Robert Wrembel · Jérôme Darmont
Stefano Rizzi (Eds.)

New Trends in Databases and Information Systems

ADBIS 2017 Short Papers and Workshops
AMSD, BigNovelTI, DAS, SW4CH, DC
Nicosia, Cyprus, September 24–27, 2017
Proceedings

 Springer

Editors
Mārīte Kirikova ⓘ
Riga Technical University
Riga
Latvia

Kjetil Nørvåg
Norwegian University of Science
and Technology
Trondheim
Norway

George A. Papadopoulos
University of Cyprus
Nicosia
Cyprus

Johann Gamper
Free University of Bozen-Bolzano
Bozen-Bolzano
Italy

Robert Wrembel
Poznan University of Technology
Poznan
Poland

Jérôme Darmont
Université de Lyon, Lyon 2, ERIC EA 3083
Lyon
France

Stefano Rizzi
University of Bologna
Bologna
Italy

ISSN 1865-0929   ISSN 1865-0937 (electronic)
Communications in Computer and Information Science
ISBN 978-3-319-67161-1   ISBN 978-3-319-67162-8   (eBook)
DOI 10.1007/978-3-319-67162-8

Library of Congress Control Number: 2017952393

Printed on acid-free paper

This Springer imprint is published by Springer Nature
The registered company is Springer International Publishing AG
The registered company address is: Gewerbestrasse 11, 6330 Cham, Switzerland

Preface

The European Conference on Advances in Databases and Information Systems (ADBIS) celebrates this year its 21st anniversary. Previous ADBIS conferences were held in St. Petersburg (1997), Poznan (1998), Maribor (1999), Prague (2000), Vilnius (2001), Bratislava (2002), Dresden (2003), Budapest (2004), Tallinn (2005), Thessaloniki (2006), Varna (2007), Pori (2008), Riga (2009), Novi Sad (2010), Vienna (2011), Poznan (2012), Genoa (2013), Ohrid (2014), Poitiers (2015), and Prague (2016). The conferences were initiated and supervised by an international Steering Committee consisting of representatives from Armenia, Austria, Bulgaria, the Czech Republic, Cyprus, Estonia, Finland, France, Germany, Greece, Hungary, Israel, Italy, Latvia, Lithuania, FYR of Macedonia, Poland, Russia, Serbia, Slovakia, Slovenia, and the Ukraine.

ADBIS can be considered as one of the most established and recognized conferences in Europe, in the broad field of databases and information systems.

The ADBIS conferences aim at: (1) providing an international forum for presenting research achievements on database theory and practice, development of advanced DBMS technologies, and their applications as well as (2) promoting interaction and collaboration between the database and information systems research communities from European countries and the rest of the world.

This volume contains short research papers, workshop papers, and doctoral consortium papers presented at the 21st European Conference on Advances in Databases and Information Systems (ADBIS), held on 24–27 September, 2017, in Nicosia, Cyprus.

The program of ADBIS 2017 included keynotes, research papers, thematic workshops, and a doctoral consortium. The main conference, workshops, and doctoral consortium had their own international Program Committees. The main conference attracted 107 paper submissions from 33 countries from all continents. After rigorous reviewing by the Program Committee (88 reviewers from 36 countries in the Program Committee and additionally by 26 external reviewers), the 26 papers included in the LNCS proceedings volume were accepted as full contributions, making an acceptance rate of 24%. In addition, 12 papers were selected as short contributions and are included in these proceedings.

The selected short papers span a wide spectrum of topics related to the ADBIS conference. Most of them are related to database and information systems technologies for advanced applications. Typical applications are text databases, streaming data, and graph processing. In addition, there are also papers covering the theory of databases.

Initially, 6 workshops were accepted to be collocated with ADBIS 2017. A workshop, to be run, had to receive at least 6 submissions. Two workshops did not reach this submission level and were canceled. Finally, the following 4 workshops were run: (1) Workshop on Novel Techniques for Integrating Big Data (BigNovelTI), (2) Workshop on Data Science: Methodologies and Use-Cases (DaS), (3) Semantic

Web for Cultural Heritage (SW4CH), and (4) Data-Driven Approaches for Analyzing and Managing Scholarly Data: Systems, Methods, and Applications (AMSD). BigNovelTI received 8 submissions, out of which 4 were accepted for publication in this volume. DaS received 20 submissions, out of which 11 were accepted. Additionally, an invited paper for DaS is included in this volume. SW4CH accepted 6 submissions out of 11. Finally, AMSD accepted 3 submissions out of 8. Thus, the overall acceptance rate for the four workshops is 51%. A summary of the content of these workshops is included in the chapter "New Trends in Databases and Information Systems: Contributions from ADBIS 2017 Workshops".

The ADBIS Doctoral Consortium (DC) 2017 was a forum where PhD students had a chance to present their research ideas to the database research community, receive inspiration from their peers and feedback from senior researchers, and tie cooperation bounds. DC papers are single-authored and aim at describing the current status of thesis research. Out of six submissions, the DC Committee selected three papers that were presented at the DC, giving an acceptance rate of 50%. Various topics were addressed, i.e., preference-based stream analysis, database reverse engineering, and conceptual object-role modeling by reasoning. The DC chairs would like to thank the DC Program Committee members for their dedicated work.

July 2017

George A. Papadopoulos
Marite Kirikova
Kjetil Nørvåg
Johann Gamper
Robert Wrembel
Jérôme Darmont
Stefano Rizzi

Organization

ADBIS Program Committee

Bader Albdaiwi	Kuwait University, Kuwait
Bernd Amann	LIP6-UPMC, France
Grigoris Antoniou	University of Huddersfield, UK
Ladjel Bellatreche	LIAS/ENSMA, France
Klaus Berberich	Max Planck Institute for Informatics, Germany
Maria Bielikova	Slovak University of Technology in Bratislava, Slovakia
Doulkifli Boukraa	Université de Jijel, Algeria
Drazen Brdjanin	University of Banja Luka, Bosnia and Herzegovina
Stephane Bressan	National University of Singapore, Singapore
Bostjan Brumen	University of Maribor, Slovenia
Albertas Caplinskas	Vilnius University, Lithuania
Barbara Catania	DIBRIS-University of Genoa, Italy
Marek Ciglan	Institute of Informatics, Slovak Academy of Sciences, Slovakia
Isabelle Comyn-Wattiau	ESSEC Business School, France
Alfredo Cuzzocrea	ICAR-CNR and University of Calabria, Italy
Ajantha Dahanayake	Georgia College and State University, USA
Christos Doulkeridis	University of Piraeus, Greece
Johann Eder	Alpen Adria Universität Klagenfurt, Austria
Erki Eessaar	Tallinn University of Technology, Estonia
Markus Endres	University of Augsburg, Germany
Werner Esswein	Technische Universität Dresden, Germany
Georgios Evangelidis	University of Macedonia, Greece
Flavio Ferrarotti	Software Competence Centre Hagenberg, Austria
Peter Forbrig	University of Rostock, Germany
Flavius Frasincar	Erasmus University Rotterdam, Netherlands
Jan Genci	Technical University of Kosice, Slovakia
Jānis Grabis	Riga Technical University, Latvia
Gunter Graefe	HTW Dresden, Germany
Franceso Guerra	University of Modena and Reggio Emilia, Italy
Hele-Mai Haav	Institute of Cybernetics at Tallinn University of Technology, Estonia
Theo Härder	TU Kaiserslautern, Germany
Katja Hose	Aalborg University, Denmark
Ekaterini Ioannou	Technical University of Crete, Greece
Mirjana Ivanovic	University of Novi Sad, Serbia
Hannu Jaakkola	Tampere University of Technology, Finland
Lili Jiang	Univeristy of Umeå, Sweden

Ahto Kalja   Tallinn University of Technology, Estonia
Dimitris Karagiannis   University of Vienna, Austria
Randi Karlsen   University of Tromsø, Norway
Panagiotis Karras   Aalborg University, Denmark
Zoubida Kedad   University of Versailles, France
Marite Kirikova   Riga Technical University, Latvia
Margita Kon-Popovska   Ss. Cyril and Methodius University, FYR of Macedonia
Michal Kopecký   Charles University, Czech Republic
Michal Kratky   VSB-Technical University of Ostrava, Czech Republic
John Krogstie   NTNU, Norway
Ulf Leser   Humboldt-Universität zu Berlin, Germany
Sebastian Link   The University of Auckland, New Zealand
Audrone Lupeikiene   Vilnius University, Lithuania
Hui Ma   Victoria University of Wellington, New Zealand
Leszek Maciaszek   Wrocław University of Economics, Poland
Federica Mandreoli   DII - University of Modena, Italy
Yannis Manolopoulos   Aristotle University of Thessaloniki, Greece
Tadeusz Morzy   Poznan University of Technology, Poland
Martin Nečaský   Charles University, Czech Republic
Kjetil Nørvåg   Norwegian University of Science and Technology,
                              Norway
Boris Novikov   St. Petersburg University, Russia
Eirini Ntoutsi   Gottfried Wilhelm Leibniz Universität Hannover, Germany
Andreas Oberweis   Karlsruhe Institute of Technology (KIT), Germany
Andreas L. Opdahl   University of Bergen, Norway
Odysseas Papapetrou   EPFL, Switzerland
Jaroslav Pokorný   Charles University in Prague, Czech Republic
Giuseppe Polese   University of Salerno, Italy
Boris Rachev   Technical University of Varna, Bulgaria
Milos Radovanovic   University of Novi Sad, Serbia
Heri Ramampiaro   Norwegian University of Science and Technology
                              (NTNU), Norway
Tore Risch   University of Uppsala, Sweden
Gunter Saake   University of Magdeburg, Germany
Petr Saloun   VSB-TU Ostrava, Czech Republic
Kai-Uwe Sattler   TU Ilmenau, Germany
Ingo Schmitt   Technical University Cottbus, Germany
Tomas Skopal   Charles University in Prague, Czech Republic
Bela Stantic   Griffith University, Australia
Kostas Stefanidis   University of Tampere, Finland
Panagiotis Symeonidis   Free University of Bozen-Bolzano, Italy
James Terwilliger   Microsoft Corporation, USA
Goce Trajcevski   Northwestern University, USA
Christoph Trattner   MODUL University Vienna, Austria
Raquel Trillo-Lado   Universidad de Zaragoza, Spain
Yannis Velegrakis   University of Trento, Italy

Goran Velinov	Ss. Cyril and Methodius University, FYR of Macedonia
Akrivi Vlachou	University of Piraeus, Greece
Gottfried Vossen	ERCIS Muenster, Germany
Robert Wrembel	Poznan University of Technology, Poland
Anna Yarygina	Saint-Petersburg University Russia
Weihai Yu	University of Tromsø, Norway
Arkady Zaslavsky	CSIRO, Australia

Additional Reviewers

Hosam Aboelfotoh	Jens Lechtenbörger
Dionysis Athanasopoulos	Jevgeni Marenkov
George Baryannis	Denis Martins
Sotiris Batsakis	Robert Moro
Panagiotis Bozanis	Ludovit Niepel
Loredana Caruccio	Wilma Penzo
Vincenzo Deufemia	Horst Pichler
Senén González	Benedikt Pittl
Sven Hartmann	Tarmo Robal
Zaid Hussain	Eliezer Souza Silva
Pavlos Kefalas	Nikolaos Tantouris
Julius Köpke	Eleftherios Tiakas
Vimal Kunnummel	H. Yahyaoui

Workshops

AMSD Program Committee

Andreas Behrend	University of Bonn, Germany
Christiane Engels	University of Bonn, Germany
Christoph Lange	Fraunhofer IAIS, Germany
Jens Lehmann	Fraunhofer IAIS, Germany
Rainer Manthey	University of Bonn, Germany
Sahar Vahdati	Fraunhofer IAIS, Germany
Hannes Voigt	Technische Universität Dresden, Germany

BigNovelTI Program Committee

Alberto Abelló	Universitat Politècnica de Catalunya, Spain
Andrea Calì	Birkbeck College London, UK
C. Maria Keet	University of Cape Town, South Africa
Diego Calvanese	Free University Bozen-Bolzano, Italy
Theodoros Chondrogiannis	Free University of Bozen-Bolzano, Italy
Martin Giese	University of Oslo, Norway
Matteo Golfarelli	University of Bologna, Italy

Gianluigi Greco	University of Calabria, Italy
Katja Hose	Aalborg University, Denmark
Petar Jovanovic	Universitat Politècnica de Catalunya, Spain
Roman Kontchakov	Birkbeck College London, UK
Antonella Poggi	Sapienza University of Rome, Italy
Oscar Romero	Universitat Politècnica de Catalunya, Spain
Mantas Simkus	TU Vienna, Austria
Sergio Tessaris	Free University of Bozen-Bolzano, Italy
Christian Thomsen	Aalborg University, Denmark
Stefano Rizzi	University of Bologna, Italy
Alejandro Vaisman	Instituto Tecnológico de Buenos Aires, Argentina
Stijn Vansummeren	Université Libre de Bruxelles, Belgium
Panos Vassiliadis	University of Ioannina, Greece
Robert Wrembel	Poznan University of Technology, Poland
Esteban Zimanyi	Université Libre de Bruxelles, Belgium

DaS Program Committee

Julien Aligon	University of Toulouse - IRIT, France
Antonio Attanasio	Istituto Superiore Mario Boella, Italy
Tania Cerquitelli	Politecnico di Torino, Italy
Agnese Chiatti	Pennsylvania State University, USA
Silvia Chiusano	Politecnico di Torino, Italy
Evelina Di Corso	Politecnico di Torino, Italy
Javier A. Espinosa-Oviedo	Barcelona Supercomputing Center, Spain
Göran Falkman	University of Skövde, Sweden
Fabio Fassetti	Università della Calabria, Italy
Juozas Gordevicius	Institute of Biotechnology, Lithuania
Natalija Kozmina	University of Latvia, Latvia
Patrick Marcel	Université François Rabelais, Tours, France
Erik Perjons	Stockholm University, Sweden
Emanuele Rabosio	Politecnico di Milano, Italy
Francesco Ventura	Politecnico di Torino, Italy
Gatis Vītols	Latvia University of Agriculture, Latvia
Xin Xiao	Huawei Technologies Co., Ltd., China
José Luis Zechinelli Martini	Universidad de las Américas Puebla, Mexico

SW4CH Program Committee

Trond Aalberg	IDI, NTNU, Norway
Carmen Brando	EHESS-CRH, Paris, France
Benjamin Cogrel	KRDB Research Centre for Knowledge and Data, Free University of Bozen-Bolzano, Italy
Donatello Conte	LI, Université François Rabelais, Tours, France
Peter Haase	Metaphacts, Walldorf, Germany

Mirian Halfeld Ferrari Alves	LIFO, University of Orléans, France
Katja Hose	Aalborg University, Denmark
Stéphane Jean	LIAS/ENSMA and University of Poitiers, France
Efstratios Kontopoulos	MKLAB, Thessaloniki, Greece
Cvetana Krstev	University of Belgrade, Serbia
Nikolaos Lagos	Xerox, Grenoble, France
Béatrice Markhoff	LI, Université François Rabelais, Tours, France
Denis Maurel	LI, Université François Rabelais, Tours, France
Carlo Meghini	CNR/ISTI, Pisa, Italy
Isabelle Mirbel	WIMMMICS, University of Nice Sophia Antipolis, France
Alessandro Mosca	SIRIS Academic, S.L., Barcelona, Spain
Dmitry Muromtsev	ITMO University, Russia
Cheikh Niang	AUF, Paris, France
Xavier Rodier	Laboratoire Archéologie et Territoires, Université François Rabelais, Tours, France
Maria Theodoridou	FORTH ICS, Heraklion, Crete, Greece
Genoveva Vargas-Solar	University of Grenoble, France
Dusko Vitas	University of Belgrade, Serbia

Doctoral Consortium Program Committee

Varunya Attasena	Kasetsart University, Kamphaeng Saen Campus, Thailand
Maria Bielikova	Slovak University of Technology in Bratislava, Slovakia
Doulkifli Boukraa	Université de Jijel, Algeria
Enrico Gallinucci	Università di Bologna, Italy
Adrien Guille	Université de Lyon, France
Sebastian Link	The University of Auckland, New Zealand
Raquel Trillo-Lado	Universidad de Zaragoza, Spain

Contents

ADBIS 2017 – Workshops

**The 1st Workshop on Data-Driven Approaches for Analyzing
and Managing Scholarly Data (AMSD 2017)**

**The 1st Workshop on Novel Techniques for Integrating Big Data
(BigNovelTI 2017)**

The 1st International Workshop on Data Science: Methodologies and Use-Cases (DaS 2017)

The 2nd International Workshop on Semantic Web for Cultural Heritage (SW4CH 2017)

ADBIS Doctoral Consortium

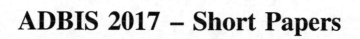

ADBIS 2017 – Short Papers

Distributing N-Gram Graphs for Classification

Ioannis Kontopoulos[1](✉), George Giannakopoulos[1], and Iraklis Varlamis[2]

[1] Institute of Informatics and Telecommunications, N.C.S.R. "Demokritos",
Agia Paraskevi, Greece
{ikon,ggianna}@iit.demokritos.gr
[2] Department of Informatics and Telematics, Harokopio University of Athens,
Kallithea, Greece
varlamis@hua.gr

Abstract. N-gram models have been an established choice for language modeling in machine translation, summarization and other tasks. Recently n-gram graphs managed to capture significant language characteristics that go beyond mere vocabulary and grammar, for tasks such as text classification. This work proposes an efficient distributed implementation of the n-gram graph framework on Apache Spark, named ARGOT. The implementation performance is evaluated on a demanding text classification task, where the n-gram graphs are used for extracting features for a supervised classifier. A provided experimental study shows the scalability of the proposed implementation to large text corpora and its ability to take advantage of a varying number of processing cores.

Keywords: Distributed processing · N-gram graphs · Text classification

1 Introduction

Text classification is a supervised machine learning technique for identifying predefined categories that are related to a specific document. Several distributed processing paradigms have been applied to handle classification in the increasing volume of generated texts in different domains (e.g. [5]). However, promising methods, such as the n-gram graph (nGG) framework [8–10] still have scalability limited to a single machine[1]. This paper proposes the $ARGOT$ framework, a distributed implementation of nGGs, tested on a text classification task. We re-engineer the time-consuming processes of feature extraction and graph merging in order to improve scalability in large corpora, while retaining the framework's state-of-the-art performance. Our experimental evaluation also studies the scalability of the solution, with respect to various parameters.

In the following, we summarize related work (Sect. 2), explain the nGG framework (Sect. 3) and ARGOT (Sect. 3). We then experimentally evaluate our approach (Sect. 4) and conclude our work (Sect. 5), summarizing the next steps.

---

[1] JInsect toolkit, found at https://github.com/ggianna/JInsect.

© Springer International Publishing AG 2017
M. Kirikova et al. (Eds.): ADBIS 2017, CCIS 767, pp. 3–11, 2017.
DOI: 10.1007/978-3-319-67162-8_1

2   Related Work

Distributed text classification in Big Data applications oftentimes employ frameworks such as MapReduce [12,15] and Spark [13] to ensure scalability. To adequately manage big data in associative classifiers (because of time complexity and memory constraints) a map-reduce solution has been presented in [2]. In [3] the efficiency of Support Vector Machine and Naive Bayes classifiers is tested using a *Word2Vec* model and algorithm on Apache Spark.

Apart from the distributed approach, graph-based techniques have been developed over the years for accurate text classification [4]. Authors in [1] have employed the nGG framework to extract rich features, usable by a text classifier. A graph-based approach has been introduced in [11] for encoding the relationships between the different terms in the text.

The primary aim – and unique positioning – of this work is to provide a parallel and distributed implementation that boosts the scalability of the original *nGG framework*. Thus, the focus is not on the classification task itself, but on the properties that affect the scalability of the classification solution. These properties drive the redesign of existing operators (graph creation, graph merging, graph similarity calculation, etc.) in the distributed setup.

The distributed implementation of nGGs is meant to empower distributed text analysis components, such as the Big Data Europe[2] project "event detection" workflow, where nGGs are used to identify real-world events from textual streams.

3   Distributed N-Gram Graphs for Classification

Character (or word) nGGs represent texts as neighborhoods of n-grams. Given a string (e.g. "Hello!"), a value n for the n-grams and a window size w, an nGG: (a) breaks the string into overlapping n-grams (e.g. for 3-grams: "Hel", "ell", "llo", ...), creating one vertex per unique n-gram; (b) connects n-grams found within a (character) distance of at most w of each other in the original string with edges (e.g. for $w = 2$: "Hel-ell", "Hel-llo", ...); maps to each edge the number of times that its connected n-grams were found to be neighbors (e.g. "Hel-llo" $\rightarrow$ 1). The extracted graph for the "Hello!" input text is shown in Fig. 1.

Given the representation, ARGOT implements and uses four different similarity operators for nGG comparison: Size Similarity (SS), expressing the ratio of graph sizes; Containment Similarity (CS) expressing the proportion of common edges between graphs, regardless of edge weights; Value Similarity (VS) which also makes use of the edge weights; Normalized Value Similarity (NVS) which is a size-ratio-independent version of VS. In our work we also apply the union (or merge) operator over m graphs, which – given m graphs – creates a

---

[2] This paper is supported by the project "Integrating Big Data, Software and Communities for Addressing Europe's Societal Challenges – BigDataEurope", which has received funding from the European Union's Horizon 2020 research and innovation programme under grant agreement No 644564. https://www.big-data-europe.eu/.

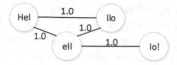

Fig. 1. The N-gram graph representation ($n = 3$, $w = 2$) of the "Hello!" string.

new graph G_U, which (a) contains as edges the union of the edges of the input graphs; (b) assigns a weight w_e to every one of its edges, where w_e is the average of the weights of all the edges e across all input graphs where e appears. For more details on the operators, please consult [8].

The first step of the classification process is to train the classifier with the training instances. In the case of nGGs (see Fig. 2), separate class graphs are composed by merging – using the union operator – the training instances (nGG representations of documents) of each class. In the classification step, each unlabeled instance (i.e. document) is first represented as an nGG. It is then compared with the N class graphs and the CS, VS and NVS similarities form the $3 \times N$ feature vector representation of the instance.

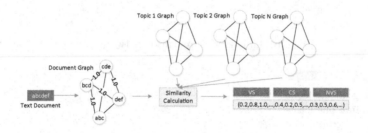

Fig. 2. Extracting the feature vector using the n-gram graph framework.

Despite its performance [10], the nGG classification requires a significant amount of time for training (i.e. for constructing the class graphs) and for creating the feature vectors (i.e. computing graph similarities). The time complexity of these processes depends both on the number of classes and the amount and size of training documents and sets a scalability challenge. In order to address this challenge, we propose a distributed implementation of the classification algorithm that uses Apache Spark, a mature, fast and general engine for large-scale data processing. In this section, the distributed algorithms for the construction of the class graphs and the computation of graph similarities are explained. The code of these implementations and the framework developed, which is called ARGOT, is publicly available and can be found on a public repository[3].

---

[3] https://github.com/ioannis-kon/ARGOT.

Distributed class graph creation: As illustrated in Fig. 3a, the process begins with the transformation of the training text documents to the respective nGGs. The documents, which can be located in a Hadoop (HDFS), Network (NFS) or any other distributed file system, are split into k partitions and each partition is processed independently. Each edge of the nGG consists of the connected vertex identifiers (ids) and the edge weight. Edges are repartitioned based on the hash of their corresponding vertex ids. As a result, edges from different documents that map to the same vertex pair are located in the same partition, increasing merging and calculation efficiency. Based on the above, a single nGG is essentially distributed across the processing nodes.

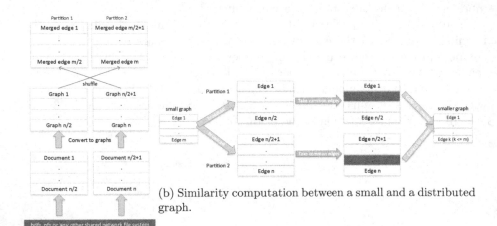

(b) Similarity computation between a small and a distributed graph.

(a) Merging of multiple documents into one class-representative graph.

Fig. 3. Distributed implementation of algorithms.

Distributed graph similarity computation: The second step, as illustrated in Fig. 3b, is the extraction of similarity features, which is based on the computation of graph similarities between the class graphs and each unclassified document. The class graph can contain millions of edges, delaying the comparison of each unclassified instance with the class graphs. To solve this problem, a graph similarity algorithm similar to the semi-join technique (used in distributed databases) has been implemented. At each comparison, the smaller document graph is broadcasted to every partition, and the partial overlap with the portion of the class graph in each partition is computed. This implementation reduces the communication overhead between nodes. Since the similarity measures are based on the overlapping edges only, we filter the partitions and keep the common edges (which are at most the edges of the smallest graph). These edges are collected by the master node, which then computes similarities locally and fast.

4   Experimental Evaluation

The distributed implementation of the document classification algorithm was evaluated for its performance and scalability on the large-scale, real-world Reuters RCV2 dataset[4], using two different hardware infrastructures. The experiments aim to study how several parameters (e.g. the total number of documents or classes) affect the algorithm performance. The processing steps of the classification task, as described in [10], were repeated: (i) on a single node with 24 cores[5] and (ii) on a cluster comprising 6 commodity machines, each with 4 cores[6]. For the implementation of ARGOT, Scala 2.11.7 and Apache Spark 2.0.1 were used.

Reuters RCV2 holds more than 480,000 news articles in 13 different languages, categorized along a class hierarchy of 104 overlapping topics and has been widely used [6,14]. In the experiments we used the top four, non overlapping, corpus categories, comprising articles in four languages (see Table 1).

Table 1. Dataset statistics.

Characteristic	Value	Characteristic	Value
Classes	4	# of edges of class graph 1	5,772,038
Documents	172,115	# of edges of class graph 2	10,134,473
Total characters ($\times10^8$)	2.07	# of edges of class graph 3	3,708,371
Characters/Document	1,205.55	# of edges of class graph 4	8,251,410
Total Tokens ($\times10^8$)	30.92	Total # of graph comparisons	688,460
Tokens/Document	179.66	Total size of dataset	201.69 MB
Av. token length	5.68	Average size of each instance	1.2 KB

Classification: In order to evaluate the performance of ARGOT implementations, we performed a 90%−10% training-test split of the dataset (as in [10]). The (four) class graphs are created by merging a randomly selected subset (90%) of the training documents. Then, we follow the process described in Sect. 3 to represent all instances and train a Naive Bayes classifier. The classification precision on the test set was 95.1%, whereas the maximum precision reported in [7] was 94% in only a subset of the same data set. This side-result highlights the value and language-independence of the nGG approach.

Scalability: In order to test the scalability of the implemented algorithms, we conducted experiments on the complete dataset described in Table 1. ARGOT

---

[4] http://trec.nist.gov/data/reuters/reuters.html.

[5] Intel(R) Xeon(R) CPU E5-2680 v3 @ 2.50 GHz) and 96 GB of RAM, running Debian 64-bit with Linux Kernel 3.16.0-4-amd64 and Java OpenJDK 1.8.

[6] Intel(R) Core(TM) i5-3330S CPU @ 2.70 GHz) and 8 GB of RAM, totaling in 24 cores and 48 GB of RAM, running OS X 10.10 (14A389) with Kernel Darwin 14.0.0 and Java OpenJDK 1.8, connected with 100-Mbit ethernet links.

scalability was tested using an increasing number of partitions (from 8 to 48) on a single node machine and on a cluster of commodity machines. We repeat the experiments 10 times and report the average times. Figure 4a shows the average merging time per class, where the distributed merging function scales better for bigger classes (i.e., class 2). Figure 4b shows the average time for extracting features from the instances. It shows great scalability capacity since the algorithm reduced the extraction time by almost 50% from 8 to 48 partitions. The performance improvement is more obvious in Fig. 4c, which shows the average time for extracting similarities per instance, which was reduced from 1.04 s to 0.58 s (i.e. 43% improvement), from 8 to 48 partitions.

Using the same dataset and partitions setup, we repeated the experiment on a distributed cluster[7], in order to test how data broadcasting and repartitioning affect performance (cf. Fig. 5a, b and c). We see that performance drops when increasing the number of partitions, since the communication overhead counters the distributed processing gain. However, in larger class graphs and datasets the system is expected to perform better with more partitions.

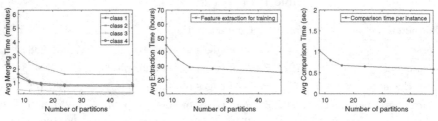

(a) Graph merging time. (b) Feature extraction time. (c) Graph comparison time.

Fig. 4. ARGOT performance in a large-scale experiment (single node).

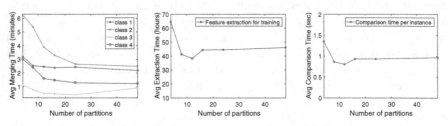

(a) Graph merging time. (b) Feature extraction time. (c) Graph comparison time.

Fig. 5. ARGOT performance in a large-scale experiment (cluster mode).

Dataset traits: We now examine how the total number of instances and classes affects performance. Two subsets of the large dataset have been created, each

---

[7] Cluster nodes are connected with 100-Mbit Ethernet links.

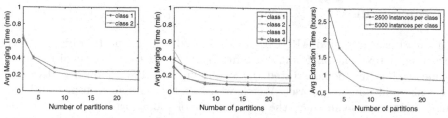

(a) Merging time - 2 topics. (b) Merging time - 4 topics. (c) Feature extraction time.

Fig. 6. Performance results of the two subset experiments.

comprising $10,000$ documents in total, from two and four classes respectively ($9,000$ training instances and $1,000$ testing instances in each subset).

We run the same experiments (10 times per setting and averaging results), but this time we used 2 to 24 partitions on the single node setup. Figure 6a and b show the average merging time per class per experiment, illustrating that the merging time depends on the number of the documents in a topic. Figure 6c depicts the average feature extraction time per experiment. While having less documents per topic and the same number of total training instances, the extraction time took longer. The difference is the number of graph comparisons, which are greater in the last case. From this we can infer that the feature extraction time depends on the number of topics. From the experiments conducted we can safely say that the more documents we have per class and the larger graphs we have, the better the algorithms scale.

Distributed vs. multi-threaded: We compared our method to the current, multi-threaded implementation of nGG algorithms, JInsect. Figures 7a, b show the average graph merging times in the aforementioned subsets, correspondingly. Figure 7c compares the ARGOT average feature extraction time to JINSECT. ARGOT is faster in graph merging but slower in feature extraction. However, JInsect failed on the full dataset, due to lack of memory scalability, which was not a problem for ARGOT. Finally, ARGOT yields the same results as JInsect in terms of classification performance, as expected.

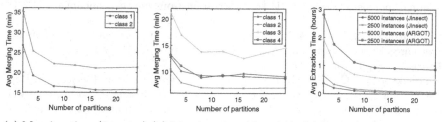

(a) Merging time (2 topics). (b) Merging time (4 topics). (c) Feature extraction time.

Fig. 7. JInsect (i.e. multi-threaded, single-machine) performance evaluation.

5 Conclusion

In this work we presented ARGOT, a distributed implementation of the nGG algorithms in text classification. We illustrated the details of the distributed implementation and demonstrated the scalability of ARGOT on a real-world, large-scale, multilingual dataset. We showed how the number of topics and the number of instances can affect the performance of the algorithms. We also presented the effect of broadcasting the data to the execution time of the experiment, when using a cluster of computers. We showed that if the number of instances per topic is large, ARGOT can benefit from the larger number of CPUs in a cluster/machine. We also compared ARGOT to the established implementation of nGGs and showed that ARGOT scales better, also exceeding inherent single-machine limitations. As a side-effect, our method demonstrated state-of-the-art performance on an established classification dataset for a single-label, multi-class, multilingual setting, highlighting the value of the approach.

References

1. Angelova, R., Weikum, G.: Graph-based text classification: learn from your neighbors. In: 29th Annual International ACM SIGIR Conference on Research and Development in Information Retrieval, pp. 485–492. ACM, New York (2006)
2. Bechini, A., Marcelloni, F., Segatori, A.: A mapreduce solution for associative classification of big data. Inf. Sci. **332**(C), 33–55 (2016)
3. Choi, M., Jin, R., Chung, T.-S.: Document classification using Word2Vec and Chi-square on apache spark. In: Park, J., Pan, Y., Yi, G., Loia, V. (eds.) CSA 2016, CUTE 2016, UCAWSN 2016. LNEE, vol. 421, pp. 867–872. Springer, Singapore (2017). doi:10.1007/978-981-10-3023-9_134
4. Das, N., Ghosh, S., Goncalves, T., Quaresma, P.: Comparison of different graph distance metrics for semantic text based classification. Polibits 51–58 (2014)
5. Fei, X., Li, X., Shen, C.: Parallelized text classification algorithm for processing large scale TCM clinical data with mapreduce. In: 2015 IEEE International Conference on Information and Automation, pp. 1983–1986, August 2015
6. Ferreira, D.C., Martins, A.F.T., Almeida, M.S.C.: Jointly learning to embed and predict with multiple languages. In: ACL (2016)
7. Fortuna, B., Shawe-Taylor, J.: The use of machine translation tools for cross-lingual text mining. In: ICML Workshop on Learning with Multiple Views (2005)
8. Giannakopoulos, G.: Automatic summarization from multiple documents. Ph.D. thesis (2009)
9. Giannakopoulos, G., Karkaletsis, V., Vouros, G., Stamatopoulos, P.: Summarization system evaluation revisited: N-gram graphs. ACM Trans. Speech Lang. Process. **5**(3), 5:1–5:39 (2008)
10. Giannakopoulos, G., Mavridi, P., Paliouras, G., Papadakis, G., Tserpes, K.: Representation models for text classification: a comparative analysis over three web document types. In: 2nd International Conference on Web Intelligence, Mining and Semantics, pp. 13:1–13:12. ACM, New York (2012)
11. Malliaros, F.D., Skianis, K.: Graph-based term weighting for text categorization. In: 2015 IEEE/ACM International Conference on Advances in Social Networks Analysis and Mining 2015, pp. 1473–1479. ACM, New York (2015)

12. Santoso, J., Yuniarno, E.M., Hariadi, M.: Large scale text classification using map reduce and naive bayes algorithm for domain specified ontology building. In: 7th International Conference on Intelligent Human-Machine Systems and Cybernetics, vol. 1, pp. 428–432 (2015)
13. Semberecki, P., Maciejewski, H.: Distributed Classification of Text Documents on Apache Spark Platform, pp. 621–630. Springer International Publishing, Cham (2016)
14. Song, Y., Upadhyay, S., Peng, H., Roth, D.: Cross-lingual dataless classification for many languages. In: Twenty-Fifth International Joint Conference on Artificial Intelligence, pp. 2901–2907. AAAI Press (2016)
15. Zhou, L., Yu, Z.: Acceleration of MapReduce Framework on a Multicore Processor, pp. 175–190. Springer International Publishing, Cham (2017)

Assessing the Quality of Spatio-Textual Datasets in the Absence of Ground Truth

Mouzhi Ge[1](✉) and Theodoros Chondrogiannis[2](✉)

[1] Masaryk University, Brno, Czech Republic
mouzhi.ge@muni.cz
[2] Free University of Bozen-Bolzano, South Tyrol, Italy
tchond@inf.unibz.it

Abstract. The increasing availability of enriched geospatial data has opened up a new domain and enables the development of more sophisticated location-based services and applications. However, this development has also given rise to various data quality problems as it is very hard to verify the data for all real-world entities contained in a dataset. In this paper, we propose ARCI, a relative quality indicator which exploits the vast availability of spatio-textual datasets, to indicate how confident a user can be in the correctness of a given dataset. ARCI operates in the absence of ground truth and aims at computing the relative quality of an input dataset by cross-referencing its entries among various similar datasets. We also present an algorithm for computing ARCI and we evaluate its performance in a preliminary experimental evaluation using real-world datasets.

Keywords: Spatio-textual data · Data quality · Relative quality

1 Introduction

The current trends in technology, such as smartphones and sensor networks, along with the proliferation of location-based social networks, such as Foursquare and Flickr, have motivated the development of new applications and services which employ spatio-textual datasets, i.e., collections of spatial objects which carry both spatial and textual information. In addition, spatio-textual queries, which combine location-based retrieval with textual information that describes the spatial objects, have recently attracted much attention [3,15].

The fact that real world entities are constantly changing though, e.g., restaurants are closing down, museums are being moved etc., makes the maintenance of spatio-textual datasets very hard. Consequently, it is quite common for spatio-textual datasets to contain inaccurate information and/or incomplete entries [7]. Furthermore, data across different datasets is not always consistent, i.e., entries of different datasets referring to the same real-world entity may provide conflicting information. Due to such data quality issues, a number of geospatial applications and initiatives have been delayed or even canceled, citing poor-quality

© Springer International Publishing AG 2017
M. Kirikova et al. (Eds.): ADBIS 2017, CCIS 767, pp. 12–20, 2017.
DOI: 10.1007/978-3-319-67162-8_2

of the available data as the main reason. Although identifying such data quality problems manually is unrealistic, it is important to provide at least an indication for the quality of spatio-textual datasets. Such indicators should operate in the absence of ground truth, i.e., when there is no available verified data.

In this paper, we propose ARCI, a novel approach to indicate the relative quality of a spatio-textual dataset by cross-referencing its entries with other spatio-textual datasets. The ARCI indicator intends to highlight the trustfulness of the dataset and confirm the confidence to use the data. Furthermore, quantifying the ARCI indicator allows us to visualize the data quality, for example using a spectrum bar in this paper. While most existing quality assessment methodologies [6,7] rely on (designed) ground truth data, our approach works in the absence of ground truth. Instead of focusing on a specific data quality dimension such as accuracy or completeness, ARCI uses data alignment as a quality indicator for the input dataset. We also present ARCI-GP, an algorithm for computing ARCI and we conduct a preliminary experimental evaluation.

The remainder of the paper is organized as follows. Section 2 overviews the related work. Section 3 presents some preliminaries on spatio-textual datasets, spatio-textual similarity search and spatio-textual similarity joins. Our approach for evaluating the relative quality of spatio-textual data is introduced in Sect. 4. Section 5 reports on the results of a preliminary experimental evaluation. Finally, Sect. 6 concludes the paper and outlines future work.

2   Related Work

A data value can be considered dirty if it does not conform to the reference data [5]. However, in reality it is usually unrealistic to define a set of comprehensive and correct reference data as the ground truth. Thus, relative data quality has been recently emerging as a research trend. Cao et al. [4] have defined the relative accuracy of data entries as the closeness of the value of different entries, and further specified that the challenge in relative accuracy is creating accuracy rules and inferring the true value. Their work is still built upon the availability of master data though.

Regarding the absence of ground-truth data, Galarus and Angryk [7] have further elaborated this challenge for cases where it is not feasible to obtain the ground-truth data at all times. In order to leverage this issue, they have developed a representative artificial dataset, which can be used as an interpolator to estimate unknown values. The focus of their approach is to mitigate the erroneous data that can possibly appear during the data processing. In their context, it is possible to construct the representative dataset from historical data, which as facts will not be changed.

Another way to detect data quality problems without using reference data is to define data quality rules. In order to discover such rules, different approaches have been proposed such as conditional functional dependencies [5], association rule learning [1] or defining quality processing requirements from users [11]. These rules are usually determined or inferred in certain context. In this work, the experiment has been conducted context-independently and the datasets used contain a limited set of attributes. Therefore, it is not suitable for this work to use rules to detect data quality problems.

Furthermore, data quality problems are usually related to certain data quality dimensions. The research of data quality dimensions can be traced back to the 90's. Wang and Strong [16] used an exploratory factor analysis to derive fifteen data quality dimensions, which are widely accepted in the subsequent data quality research. Afterwards, different data quality dimensions have been further studied and refined such as consistency, accuracy [6] or completeness [13]. This paper does not contribute to a specific data quality dimension, but instead introduces an overall quality indicator to show the trust level of the data.

3   Spatio-Textual Similarity Search

Building upon the definition of spatio-textual objects in [3], a spatio-textual dataset $D = \{x_1, \ldots, x_k\}$ is a collection of spatio-textual objects $x_i = \langle x.id, x.txt, x.loc, x.attr \rangle$ where $x.id$ is the id of the object within the dataset, $x.txt$ is a textual attribute of x (i.e. the name), $x.loc$ is the location of x in the two dimensional geographical space and $x.attr$ is the attribute which is to be assessed.

To determine whether two spatio-textual objects refer to the same real-world entity, one can aim either for exact or for approximate matching. Given two spatio-textual objects x_1 and x_2, for exact matching $x_1.txt$ must be equal to $x_2.txt$, and $x_1.loc$ must also be equal to $x_2.loc$. Clearly, such an approach is not practical as tiny differences, especially in the spatial attribute (i.e. the locations of the objects are few centimeters apart) will prevent a matching. For approximate matching, we employ a similarity function $sim(x_1.txt, x_2.txt)$ for the textual attribute and a distance function $dist(x_1.loc, x_2.loc)$ for the spatial attribute. Hence we declare that x_1 and x_2 refer to the same real-world entity if $sim(x_1.txt, x_2.txt) > \theta$ and $dist(x_1.loc, x_2.loc) \leq \epsilon$, where θ is a textual similarity threshold, and ϵ is the maximum distance that two locations can be apart. Since attributes $x.txt$ and $x.loc$ are used to identify one spatio-textual object from another, throughout this paper, we refer to these two attributes as the identifiers of x.

To perform the spatio-textual similarity search, first we need to decide on a similarity metric for textual attributes, and a distance metric for spatial attributes. For textual similarity in particular, a variety of metrics has been proposed [14]. In this paper, we employ the Levenshtein [10] distance to compute the spatio-textual similarity and we also run experiments using the N-Gram textual similarity [8]. However, choosing the most suitable similarity metric is out of the scope of this paper. Regarding the spatial distance between two objects, it is given by their geodetic distance.

3.1   Spatio-Textual Similarity Joins

An efficient approach to achieve approximate matching between objects from different datasets is to perform a *Spatio-textual similarity join*(STJoin) [2,3,12]. STJoin queries aim at identifying similar spatio-textual objects across different datasets that are close to each other and have similar textual descriptions. More specifically, given two datasets D_i and D_j, a textual similarity threshold ϵ and a spatial distance threshold θ, a spatio-textual similarity join query retrieves all pairs of objects $\{x_i, x_j\}$ where $x_i \in D_i$ and $x_j \in D_j$, the spatial distance of x_i and x_j is $dist(x_i.loc, x_j.loc) \leq \epsilon$ and the textual similarity is $sim(x_i.txt, x_j.txt) > \theta$.

Various algorithms for processing STJoin queries have been proposed. According to the results of the experimental evaluation in [3], the most efficient algorithm for evaluating STJoin queries is PPJ-C algorithm. Given a spatio-textual dataset, PPJ-C first defines a dynamic grid partitioning such that the length of the side of each grid cell is equal to the distance threshold ϵ. Next, for each cell C, PPJ-C identifies the set of join cells A_C, i.e., cell C along with its adjacent cells. Finally, for each pair of join cells, the algorithm compares the elements in the two cells and adds to the result set all pairs of objects that satisfy both the textual similarity and the spatial distance constraint.

4   Spatio-Textual Data Quality Assessment

In this section, we describe our methodology for assessing the relative quality of a spatio-textual dataset. Let a spatio-textual dataset D be the dataset the quality of which we will evaluate. Since we operate in the absence of ground truth, the main idea behind our approach is to cross-reference the entries of D with other similar spatio-textual datasets. For example, given a dataset D_e the quality of which we want to assess, and another dataset D_r, we first match each spatio-textual object $x \in D_e$ with some spatio-textual object $y \in D_r$ such that x and y refer to the same real-world entity. Then, for each attribute of x we compute a relative correctness indicator by comparing the attributes of x with the attributes of y.

4.1   Attribute Relative Correctness Indicator

We propose the *Attribute Relative Correctness Indicator* (ARCI) to assess the quality of a spatio-textual dataset. Let D_e be the spatio-textual dataset the quality of which we want to assess and D_r be another spatio-textual dataset. Each dataset containts spatio-textual objects $x = \langle x.id, x.txt, x.loc, x.attr \rangle$, where $x.txt$ and $x.loc$ are the identifiers of x. Given a textual similarity threshold θ, and a spatial distance threshold ϵ, for every spatio-textual object $x_i \in D_e$ we compute the spatio-textual object $y_j \in D_r$ which is the most similar to x_i and does not violate the θ and ϵ thresholds (it is possible that no matching spatial-object is found). Having matched all possible spatial objects, the ARCI of $x_i.attr$ is

$$ARCI(x_i.attr, y_j.attr) = \begin{cases} Sim(x_i.attr, y_j.attr), & \text{if } y_j \text{ is the best match to } x_i \\ 0.5, & \text{if } \nexists y_j \text{ that matches } x_i. \end{cases}$$

Figure 1 illustrates two datasets: Dataset 1 (D_e), the dataset the quality of which we evalute, and Dataset 2, (D_r) another dataset we use as reference. The identifiers of both D_e and D_r are the *name* and the *location*, while the attribute for which we want to compute the ARCI is the *type*. The textual similarity is given by the Levenshtein distance, while the spatial distance is given by the Euclidean distance. Finally, we set the textual similarity threshold $\epsilon = 0.6$ and the spatial distance threshold $\theta = 1$.

Dataset 1 (D_{eval})			
id	name	type	loc.
x_1	Loacker	Café	$(1,1)$
x_2	Marilyn	Café	$(3,2)$
x_3	Cavalino Bianco	Restaurant	$(1,6)$
x_4	Dai Carrettai	Restaurant	$(5,5)$
x_5	Enovit Bar	Bar	$(8,9)$
x_6	Hopfen	Brewery	$(7,5)$

Dataset 2 (D_{ref})			
id	name	type	loc.
y_1	Da Pichio	Bar	$(4,5)$
y_2	Loacker	Café	$(1,1)$
y_3	Nadamas	Bar	$(5,1)$
y_4	Marylin	Café	$(3,3)$
y_5	Enovit Wine Bar	Wine Bar	$(9,8)$
y_6	Stadt Café Città	Café	$(8,2)$
y_7	Nussbaumer	Restaurant	$(3,7)$
y_8	Hopfen & Co.	Restaurant	$(7,5)$
y_9	Carrettai	Restaurant	$(7,7)$

Fig. 1. Sample spatio-textual datasets.

First, we attempt to match each entry $x_i \in D_e$ with an entry $y_i \in D_r$. The result of this operation is illustrated in Fig. 2. Having computed the matching spatio-textual objects, we compute the ARCI for the attribute "type" for each spatio-textual object in D_e. For x_1 and x_2 we have an exact match, hence $ARCI(x_1.attr) = 1$ and $ARCI(x_2.attr) = 1$. For x_1 and x_2 the ARCI indicates that both $x_1.attr$ and $x_2.attr$ are correct with regard to D_r. For, x_3 and x_4 no matching object in D_r was found; thus $ARCI(x_3.attr) = 0.5$ and $ARCI(x_4.attr) = 0.5$ meaning that it was not possible to verify the correctness of the attributes. Finally, for elements x_5 and x_6 the respective ARCIs are $ARCI(x_5.attr) = 0.36$ and $ARCI(x_6.attr) = 0.1$ which are relatively low and, therefore, our approach indicates that $x_5.attr$ and $x_6.attr$ might be wrong.

$x_i \in D_{eval}$	$y_j \in D_{ref}$	$Sim(x_i.txt, y_j.txt)$	$dist(x_i.loc, y_i.loc)$	ARCI
x_1	y_2	1	0	1
x_2	y_4	0	1	1
x_3	-	-	-	0.5
x_4	-	-	-	0.5
x_5	y_5	0.667	1.414	0.36
x_6	y_8	0.667	0	0.1

0	0.5	1

Fig. 2. Matching entries of D_e and D_r of Fig. 1.

Note that ARCI cannot be computed for the identifiers of a spatio-textual object. Apparently, the selection of the identifiers (especially the textual one) is crucial and can possibly affect the efficiency of our approach. However, as determining the best possible identifiers of a spatio-textual object is out of the scope of this paper, we work under the assumption that each spatio-textual object comes with the proper identifiers.

4.2   The ARCI-GP Algorithm

To compute the ARCI for each spatio-textual object in a dataset we propose ARCI-GP, an algorithm which is inspired by the PPJ-C [3] algorithm for spatio-textual similarity joins. Let D_e be the spatio-textual dataset the quality of which we want to assess and D_r be the reference dataset. First, ARCI-GP defines a dynamic grid partitioning and organizes the spatio-textual objects of both D_e and D_r into cells. The grid is defined such that the length of the side of every cell equals the spatial distance threshold ϵ. For each cell C, ARCI-GP identifies a set of join cells A_C, i.e., the set of the adjacent cells of C along with C itself. For every element $x \in C$ of D_e, A_C contains the objects $y \in D_r$ which can possibly match with x. Due to the properties of the grid, the distance of all other elements $y' \notin A_C$ from x is more than ϵ, i.e., they violate the spatial distance constraint. Next, ARCI-GP computes for each object $x \in C$ of D_e the object $y \in A_C$ of D_r which is the most similar to x. To determine the best match of x, the algorithm examines only the objects in A_C. During this process, it is possible that the algorithm does not find any matching object to x. The same process is executed over all the cells that contain at least one object $x \in D_e$.

Algorithm 1 illustrates the pseudocode of our ARCI-GP algorithm. First, the result set R is initialized in Line 1 and a grid partition G_R is constructed for the spatio-textual objects in both input datasets D_e and D_r. In Lines 3–16 the algorithm iterates over the cells of the partition that contain at least one element of D_e and, for each cell C, it computes the set of join cells A_C. Then, for each cell $C' \in A_C$, the algorithm attempts to find a match between the element $o_i \in C$ that is an element of D_e, and the element $o_j \in C'$ that is an element of D_r (Lines 6–16). A new entry r is initialized to *null* in Line 7. Each element o_i is matched with only one element o_j for which the textual similarity $Sim(o_i.txt, o_j.txt)$ is maximum and the spatial distance $dist(o_i.loc, o_j.loc)$ is minimum (Lines 8–12). If a match is found, i.e. r is not *null*, then r is added to the result set with the computed ARCI value in Line 14. Otherwise, r is added to the result set with the default 0.5 ARCI value. Finally, the result set is returned in Line 17.

Algorithm 1. ARCI-GP $(D_e, D_r, \epsilon, \theta)$

Input: Collection of spatio-textual objects D_e; collection of spatio-textual
      objects D_r; spatial distance threshold ϵ; textual similarity threshold θ
Output: Result set R

1   $R \leftarrow \emptyset$;
2   $G_R \leftarrow$ ConstructGridPartitioning $(D_e \cup D_r, \epsilon)$;
3   **foreach** *cell* $C \in G_R$ **do**
4      $A_C \leftarrow$ IdentifyJoinCells (G_R, C);
5   **foreach** *cell* $C' \in A_C$ **do**
6   **foreach** *object* $o_i \in C \cap D_e$ **do**
7   intitialize entry $r \leftarrow$ null;
8             $sim_{max} = \theta$;
9   **foreach** *object* $o_j \in C \cap D_r$ **do**
10   **if** $Sim(o_i.txt, o_j.txt) > sim_{max}$ *and* $dist(o_i.loc, o_j.loc) \leq \epsilon$ **then**
11                    $r \leftarrow \langle o_i, ARCI(o_i, o_j) \rangle$;
12                    $sim_{max} \leftarrow Sim(o_i.txt, o_j.txt)$;

13   **if** $r \neq null$ **then**
14                $R \leftarrow R \cup r$;   ▷ Best match added to result
15   **else**
16                $R \leftarrow R \cup \langle o, 0.5 \rangle$;   ▷ No match was found

17 **return** R;

---

5   Preliminary Experimental Evaluation

In this section, we report the results of a preliminary experimental evaluation
and compare two different implementations of our ARCI-GP algorithm. We use
two different spatio-textual datasets in our experiments. The first dataset (D_e)
contains $500,000$ spatio-textual objects and was compiled by combining datasets
obtained from Tourpedia[1]. The second dataset (D_r) contains $1,000,000$ spatio-
textual objects and was obtained by querying the Foursquare API[2] using *ids*
obtained from [9]. To observe the effect of the textual similarity metric, we
measure the runtime of ARCI-GP using two different metrics: the Levenshtein
similarity [10] and the NGram similarity [8]. We implemented our algorithm with
Java 1.8 and the tests run on a machine with 4 Intel Xeon X5550 (2.67 GHz)
processors and 48 GB main memory running Ubuntu Linux.

    Figure 3 shows our measurement results on the runtime of our ARCI-GP
algorithm by varying the sizes of D_e and D_r. More specifically, Fig. 3a shows
the runtime of ARCI-GP varying the size of D_e from 100,000 to 500,000 entries
using the entire D_r dataset, and Fig. 3b shows the runtime of ARCI-GP varying
the size of D_r from 200,000 to 1,000,000 entries using the entire D_r dataset. In
both figures we observe that the runtime of ARCI-GP increases with the size of

---

[1] http://tour-pedia.org/about/datasets.html.
[2] https://developer.foursquare.com.

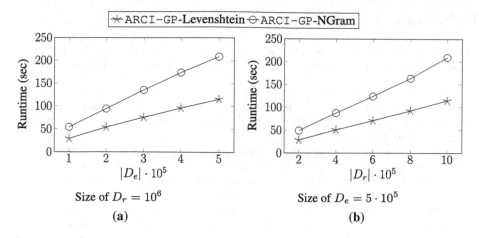

Fig. 3. Performance of PPLJ.

D_e and D_r. However, we can observe that the runtime increases faster with the size of D_e. For example, the runtime for $D_e = 2 \cdot 10^5$ and $D_r = 10^6$ (Fig. 3a) and the runtime for $D_e = 4 \cdot 10^5$ and $D_r = 5 \cdot 10^5$ (Fig. 3b). Although in the first case the total number of involved entries is higher by 300,000 entries, the runtime is approximately the same. Apparently, the size of D_e influences the runtime of ARCI-GP much more than the size of D_r. Such a result is to be expected as ARCI-GP considers all the elements in D_e, while many of the elements in D_r are filtered out by the grid partitioning.

Finally, with regard to the textual similarity metric, we observe that our algorithm is almost two time faster when using the Levenshtein similarity than when using the NGram similarity. Since the algorithm requires the execution of many textual similarity operations, the computational cost of the similarity metric has a great influence on the total runtime of the ARCI-GP.

6   Conclusion

In this paper, we proposed ARCI, an indicator which operates in the absence of ground truth and shows the relative quality of a spatio-textual dataset by cross-referencing it with similar datasets. ARCI is computed for the attributes of a data entry in a spatio-textual dataset to indicate its relative correctness. We have also shown that ARCI can be used directly to provide visual information on the quality of the data, e.g., using a spectrum bar. Furthemore, we proposed an algorithm for computing ARCI and evaluated its performance using real-world spatio-textual datasets.

As future work, we will explore different strategies to develop efficient algorithm for computing ARCI such as executing textual-matches first. We will also consider utilizing indexing structures to improve performance even further. Finally, we plan to investigate alternative metrics for computing textual similarity such as semantic similarity metrics.

References

1. Abedjan, Z., Akcora, C.G., Ouzzani, M., Papotti, P., Stonebraker, M.: Temporal rules discovery for web data cleaning. Proc. VLDB Endowment **9**(4), 336–347 (2015)
2. Ballesteros, J., Cary, A., Rishe, N.: Spsjoin: parallel spatial similarity joins. In: Proceedings of the 19th ACM SIGSPATIAL GIS Conference, pp. 481–484 (2011)
3. Bouros, P., Ge, S., Mamoulis, N.: Spatio-textual similarity joins. Proc. VLDB Endowment **6**(1), 1–12 (2012)
4. Cao, Y., Fan, W., Yu, W.: Determining the relative accuracy of attributes. In: Proceedings of the 2013 ACM SIGMOD Conference, pp. 565–576 (2013)
5. Chiang, F., Miller, R.J.: Discovering data quality rules. Proc. VLDB Endowment **1**(1), 1166–1177 (2008)
6. Cong, G., Fan, W., Geerts, F., Jia, X., Ma, S.: Improving data quality: consistency and accuracy. In: Proceedings of the 33rd VLDB Conference, pp. 315–326 (2007)
7. Galarus, D., Angryk, R.: A smart approach to quality assessment of site-based spatio-temporal data. In: Proceedings of the 24th ACM SIGSPATIAL GIS Conference, pp. 55:1–55:4 (2016)
8. Kondrak, G.: N-gram similarity and distance. In: Consens, M., Navarro, G. (eds.) SPIRE 2005. LNCS, vol. 3772, pp. 115–126. Springer, Heidelberg (2005). doi:10. 1007/11575832_13
9. Levandoski, J.J., Sarwat, M., Eldawy, A., Mokbel, M.F.: Lars: a location-aware recommender system. In: Proceedings of the 28th IEEE ICDE, pp. 450–461 (2012)
10. Levenshtein, V.: Binary codes capable of correcting deletions, insertions, and reversals. Soviet Phys. Doklady **10**, 707–710 (1965)
11. Missier, P., Embury, S., Greenwood, M., Preece, A., Jin, B.: Quality views: capturing and exploiting the user perspective on data quality. In Proceedings of the 32nd VLDB Conference, pp. 977–988 (2006)
12. Rao, J., Lin, J., Samet, H.: Partitioning strategies for spatio-textual similarity join. In: Proceedings of the 3rd ACM International Workshop on Analytics for Big Geospatial Data, pp. 40–49 (2014)
13. Razniewski, S., Nutt, W.: Completeness of queries over incomplete databases. Proc. VLDB Endowment **4**(11), 749–760 (2011)
14. Recchia, G., Louwerse, M.: A comparison of string similarity measures for toponym matching. In: Proceedings of The 1st ACM International COMP Workshop, pp. 54:54–54:61 (2013)
15. Tsatsanifos, G., Vlachou, A.: On processing top-k spatio-textual preference queries. In: Proceedings of the 18th EDBT Confernce, pp. 433–444 (2015)
16. Wang, R.Y., Strong, D.M.: Beyond accuracy: what data quality means to data consumers. J. Manage. Inf. Syst. **12**(4), 5–33 (1996)

T²K²: The Twitter Top-K Keywords Benchmark

Ciprian-Octavian Truică[1](✉) and Jérôme Darmont[2]

[1] Computer Science and Engineering Department, Faculty of Automatic Control
and Computers, University Politehnica of Bucharest, Bucharest, Romania
ciprian.truica@cs.pub.ro
[2] Université de Lyon, Lyon 2, ERIC EA 3083, Lyon, France
jerome.darmont@univ-lyon2.fr

Abstract. Information retrieval from textual data focuses on the construction of vocabularies that contain weighted term tuples. Such vocabularies can then be exploited by various text analysis algorithms to
extract new knowledge, e.g., top-k keywords, top-k documents, etc. Top-
k keywords are casually used for various purposes, are often computed
on-the-fly, and thus must be efficiently computed. To compare competing
weighting schemes and database implementations, benchmarking is customary. To the best of our knowledge, no benchmark currently addresses
these problems. Hence, in this paper, we present a top-k keywords benchmark, T²K², which features a real tweet dataset and queries with various
complexities and selectivities. T²K² helps evaluate weighting schemes
and database implementations in terms of computing performance. To
illustrate T²K²'s relevance and genericity, we show how to implement the
TF-IDF and Okapi BM25 weighting schemes, on one hand, and relational
and document-oriented database instantiations, on the other hand.

Keywords: Top-k keywords · Benchmark · Term weighting · Database
systems

1 Introduction

Analyzing textual data is a current challenge, notably due to the vast amount
of text generated daily by social media. One approach for extracting knowledge
is to infer from texts the top-k keywords to determine trends [1,14], or to detect
anomalies or more generally events [7]. Computing top-k keywords requires building a weighted vocabulary, which can also be used for many other purposes such
as topic modeling and clustering. Term weights can be computed at the application level, which is inefficient when working with large data volumes because
all information must be queried and processed at a layer different from storage.
A presumably better approach is to process information at the storage layer
using aggregation functions, and then return the result to the application layer.
Yet, the term weighting process remains very costly, because each time a query
is issued, at least one pass through all documents is needed.

© Springer International Publishing AG 2017
M. Kirikova et al. (Eds.): ADBIS 2017, CCIS 767, pp. 21–28, 2017.
DOI: 10.1007/978-3-319-67162-8_3

To compare combinations of weighting schemes, computing strategies and physical implementations, benchmarking is customary. However, to the best of our knowledge, there exists no benchmark for this purpose. Hence, we propose in this paper the Twitter Top-K Keywords Benchmark (T^2K^2), which features a real tweet dataset and queries with various complexities and selectivities. We designed T^2K^2 to be somewhat generic, i.e., it can compare various weighting schemes, database logical and physical implementations and even text analytics platforms [18] in terms of computing efficiency. As a proof of concept of T^2K^2's relevance and genericity, we show how to implement the TF-IDF and Okapi BM25 weighting schemes, on one hand, and relational and document-oriented database instantiations, on the other hand.

The remainder of this paper is organized as follows. Section 2 reviews text-oriented benchmarks. Section 3 provides T^2K^2's generic specification. Section 4 details T^2K^2's proof of concept, i.e., its instantiation for several weighting schemes and database implementations. Finally, Sect. 5 concludes this paper and hints at future research.

2 Related Work

Term weighting schemes are extensively benchmarked in sentiment analysis [15], semantic similarity [11], text classification and categorization [8,9,11,13], and textual corpus generation [19]. Benchmarks for text analysis focus mainly on algorithm accuracy, while either term weights are known before the algorithm is applied, or their computation is incorporated with preprocessing. Thus, such benchmarks do not evaluate weighting scheme construction efficiency as we do.

Other benchmarks evaluate parallel text processing in big data applications in the cloud [4,5]. PRIMEBALL notably specifies several relevant properties characterizing cloud platforms [4], such as scale-up, elastic speedup, horizontal scalability, latency, durability, consistency and version handling, availability, concurrency and other data and information retrieval properties. However, PRIMEBALL is only a specification; it is not implemented.

3 T^2K^2 Specification

Typically, a benchmark is constituted of a data model (conceptual schema and extension), a workload model (set of operations) to apply on the dataset, an execution protocol and performance metrics [3]. In this section, we provide a conceptual description of T^2K^2, so that it is generic and can cope with various weighting schemes and database logical and physical implementations.

3.1 Data Model

The base dataset we use is a corpus of 2 500 000 tweets that was collected using Twitter's REST API to read and gather data. Moreover, we applied preprocessing steps to the raw corpus to extract the additional information needed to build a weighted vocabulary: (1) extract all tags and remove links; (2) expand contractions, i.e., shortened versions of the written and spoken forms of a word, syllable, or word group, created by omission of internal letters and sounds [2], e.g., "it's" becomes "it is"; (3) extract sentences and remove punctuation in each sentence, creating a clean text; (4) for each sentence, extract lemmas and create a lemma text; (5) for each lemma t in tweet d, compute the number of co-occurrences $f_{t,d}$ and term frequency $TF(t, d)$, which normalizes $f_{t,d}$.

T²K² database's conceptual model (Fig. 1) represents all the information extracted after the text preprocessing steps. Information about tweet *Author* are a unique identifier, first name, last name and age. Information about author *Gender* is stored in a different entity to minimize the number of duplicates of gender type. *Documents* are identified by the tweet's unique identifier and store the raw tweet text, clean text, lemma text, and the tweet's creation date. *Writes* is the relationship that associates a tweet to its author. Tweet location is stored in the *Geo_Location* entity to avoid duplicates again. *Word* bears a unique identifier and the actual lemma. Finally, weights $f_{t,d}$ and $TF(t, d)$ for each lemma and each document are stored in the *Vocabulary* relationship.

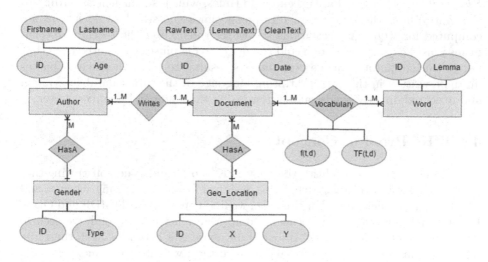

Fig. 1. T²K² conceptual data model

The initial 2 500 000 tweet corpus is split into 5 different datasets that all keep an equal balance between the number of tweets for both genders, location and date. These datasets contain 500 000, 1 000 000, 1 500 000, 2 000 000 and 2 500 000 tweets, respectively. They allow scaling experiments and are associated to a scale factor (SF) parameter, where $SF \in \{0.5, 1, 1.5, 2, 2.5\}$, for conciseness sake.

3.2   Workload Model

The queries used in T²K² are designed to achieve two goals: (1) compute different term weighting schemes using aggregation functions and return the top-k keywords; (2) test the performance of different database management systems. T^2K^2 queries are sufficient for achieving these goals, because they test the query execution plan, internal caching and the way they deal with aggregation. More precisely, they take different group by attributes into account and aggregate the information to compute weighting schemes for top-k keywords.

T^2K^2 features four queries $Q1$ to $Q4$ that compute top-k keywords w.r.t. constraint(s): $c_1(Q1)$, $c_1 \wedge c_2(Q2)$, $c_1 \wedge c_3(Q3)$, $c_1 \wedge c_2 \wedge c_3(Q4)$. c_1 is $Gender.Type = pGender$, where parameter $pGender \in \{male, female\}$. c_2 is $Document.Date \in [pStartDate, pEndDate]$, where $pStartDate, pEndDate \in [2015\text{-}09\text{-}17\ 20{:}41{:}35, 2015\text{-}09\text{-}19\ 04{:}05{:}45]$ and $pStartDate < pEndDate$. c_3 is $Geo\_location.X \in [pStartX, pEndX]$ and $Geo\_location.Y \in [pStartY, pEndY]$, where $pStartX, pEndX \in [15, 50]$, $pStartX < pEndX$, $pStartY, pEndY \in [-124, 120]$ and $pStartY < pEndY$. Queries bear different levels of complexity and selectivity.

3.3   Performance Metrics and Execution Protocol

We use each query's response time $t(Q_i)$ as metrics in T^2K^2. Given scale factor SF, all queries $Q1$ to $Q4$ are executed 40 times, which is sufficient according to the central limit theorem. Average response times and standard deviations are computed for $t(Q_i)$. All executions are warm runs, i.e., either caching mechanisms must be deactivated, or a cold run of $Q1$ to $Q4$ must be executed once (but not taken into account in the benchmark's results) to fill in the cache. Queries must be written in the native scripting language of the target database system and executed directly inside said system using the command line interpreter.

4 T^2K^2 Proof of Concept

In this section, we aim at illustrating how T^2K^2 works and at demonstrating that it can adequately benchmark what it is designed for, i.e., weighting schemes and database implementations. For this sake, we first compare the TF-IDF and Okapi BM25 weighting schemes in terms of computing efficiency. Second, we seek to determine whether a document-oriented database is a better solution than in a relational databases when computing a given term weighting scheme.

4.1   Weighting Schemes

Let D be the corpus of tweets, $N = |D|$ the total number of documents (tweets) in D and n the number of documents where some term t appears. The TF-IDF weight is computed by multiplying the augmented term frequency $TF(t, d) = K + (1 - K) \cdot \frac{f_{t,d}}{\max_{t' \in d}(f_{t',d})})$ by the inverted document frequency $IDF(t, D) = 1 + \log \frac{N}{n}$, i.e., $TFIDF(t, d, D) = TF(t, d) \cdot IDF(t, D)$. The augmented form of TF prevents a bias towards long tweets when the free parameter K is set to 0.5 [12]. It uses the number of co-occurrences $f_{t,d}$ of a word in a document, normalized with the frequency of the most frequent term t', i.e., $\max_{t' \in d}(f_{t',d})$.

The Okapi BM25 weight is given in Eq. (1), where $||d||$ is d's length, i.e., the number of terms appearing in d. Average document length $avg_{d' \in D}(||d'||)$ is used to remove any bias towards long documents. The values of free parameters k_1 and b are usually chosen, in absence of advanced optimization, as $k_1 \in [1.2, 2.0]$ and $b = 0.75$ [10, 16, 17].

$$Okapi(t, d, D) = \frac{TFIDF(t, d, D) \cdot (k_1 + 1)}{TF(t, d) + k_1 \cdot (1 - b + b \cdot \frac{||d||}{avg_{d' \in D}(||d'||)})} \tag{1}$$

The sum $S\_TFIDF(t, d, D) = \sum_{i=1}^{N} TFIDF(t, d_i, D)$ of all TF-IDFs and the sum $S\_Okapi(t, d, D) = \sum_{i=1}^{N} Okapi(t, d_i, D)$ of all Okapi BM25 weights constitute the term's weights that are used to construct the list of top-k keywords.

4.2 Relational Implementation

Database. The logical relational schema used in both relational databases management systems (Fig. 2) directly translates the conceptual schema from Fig. 1.

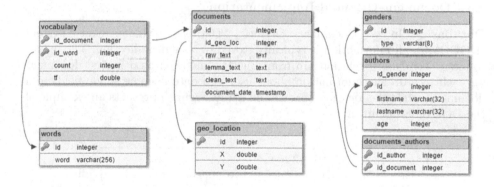

Fig. 2. T²K² relational logical schema

Queries. Text analysis deals with discovering hidden patterns from texts. In most cases, it is useful to determine such patterns for given groups, e.g., males and females, because they have different interests and talk about disjunct subjects. Moreover, if new events appear, depending on the location and time of day, these subject can change for the same group of people. The queries we propose aim to determine such hidden patterns and improve text analysis and anomaly detection.

Let us express T²K²'s queries in relational algebra. c_1, c_2 and c_3 are the constraints defined in Sect. 3.2, adapted to the relational schema.

$Q1 = \gamma_L \left(\pi_{documents.id,words.word,f_w(vocabulary.count,vocabulary.tf)} \left(\sigma_{c_1} \left(documents \bowtie_{c_4} documents\_authors \bowtie_{c_5} authors \bowtie_{c_6} genders \bowtie_{c_7} vocabulary \bowtie_{c_8} words\right)\right)\right)$, where c_4 to c_8 are join conditions; f_w is the weighting function that computes TF-IDF or Okapi BM25, which takes two parameters: $vocabulary.count = f_{t,d}$ and $vocabulary.tf = TF(t,d)$; γ_L is the aggregation operator, where $L = (F,G)$, with $F = $ sum$(f_w(vocabulary.count, vocabulary.tf))$ and G is the $words.word$ attribute that appears in the group by clause.

$Q2 = \gamma_L \left(\pi_{documents.id,words.word,f_w(vocabulary.count,vocabulary.tf)} \left(\sigma_{c_1 \wedge c_2} \left(documents \bowtie_{c_4} documents\_authors \bowtie_{c_5} authors \bowtie_{c_6} genders \bowtie_{c_7} vocabulary \bowtie_{c_8} words\right)\right)\right)$.

$Q3 = \gamma_L \left(\pi_{documents.id,words.word,f_w(vocabulary.count,vocabulary.tf)} \left(\sigma_{c_1 \wedge c_3} \left(documents \bowtie_{c_4} documents\_authors \bowtie_{c_5} authors \bowtie_{c_6} genders \bowtie_{c_7} vocabulary \bowtie_{c_8} words \bowtie_{c_9} geo\_location\right)\right)\right)$, where c_9 is the join condition between $documents$ and $geo\_location$.

$Q4 = \gamma_L \left(\pi_{documents.id,words.word,f_w(vocabulary.count,vocabulary.tf)} \left(\sigma_{c_1 \wedge c_2 \wedge c_3} \left(documents \bowtie_{c_4} documents\_authors \bowtie_{c_5} authors \bowtie_{c_6} genders \bowtie_{c_7} vocabulary \bowtie_{c_8} words \bowtie_{c_9} geo\_location\right)\right)\right)$.

4.3 Document-Oriented Implementation

Database. In a Document Oriented Database Management System (DODBMS), all information is typically stored in a single collection. The many-to-many *Vocabulary* relationship from Fig. 1 is modeled as a nested document for each record. The information about user and date become single fields in a document, while the location becomes an array. Figure 3 presents an example of the DODBMS document.

```
{    _id : 644626677310603264,
     rawText : "Amanda's car is too much for my headache",
     cleanText : "Amanda is car is too much for my headache",
     lemmaText : "amanda car headache",
     author : 970993142,
     geoLocation : [ 32, 79 ],
     gender : "male",
     age : 23,
     lemmaTextLength : 3,
     words : [ { "tf" : 1, "count" : 1, "word" : "amanda"},
             { "tf" : 1, "count" : 1, "word" : "car" },
             { "tf" : 1, "count" : 1, "word" : "headache"} ],
     date : ISODate("2015-09-17T23:39:11Z") }
```

Fig. 3. Sample DODBMS document

Queries. In DODBMSs, user-defined (e.g., JavaScript) functions are used to compute top-k keywords. The TF-IDF weight can take advantage of both native database aggregation (NA) and MapReduce (MR). However, due to the multitude of parameters involved and the calculations needed for the Okapi BM25 weighting scheme, the NA method is usually difficult to develop. Thus, we recommend to only use MR in benchmark runs.

5 Conclusion

Jim Gray defined four primary criteria to specify a "good" benchmark [6]. *Relevance:* The benchmark must deal with aspects of performance that appeal to the largest number of users. Considering the wide usage of top-k queries in various text analytics tasks, we think T^{2}K^2 fulfills this criterion. We also show in Sect. 4 that our benchmark achieves what it is designed for.

Portability: The benchmark must be reusable to test the performances of different database systems. We successfully instantiated T^{2}K^2 within two types of database systems, namely relational and document-oriented systems.

Simplicity: The benchmark must be feasible and must not require too many resources. We designed T^{2}K^2 with this criterion in mind (Sect. 3), which is particularly important for reproducibility. We notably made up parameters that are easy to setup.

Scalability: The benchmark must adapt to small or large computer architectures. By introducing scale factor SF, we allow users to simply parameterize T^{2}K^2 and achieve some scaling, though it could be pushed further in terms of data volume.

In future work, we plan to expand T^{2}K^2's dataset significantly to aim at big data-scale volume. We also intend to further our proof of concept and validation efforts by benchmarking other NoSQL database systems and gain insight regarding their capabilities and shortcomings. We also plan to adapt T^{2}K^2 so that it runs in the Hadoop and Spark environments.

References

1. Bringay, S., Béchet, N., Bouillot, F., Poncelet, P., Roche, M., Teisseire, M.: Towards an on-line analysis of tweets processing. In: Hameurlain, A., Liddle, S.W., Schewe, K.-D., Zhou, X. (eds.) DEXA 2011. LNCS, vol. 6861, pp. 154–161. Springer, Heidelberg (2011). doi:10.1007/978-3-642-23091-2_15
2. Cooper, J.D., Robinson, M.D., Slansky, J.A., Kiger, N.D.: Literacy: Helping Students Construct Meaning. Cengage Learning, Boston (2014)
3. Darmont, J.: Data processing benchmarks. In: Khosrow, M. (ed.) Encyclopedia of Information Science and Technology, 3rd edn., pp. 146–152. IGI Global, Hershey (2014)

4. Ferrarons, J., Adhana, M., Colmenares, C., Pietrowska, S., Bentayeb, F., Darmont, J.: PRIMEBALL: a parallel processing framework benchmark for big data applications in the cloud. In: Nambiar, R., Poess, M. (eds.) TPCTC 2013. LNCS, vol. 8391, pp. 109–124. Springer, Cham (2014). doi:10.1007/978-3-319-04936-6_8

5. Gattiker, A.E., Gebara, F.H., Hofstee, H.P., Hayes, J.D., Hylick, A.: Big data text-oriented benchmark creation for Hadoop. IBM J. Res. Dev. **57**(3/4), 10: 1–10: 6 (2013)

6. Gray, J.: The Benchmark Handbook for Database and Transaction Systems, 2nd edn. Morgan Kaufmann, Burlington (1993)

7. Guille, A., Favre, C.: Event detection, tracking, and visualization in twitter: a mention-anomaly-based approach. Soc. Netw. Anal. Min. **5**(1), 18 (2015)

8. Kılınç, D., Özçift, A., Bozyigit, F., Yildirim, P., Yücalar, F., Borandag, E.: TTC-3600: A new benchmark dataset for turkish text categorization. J. Inf. Sci. **43**(2), 174–185 (2017)

9. Lewis, D.D., Yang, Y., Rose, T.G., Li, F.: RCV1: A new benchmark collection for text categorization research. J. Mach. Learn. Res. **5**, 361–397 (2004)

10. Manning, C.D., Raghavan, P., Schütze, H.: Introduction to Information Retrieval. Cambridge University Press, New York (2008)

11. O'Shea, J., Bandar, Z., Crockett, K.A., McLean, D.: Benchmarking short text semantic similarity. Int. J. Intell. Inf. Database Syst. **4**(2), 103–120 (2010)

12. Paltoglou, G., Thelwall, M.: A study of information retrieval weighting schemes for sentiment analysis. In: 48th Annual Meeting of the Association for Computational Linguistics, pp. 1386–1395 (2010)

13. Partalas, I., Kosmopoulos, A., Baskiotis, N., Artières, T., Paliouras, G., Gaussier, É., Androutsopoulos, I., Amini, M., Gallinari, P.: LSHTC: a benchmark for large-scale text classification. CoRR abs/1503.08581 (2015)

14. Ravat, F., Teste, O., Tournier, R., Zurfluh, G.: Top_Keyword: an aggregation function for textual document OLAP. In: Song, I.-Y., Eder, J., Nguyen, T.M. (eds.) DaWaK 2008. LNCS, vol. 5182, pp. 55–64. Springer, Heidelberg (2008). doi:10.1007/978-3-540-85836-2_6

15. Reagan, A.J., Tivnan, B.F., Williams, J.R., Danforth, C.M., Dodds, P.S.: Benchmarking sentiment analysis methods for large-scale texts: a case for using continuum-scored words and word shift graphs. CoRR abs/1512.00531 (2015)

16. Spärck Jones, K., Walker, S., Robertson, S.E.: A probabilistic model of information retrieval: development and comparative experiments: part 1. Inf. Process. Manage. **36**(6), 779–808 (2000)

17. Spärck Jones, K., Walker, S., Robertson, S.E.: A probabilistic model of information retrieval: development and comparative experiments: part 2. Inf. Process. Manage. **36**(6), 809–840 (2000)

18. Truică, C.O., Darmont, J., Velcin, J.: A scalable document-based architecture for text analysis. In: International Conference on Advanced Data Mining and Applications (ADMA), pp. 481–494 (2016)

19. Wang, L., Dong, X., Zhang, X., Wang, Y., Ju, T., Feng, G.: TextGen: a realistic text data content generation method for modern storage system benchmarks. Front. Inf. Technol. Electron. Eng. **17**(10), 982–993 (2016)

Outlier Detection in Data Streams Using OLAP Cubes

Felix Heine[(✉)]

Department of Computer Science, Faculty IV,
Hannover University of Applied Sciences and Arts,
Ricklinger Stadtweg 120, 30459 Hannover, Germany
felix.heine@hs-hannover.de

Abstract. Outlier detection is an important tool for many application areas. Often, data has some multidimensional structure so that it can be viewed as OLAP cubes. Exploiting this structure systematically helps to find outliers otherwise undetectable. In this paper, we propose an approach that treats streaming data as a series of OLAP cubes. We then use an offline calculated model of the cube's expected behavior to find outliers in the data stream. Furthermore, we aggregate multiple outliers found concurrently at different cells of the cube to some user-defined level in the cube. We apply our method to network data to find attacks in the data stream to show its usefulness.

1 Introduction

Outlier detection is an important basic technique used in many application areas including data quality, fraud detection, or intrusion detection in networks. In this paper, we look at outlier detection in multidimensional data. We present a novel method for outlier detection that exploits the multidimensional nature of data. It is a one-class anomaly detection approach that includes a training phase using only-normal data. The training is done offline, the detection phase can be run either online or offline. Exploiting the multidimensional structure of data helps to find outliers that are undetectable at the highest aggregation level or to collect more evidence for suspicious data by looking at various multidimensional views.

This also helps to find contextual and collective outliers (see [2]). First, aggregating the data in different ways allows for contextual models by assigning different models to different groupings. Second, collective outliers can be spotted more easily when looking at the grouping where the outlier becomes most evident. In order to define and search the groupings in a systematic way, we treat all data sources as *multidimensional data cubes*. For most data, this is a natural way to model the data. For example, network data can be grouped and aggregated e.g. using target IP ranges, source IP ranges, and protocols at different layers.

This work has partly been developed in the project IQM4HD (reference number: 01IS15053B). IQM4HD is partly funded by the German ministry of education and research (BMBF) within the research program KMU Innovativ.

M. Kirikova et al. (Eds.): ADBIS 2017, CCIS 767, pp. 29–36, 2017.
DOI: 10.1007/978-3-319-67162-8_4

The advantage of this approach is the natural treatment of collective and contextual outliers. We integrate the anomaly detection results coming from different views of the data to *collect evidence* and to build a final score for each entity of interest. The method itself is general. Domain knowledge is only needed to properly define the cube's dimensions and metrics.

The main contribution of this paper is a one-class method to find anomalies in multidimensional data cubes using models that describe the normal behavior of the data at a fine-grained level. This method is complemented by a second method that combines anomaly scores that were calculated for different data views and to relate these scores to relevant *units of inspection*. These units are application dependent and define the granularity of the generated anomaly events. For network data, a useful unit of inspection is a single network connection. As an application example, the method is applied to network data. We provide an evaluation using data from the 1998 DARPA IDS evaluation.

The next section describes the proposed method in detail on a generic level that is independent from a specific application. Transfer to a network example scenario is done in the evaluation Sect. 3. Before concluding, related work is described. Please note that we assume that the reader is familiar with basic OLAP cube terminology and iceberg cubes. For an introduction, see [4, 14].

2   Anomaly Detection in Multidimensional Data

In this section, we describe our approach. The basis is time-related multidimensional data, i.e. a stream of tuples that each contain a time-stamp, multiple dimensional attributes (categorical) and multiple metric attributes (continuous). We furthermore assume to have recorded old data that well reflects normal behavior and can be used to train the system.

We split both training and test data into a sequence of cubes by accumulating data over an interval ΔT. In the training phase, this is done offline. Here, we build the models that describe the normal behavior of the data. During the online operation, we collect data until the interval is completed and then build the cube for the past interval. We then use the models to find outliers in the current time slice. As an outlier will be visible in multiple aggregation views, we furthermore integrate multiple outlier scores in a single score per unit. We now describe details for both phases.

2.1   Training Phase

The basic idea is to build a separate model for each cell of the cube, and to check online data against this model. The training data must not contain anomalies. As a basic model, we start with a Gaussian model, however, more complex models are possible.

The model should be able to assess the outlier score for the metric values in a single cell as well as the correlation between different cells. Our basic model captures the distribution of each metric in individual cells without inter-relationships between the metrics as a univariate Gaussian model. Thus it is easy to calculate model parameters (estimated standard deviation and estimated mean), and easy to apply (calculation of Z-scores). More complex models could fit a time series model to a cell, e.g. an autoregressive model. However, already this simple model leads to a huge number of parameters, when looking at the whole cube, thus we aim to limit the number cells where a model is created. The number of underlying data samples (size of a cell) is a good indicator to do the pruning. We use a system-wide parameter T_m as a threshold for the minimal number of time steps where the data must be present for a cell in order to build an individual model for this cell. This avoids models that are not backed by enough data and limits the size of the overall model.

Beside the behavior of individual cells, we also aim to model the relationship between cells. We capture the relationship between each cell and all of its parent cells by providing a model for every direct connection in the cube lattice. In the initial implementation, this model uses the ratio between the parent cell's metrics and the child cell's metrics. As long as we restrict the aggregation functions to summation over non-negative values, we know that the parent's metric is always larger or equal to the child's metric. This leads to a ratio between 1 and 0, that we then model using a univariate Gaussian.

2.2 Online Operation

During online operation, we collect data for each time interval (again of size ΔT) and build an iceberg cube as soon as the interval is complete. Here, the iceberg condition is a cell with a minimal count of T_i underlying base records. For each cell that appears in the iceberg cube, a model is looked up. If one is found, we calculate an outlier score and store it to an *outlier cube*. Also for all upper level cells in the cube lattice, the scores are calculated and stored.

There will be lots of cases where no fitting model is available. This happens either when the training data did not contain the test cell, or when the training data contained fewer than T_m time intervals with data for the test cell. In this case, the cell model was pruned from the model. So we cannot compute any score for this cell. However, due to the collection of scores from related cells, we are still able to compute a score for each unit of interest.

The result of the first step is an outlier cube, that contains anomaly score facts corresponding to the facts in the base cube. In the second step, we relate the anomaly scores to the units of inspection. In the case of network data, the unit of inspection is a connection, corresponding to a base cell of the cube. However, this does not need to be the case.

For each cell that corresponds to a unit of inspection we collect all related scores. These are all scores from either ancestor cells or descendent cells of the inspected cell in the outlier score cube. We store all scores in a single record that integrates all evidence that is related to the inspected unit. In this record, we have scores related to the correlation models, and there is one score per metric, both for the basic models and the correlation models.

This results in a large, however fixed number of scores for each unit. The number is fixed as it only depends on the cube model and not on the current data instances. For our sample cube, there are more than 1000 individual scores per unit. Each additional dimension would multiply this number. However, due to iceberg conditions, not every score might be present for every unit.

We have multiple options to structure this record. In order to lose as few information as possible, we could reserve one column for each potential ancestor or descendent, resulting in high-dimensional records. To make the dimensionality smaller, we map outlier scores to the distance in the cube lattice. By this, we mean the length of the shortest path from the inspected cell to the outlier score cell. Thus the cell itself has distance zero, the parent cells have distance one, direct children have distance -1, and so on. In the case of our example network cube, we have 11 different distances.

In case a unit lives for more than one time unit, we have multiple such records. For the final result, we build both the average score per distance over all time units and the maximum score over all time units. So the final score record for each unit looks as follows. By mind and maxd, we mean the minimum and maximum distance of any cell in the cube to the units of interest. These values only depend on the cube structure and are thus fixed for a given cube. In our example, mind is 0 and maxd is 10.

$$(\text{avgscr}_{\text{mind}}, \ldots, \text{avgscr}_{\text{maxd}}, \text{maxscr}_{\text{mind}}, \ldots, \text{maxscr}_{\text{maxd}})$$

In the final step, all scores contained in the records that were build in the last step must be integrated into a single score that is used to decide whether the specific unit is regarded to be an outlier or not. Each individual attribute of the record describes some evidence. When there is a huge score in the cell itself (distance 0), this is a clear indicator for an outlier. However, not all outliers will be visible at this layer. Some will only be visible when looking at more distant cells. In a denial of service attack, for example, the individual connection itself does not look suspicious, while the aggregated number of connections from various sources to a single target might generate an outlier score. In the other extreme, an outlier score at the apex cell is very unspecific, so it should not make every cell at the same time an outlier. So we propose to weight the scores using their distance to the relevant cell:

$$\text{score} = \sum_{d=\text{mind}}^{\text{maxd}} \frac{1}{|d| + 1}(\text{avgscr}_d + \text{maxscr}_d)$$

3   Evaluation

In this section, we evaluate the method using data from the network domain. We use the well-known dataset from the 1998 DARPA IDS evaluation.[1] The goal of the evaluation is to gain insights in the quality of the solution and to explore the effects of different parameters with respect to the detection capability. Here, the DARPA data is very useful as we have ground truth available.

---

[1] See https://www.ll.mit.edu/ideval/data/1998data.html.

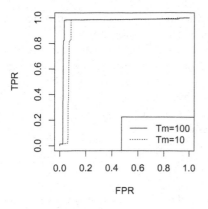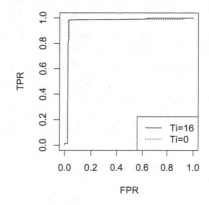

Fig. 1. Evaluation results.

First, we preprocess the training data by removing all attacks. Thus we start with training data that only contains normal data, while the test data still contains attacks *and* normal connections. We model a cube from the data including different metrics like packet count, data volume and specific metrics e.g. for capturing specific flags. The system also adds by default another fact that counts the number of base cells below the current cell in the lattice that have at least one record. For the dimensions, we first use the IP source and destination address. We build a hierarchy for these dimensions based on the /24, /16, and /8 address prefixes. IP communication will have two additional dimensions that indicate the layer 4 and layer 7 protocol fields. In layer 4, this field differentiates between UDP, TCP, ICMP, or other protocols. In layer 7, the initial target port (well known port) for a UDP or TCP connection will be used. For ICMP, the command type is used. These two protocol attributes build a protocol hierarchy.

The DARPA data contains different types of attacks. We will focus on DoS attacks, as they result in typical outlier patterns in the data that we aim to detect with our method. The main parameters are the threshold used to build the models (T_m), and the threshold used to build the current time frame's iceberg cube (T_i). First, we compare the results for model thresholds $T_m = 10$ and $T_m = 100$; see the left side of Fig. 1. The detection is much better at a larger model threshold value. This is an indicator that the models build with only a few sample time frames in the training data are too volatile. They seem to introduce too specific normal behavior so that later on different behavior is flagged wrongly as anomalous behavior. For both curves, we set $T_i = 0$.

Now we look at what happens when we restrict our attention to cells that fulfill a certain iceberg condition. In the right graph of Fig. 1, the detection capability for DoS attacks is shown with an iceberg threshold of $T_i = 0$ to $T_i = 16$ during online operation, meaning we exclude cells with fewer than 16 packets for a given time frame. Here, the detection capability is nearly unchanged. This is quite intuitive, as outliers consist of a larger data volume and thus are typically part of the iceberg cube. This means that restricting the computation to iceberg cubes is a good way to improve performance without losing quality.

Finally, we provide *preliminary* performance insights for the online phase. We processed two weeks of network data. The overall throughput is averaged to 30 MBit/sec. The results where measured on a commodity PC with Intel i7 processor. However, this includes heavy I/O overhead as we are currently logging many intermediate results. Thus we think it is possible to analyze data streams with much higher data rates in an online fashion.

4   Related Work

In this section, we compare our work to similar work in the area of outlier detection (see e.g. [1,2]). In [5], Le and Han describe the SUITS algorithm to find anomalies in OLAP cubes viewed as multi-dimensional time-series data. However, they focus on unsupervised outlier detection. Their notion of outliers assumes that time series from child cells normally have the same shape compared to the parent cell at another scale. Any cell that deviates from this assumption is an outlier. This is fundamentally different to our approach of taking past data as a normal reference and is only useful for offline processing. For an alternative approach to unsupervised outlier detection in OLAP cubes, see [6]. A similar outlier definition based on unsupervised analysis of an OLAP cube is given by Sarawagi et al. [11]. For each cell, the outlier score stems from the relative deviation to an "expected value", which is computed from related cells in various ways. They do not have a special processing for the time dimension. The outlier score is used in an interactive environment to guide the user to interesting cells of the cube. Also Dunstan et al. [3] describe the idea to present the outliers in reports using various groupings. This resembles our idea of collecting evidence from different related cells. Palpanas et al. [9] give a method to reconstruct a cube's base data based on marginal distributions (i.e. aggregate values) using maximum entropy estimation. By comparing the reconstructed values and the actual values, outliers can be identified.

There is also work that looks at OLAP cubes in data streams. In [4], stream cubes are defined. However, in order to reduce the number of cells, the authors use a minimal interest layer. This contradicts our idea to look also at very detailed levels when there is enough data present. Restricting the cube to a minimal interest layer might lead to overlooking important hints for outliers. In Rettig et al. [10], an infrastructure is proposed to find outliers in streaming data from the telco domain in real time. There are some similar ideas, as e.g. the cell[2] usage data has multidimensional structure including a region and an event type dimension. However, the aggregation is specific to the application example in contrast to our generic method. Furthermore, data is compared to older time steps for outlier detection and not to any model.

In [12], the authors follow an approach using OLAP cubes to detect outliers in wireless networks. Apart from a different outlier score definition,

---

[2] A mobile network cell, not a cube cell.

the main difference to our work is that they use domain expertise to prede-
fine those cuboids that are explored in search for outliers. Our method explores
all cuboids dynamically that contain enough data (iceberg).

There is a large body of work about outlier detection in high dimensional
data, see e.g. Chap. 5 of [1]. However, as pointed out in [5], the data model of the
OLAP approach is different. Outliers are defined in terms of distance to other
data points in the high-dimensional space, and an important approach to finding
outliers is to look at the data using different projections. This is in contrast to our
method. We look at each metric (i.e. feature) individually, which means that we
use one-dimensional projections in the feature space. However, we use different
subsets of the data, with aggregated values of the individual metric values based
on the OLAP dimensions. The only similarity is on a very high level: to find
outliers, it is beneficial to look at the data from multiple different perspectives.
Müller et al. [8], as an example for this line of work, explore subspace clusters
and collect outlier indicators from these subspaces. However, a single outlier is
always a single data row, no collective outliers can be found with this method.

5   Conclusion

This paper presents a new approach to outlier detection in a stream of multidimensional data records. The basic idea is to calculate models specific to various
groupings of the data, using an offline training phase and normal-only training
data. This makes the approach a one-class anomaly detection method. In the
online detection phase, a cube is built from the data for a series of small time
intervals. The models are used to calculate multiple outlier scores at different
cells of the cube.

The outlier scores in these cubes are then related to interesting real-world
entities in order to collect evidence for the outlierness of these entities. If entities
live for multiple time steps, further evidence is collected from subsequent cubes.
The collected scores are condensed into a single outlier score for each entity.

To make the computation efficient and to avoid volatile models, only cube
cells with enough training data will be included in the cube model. During online
operation, a cube cell without a model is ignored. However, due to evidence from
other cells, outliers can still be found. Furthermore, test data cube cells with too
few data can also be ignored, using an iceberg cube approach.

An evaluation based on an network data scenario shows that the method is
able to distinguish between normal data and attacks very well. The evaluation
also shows the effects of different parameter settings, with the conclusion that
using appropriate thresholds for the model is indeed important to avoid spurious
models. Restricting the attention to the iceberg cube during processing does not
lower the detection quality.

References

1. Aggarwal, C.C.: Outlier Analysis, 1st edn. Springer, New York (2013). doi:10.1007/978-1-4614-6396-2
2. Chandola, V., Banerjee, A., Kumar, V.: Anomaly detection: a survey. ACM Comput. Surv. **41**(3), 15:1–15:58 (2009). http://doi.acm.org/10.1145/1541880.1541882
3. Dunstan, N., Despi, I., Watson, C.: Anomalies in multidimensional contexts. WIT Transa. Inform. Commun. Technol. **42**, 173 (2009). http://www.witpress.com/elibrary/wit-transactions-on-information-and-communication-technologies/42/19978
4. Han, J., Chen, Y., Dong, G., Pei, J., Wah, B.W., Wang, J., Cai, Y.D.: Stream cube: an architecture for multi-dimensional analysis of data streams. Distrib. Parallel Databases **18**(2), 173–197 (2005). http://link.springer.com/article/10.1007/s10619-005-3296-1
5. Li, X., Han, J.: Mining approximate top-k subspace anomalies in multi-dimensional time-series data. In: Proceedings of the 33rd International Conference on Very Large Data Bases, pp. 447–458. VLDB Endowment (2007)
6. Lin, S., Brown, D.E.: Outlier-based Data Association: Combining OLAP and Data Mining. Department of Systems and Information Engineering University of Virginia, Charlottesville, VA 22904 (2002). http://web.sys.virginia.edu/files/tech_papers/2002/sie-020011.pdf
7. Lippmann, R., Fried, D., Graf, I., Haines, J., Kendall, K., McClung, D., Weber, D., Webster, S., Wyschogrod, D., Cunningham, R., Zissman, M.: Evaluating intrusion detection systems: the 1998 DARPA off-line intrusion detection evaluation. In: DARPA Information Survivability Conference and Exposition, DISCEX 2000, Proceedings, vol. 2, pp. 12–26 (2000)
8. Müller, E., Assent, I., Iglesias, P., Mülle, Y., Böhm, K.: Outlier ranking via subspace analysis in multiple views of the data. In: 2012 IEEE 12th International Conference on Data Mining, pp. 529–538. IEEE (2012). http://ieeexplore.ieee.org/xpls/abs_all.jsp?arnumber=6413873
9. Palpanas, T., Koudas, N., Mendelzon, A.: Using datacube aggregates for approximate querying and deviation detection. IEEE Trans. Knowl. Data Eng. **17**(11), 1465–1477 (2005)
10. Rettig, L., Khayati, M., Cudré-Mauroux, P., Piórkowski, M.: Online anomaly detection over Big Data streams. In: 2015 IEEE International Conference on Big Data (Big Data), pp. 1113–1122 (2015)
11. Sarawagi, S., Agrawal, R., Megiddo, N.: Discovery-driven exploration of OLAP data cubes. In: Schek, H.-J., Alonso, G., Saltor, F., Ramos, I. (eds.) EDBT 1998. LNCS, vol. 1377, pp. 168–182. Springer, Heidelberg (1998). doi:10.1007/BFb0100984
12. Sithirasenan, E., Muthukkumarasamy, V.: Substantiating anomalies in wireless networks using group outlier scores. J. Softw. **6**(4), 678–689 (2011)
13. Thatte, G., Mitra, U., Heidemann, J.: Parametric methods for anomaly detection in aggregate traffic. IEEE/ACM Trans. Networking **19**(2), 512–525 (2011)
14. Xin, D., Han, J., Li, X., Wah, B.W.: Star-cubing: Computing iceberg cubes by top-down and bottom-up integration. In: Proceedings of the 29th International Conference on Very Large Data Bases, vol. 29, pp. 476–487. VLDB Endowment (2003)

Balancing Performance and Energy
for Lightweight Data Compression Algorithms

Annett Ungethüm$^{(\boxtimes)}$, Patrick Damme, Johannes Pietrzyk, Alexander Krause,
Dirk Habich, and Wolfgang Lehner

Database Systems Group, Technische Universität Dresden, Dresden, Germany
{annett.ungethuem,patrick.damme,johannes.pietrzyk,alexander.krause,
dirk.habich,wolfgang.lehner}@tu-dresden.de

Abstract. Energy consumption becomes more and more a critical
design factor, whereby performance is still an important requirement.
Thus, a balance between performance and energy has to be established.
To tackle that issue for database systems, we proposed the concept of
work-energy profiles. However, generating such profiles requires extensive benchmarking. To overcome that, we propose to approximate work-
energy-profiles for complex operations based on the profiles of low-level
operations in this paper. To show the feasibility of our approach, we use
lightweight data compression algorithms as complex operations, since
compression as well as decompression are heavily used in in-memory
database systems, where data is always managed in a compressed representation. Furthermore, we evaluate our approach on a concrete hardware
system.

Keywords: Energy efficiency · In-memory databases · Compression

1 Introduction

Database systems constantly adapt to new hardware features to satisfy performance demands [7,8,14]. However, these features do not only influence the performance, but also the energy consumption [3]. As energy consumption becomes
more and more a critical factor [5], a balance between performance and energy
has to be established [12]. To tackle this balancing challenge in a fine-grained way
for in-memory database systems, we proposed the concept of *work-energy profiles*
in a previous work [13]. A *work-energy profile* exposes a relation between reach-
able performances and the resulting energy-efficiency, thereby a profile covers a
wide range of CPU features, such as different vector extensions, multithreading,
CPU pinning, and frequency scaling [13]. Thus, these *work-energy profiles* can
be used to determine the most energy-efficient hardware configuration for any
required performance demand. However, generating *work-energy profiles* requires
extensive benchmarking and must be done once for every database operator and
every possible hardware configuration [12,13]. Therefore, the number of necessary benchmark tests can quickly add up to thousands or millions. Even if every

© Springer International Publishing AG 2017
M. Kirikova et al. (Eds.): ADBIS 2017, CCIS 767, pp. 37–44, 2017.
DOI: 10.1007/978-3-319-67162-8_5

test needs only a one second micro benchmark, a full benchmark for a database system on a specific hardware system would need hours or days.

Our Contributions and Outline. To overcome this benchmarking overhead, we present a novel approach to approximate work-energy profiles of *complex operations* from work-energy profiles of *low-level operations* in this paper. In detail, our contributions are: (i) We briefly summarize the core concept of our *work-energy profiles* in Sect. 2. In particular, we state which low-level profiles are necessary for our approximation approach. (ii) To illustrate our approximation approach, we use a lightweight data compression algorithm and we describe the concrete algorithm in Sect. 3. (iii) Then, our approximation approach for complex operations is introduced in Sect. 4. Thereby the approximation is based on linear combination of work-energy profiles of low-level operations. (iv) We evaluate our approach on a concrete hardware system to show the feasibility in Sect. 5. (v) We conclude the paper with related work and a summary in Sects. 6 and 7.

2   Work-Energy-Profiles

Modern hardware, especially CPUs, offers a lot of features like vectorization [3], multithreading, or frequency scaling [9]. These features usually have an influence on performance as well as energy consumption. However, the mapping between hardware configurations – meaning which features to which extend should be used in which way –, performance, and energy-efficiency is not trivial [13]. To capture these effects for all possible hardware configurations, we proposed the concept of *work-energy profiles* in [13].

A *work-energy profile* is a set of the useful work done during a fixed time span and the required energy for this work for all possible hardware configurations [13]. Thus, the *work-energy profile* for a specific application has to be benchmarked on a concrete hardware system [13]. Figure 1 shows three different example *work-energy profiles*, which we measured on a concrete hardware. While the performance is plotted on the x-axis, the y-axis shows the energy-efficiency. Each dot in this graph represents a specific hardware configuration. We measured the performance as *work done* per *second* and the energy-efficiency as *work done* per *Joule* (work energy quotient – WEQ). Hence, a work-energy-profile is a set of (performance,WEQ)-tuples, each of them representing one specific hardware configuration. As we can see in Fig. 1, different hardware configurations offer the same performance range with a high variance in the energy-efficiency. To balance performance and energy, the hardware configuration with the highest *energy-efficiency* within a desired performance range should be used for application execution. This hardware configuration can be extracted from our *work-energy profile*.

Generally, our main focus is on energy-efficient in-memory database systems [12]. Here, the performance and energy-efficiency of a hardware configuration depends on a multitude of factors (e.g., data characteristics and size, operator types, etc.). Moreover, main memory bandwidth and latency are limiting factors that could cause a non trivially predictable hardware behavior.

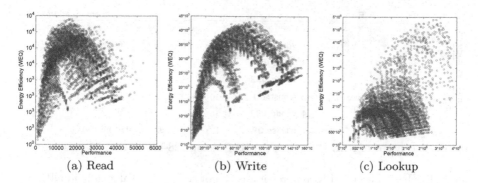

(a) Read   (b) Write   (c) Lookup

Fig. 1. ODROID-XU3-based work-energy profiles for different memory access patterns.

To get a deeper understanding and a specific foundation, we propose to benchmark *work-energy profiles* only for fine-grained memory access patterns – database primitives – which are highly utilized in in-memory database systems: (i) read-primitive, (ii) write-primitive, (iii) lookup-primitive, (iv) compute-primitive, and (v) processing-primitive, whereas the processing is a combination of read and compute, since for data processing a set of data has to be read first and then some kind of computation is triggered. The ratio between read and compute can vary, which we consider in our benchmark.

Figure 1 depicts the resulting *work-energy profiles* for these primitives on an ARM big.LITTLE hardware system. Concretely, we used an ODROID-XU3, which consists of a big and a little cluster, each of them featuring 4 cores. Additionally, the ODROID-XU3 is equipped with on-board power sensors allowing us to measure the power level of individual core clusters and the main memory separately. The different combinations of cores and their frequencies add up to roughly 6000 different configurations [13] and each dot in Fig. 1 represents a specific configuration. As we can see, the shapes of the *work-energy profiles* of our primitives are different.

3 Example Operation RLE

To describe our approximation approach, the physical operation must be precisely known. In this paper, we focus on run-length encoding (RLE), since RLE is a heavily applied compression technique in in-memory database systems [1,6]. RLE tackles uninterrupted sequences of occurrences of the same value, so called runs. In its compressed format, each run is represented by its value and length. Therefore, the compressed data is a sequence of such pairs as illustrated in Fig. 2(a). In addition to the simplicity of RLE, there are two other advantages: (i) RLE can be easily parallelized by data partitioning, so that the parallelization itself does not produce any mentionable communication overhead between the processing cores and (ii) RLE can be vectorized [2].

Our vectorized implementation consists of four steps as shown in Fig. 2(b) using the ARM NEON implementation of SIMD. In step one, four copies of

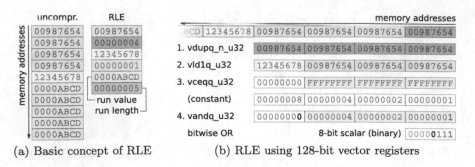

(a) Basic concept of RLE   (b) RLE using 128-bit vector registers

Fig. 2. The basic idea of RLE compression is to not store every value individually, but only once followed by the number of sequential elements with the same value.

the current input element are loaded into one 128-bit vector register using the vdupq_n_u32() NEON operation. In Step two, the next four input elements are loaded into a second 128-bit vector register using the vld1q_u32() intrinsic. In step three, these four values are compared in parallel with the current run value using vceqq_u32(). The result is stored in a third vector register. In each 32-bit element of this vector register, either all bits are set or all bits are not set, depending on whether the corresponding elements were equal or not. In step four, from this register we extract a bit mask by ANDing it with the constant vector $(1, 2, 4, 8)$ using vandq_u32(), storing the result to memory using vst1q_u32(), and sequentially ORing the lowest bytes of all four elements. The number of trailing ones in this bit mask tells us for how many elements the current run continues. We look up this number in a table created offline and indexed with the 16 possible masks. If the obtained number is four, then we have not seen the run's end yet, and continue at step 2. Otherwise, we have reached the run's end and append the run value and run length to the output and continue with step 1 at the next element after the run's end.

4 Work-Energy-Profile Approximation

Unfortunately, the benchmarking of a *work-energy profile* for a specific primitive on the ODROID-XU3 takes about eight hours. To do this for all physical database operators or even queries would be way too much overhead. However, the profiles are necessary to determine the most energy-efficient hardware configuration for a demanded performance. To tackle this challenge, our idea is to approximate the *work-energy profiles* of complex operations from the profiles of these low-level primitives. In this section, we want to demonstrate that our idea is feasible using RLE as an example.

4.1 RLE and Low-Level Operations

As just described, our vectorized RLE algorithm is fixed, but the input data determines the execution behavior, thereby two extremes arise. One extreme is obtained if there is no run in the input data at all (average run length equals 1). In this case, RLE-compressed data is twice as large as the original data, because for each array element, a run length of 1 is additionally stored. This means for the processing, that our vectorized compression algorithm performs essentially random reads and random writes with a ratio of 1:1. Random reads, since we always read 4 elements in each iteration, with 3 of them already being read in the previous iteration, producing overlapping reads. The second extreme occurs when each element in the uncompressed data array equals a single value (average run length equals number of elements). In this case, RLE-compressed data consists of two values, the single value and the number of elements as run length and these two values are written once by the compression algorithm. Thus, the read/write ratio approaches 1:0, while the read accesses are still random. Furthermore, between the reads and writes, there is also the actual compression which is a computation bound work. If this computation is slower than the I/O accesses, the memory access pattern is not the bottleneck for the performance anymore, but the computation itself.

Therefore, three low-level operations are used within our vectorized RLE compression as well as decompression algorithms: (i) read, (ii) write, and (iii) compute. Depending on the input data, these operations are composed differently. While the number of compute operations per read operation is constant, the ratio between read and write operations changes depending on the average run length. Hence, the profile for a specific average run length is within the spectrum between read-bound and write-bound operations.

4.2   Approximation Using Linear Combination

To approximate *work-energy profiles*, we propose to combine low-level primitives in a linear way. The new profile, containing (performance, WEQ)-tuples for i different configurations, is obtained from j low-level profiles and the tuples of the new profile are determined by

$$(Performance, WEQ)_{i,new} = f^{-1}(\sum_{0}^{j-1} w_j * f(Performance, WEQ)_{i,j}) \quad (1)$$

where w_j is a weighting factor which describes the influence of the profile j. The adjustment function f modifies the performance- and WEQ-values for the combination. This is necessary if the limiting factors, i.e. the low-level-profiles, do not scale linearly when the parameters of the operation are changed.

For RLE compression, the only available parameter is the average run length. This parameter defines the number of read and write accesses, which define the scaling of the *work-energy profile*, i.e. the adjustment function f. Whereas the ratio between the read and write accesses defines the weighting factors w_j. The number of read or write accesses as a function of the run length rl, can be extracted from the vectorized algorithm presented in Sect. 3. As shown in Eqs. 2 and 3, we denote to the number of reads and writes as $count_{reads}$ and $count_{writes}$ respectively. A sequence of run lengths is described as $RL = [rl_0, rl_1, rl_{|RL|-1}]$

with $count_{runs} = |RL|$ runs, and k is the vector width. For a constant run length, $count_{runs}$ can be computed by $count_{elements}/rl$.

$$count_{reads} = \sum_{i=0}^{|RL|-1} (2 + \lfloor \frac{rl_i - 1}{k} \rfloor) \tag{2}$$

$$count_{writes} = 2 * count_{runs} \tag{3}$$

Both functions are rational. Hence, our function f must be rational, too. Since there are no polynoms or exponential parts in either $count_{reads}$ or $count_{writes}$, it is safe to use the most simple rational function $f(Performance, WEQ) = (1/Performance, 1/WEQ)$ for Eq. 1.

The ratio between the read and write operations, and therefore the weighting factors w_j, follow from the Eqs. 2 and 3 as well:

$$w_0 : w_1 = 2 * count_{runs} : \sum_{i=0}^{|RL|-1} (2 + \lfloor \frac{rl_i - 1}{k} \rfloor).$$

5   Evaluation

To validate the results of our profile approximation approach, we also benchmarked the profiles on the ODROID-XU3. For comparing the quality of an approximated profile to a benchmarked profile, we filtered the configurations which are part of the Pareto front of the measured and the approximated profile. Then, we divided the profile into ten performance ranges and compared the approximated and measured configurations which are the closest to the middle of these performance ranges. As a result, there are two measures to quantify the quality of the approximated profile: (1) the chance that the approximated configuration actually is in the performance range, in which we expect it to be, and (2) the mean deviation of the WEQ from the measured configuration in the middle of this performance range.

Figure 3 shows the results for our RLE compression algorithm applied on synthetic data containing values with an average run length of 45. The figure shows that the match of the approximation and the benchmarked profile is not exact but close to the optimal solution. To quantify the difference, we calculated the two measures as described above: (1) The chance that a configuration, which we expect to be close to the middle of a specified performance range, is actually in this performance range, is 100% in our example. This means, that all ten configurations, we filtered for the ten performance ranges were actually within a 10% radius of this performance range. (2) The mean deviation of the WEQ from the measured configuration was only 3%.

We conducted the evaluation on different data sets as well as on different hardware systems. In almost all cases, we observed a similar behavior, so that we are able to conclude that our approximation approach is well-suited for the considered operations of RLE compression and decompression.

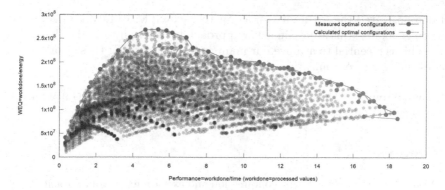

Fig. 3. A benchmarked work-energy profile for RLE compression on the ODROID-XU3 with average run length of 45. The approximated optimal configurations are highlighted in orange, the actual optimum is highlighted in green. (Color Figure Online)

6   Related Work

Lossless compression techniques play an important role in in-memory database systems [1,6]. They reduce not only the amount of needed space, but also the time spent on i/o instructions. Thus, they have already been investigated for query performance [1]. Further works on vectorized compression techniques also focus on performance [10], but not on energy efficiency. An extensive experimental evaluation on lightweight compression techniques has been done by Damme et al. [2]. However, to the best of our knowledge, none of these works explicitly regards energy efficiency. Vice versa, the works regarding energy-efficiency (1) focus on query execution, usually evaluated by running a TPC benchmark, rather than on compression itself [11], and (2) treat the performance-energy-tradeoff as a binary decision between energy-efficiency and performance [15]. In general, there are analytical models to estimate and reduce the energy consumption [15] and benchmark-based approaches [4]. The latter has only been applied for homogeneous systems and analytical models become more complex the more complex the hardware becomes. Regarding the ever-growing world of heterogeneous hardware, we developed an approach which is mostly benchmark based and applicable to various systems.

7   Conclusion and Outlook

As energy consumption becomes more and more a critical design factor, a balance between performance and energy has to be established. To tackle this challenge for in-memory database systems, we proposed the concept of work-energy profiles in [13]. A *work-energy profile* is a set of the useful work done during a fixed time span and the required energy for this work for all possible hardware configurations [13]. In this paper, we proposed to approximate work-energy-profiles for complex operations based on the work-energy profiles of low-level operations and demonstrated the feasibility on a concrete example. In future work, we are going to generalize this approximation approach to other physical database operators. This is essential to achieve our main goal of integrating these work-energy profiles into query optimization.

Acknowledgments. This work is partly funded within the DFG-CRC 912 (HAEC) and by the DFG-project LE-1416/26.

References

1. Abadi, D.J., et al.: Integrating compression and execution in column-oriented database systems. In: SIGMOD (2006)
2. Damme, P., et al.: Lightweight data compression algorithms: an experimental survey (experiments and analyses). In: EDBT (2017)
3. Firasta, N., et al.: Intel AVX: new frontiers in performance improvements and energy efficiency. Intel White Paper (2008)
4. Götz, S., et al.: Energy-efficient databases using sweet spot frequencies. In: UCC 2014 (2014)
5. Harizopoulos, S., et al.: Energy efficiency: the new holy grail of data management systems research. In: CIDR (2009)
6. Hildebrandt, J., Habich, D., Damme, P., Lehner, W.: Compression-aware in-memory query processing: vision, system design and beyond. In: Blanas, S., Bordawekar, R., Lahiri, T., Levandoski, J., Pavlo, A. (eds.) IMDM/ADMS - 2016. LNCS, vol. 10195, pp. 40–56. Springer, Cham (2017). doi:10.1007/978-3-319-56111-0_3
7. Karnagel, T., et al.: Adaptive work placement for query processing on heterogeneous computing resources. PVLDB **10**(7), 733–744 (2017)
8. Kissinger, T., et al.: ERIS: a numa-aware in-memory storage engine for analytical workload. In: ADMS@VLDB, pp. 74–85 (2014)
9. Le Sueur, E., et al.: Dynamic voltage and frequency scaling: the laws of diminishing returns. In: Proceedings of the 2010 International Conference on Power Aware Computing and Systems, pp. 1–8 (2010)
10. Lemire, D., Boytsov, L.: Decoding billions of integers per second through vectorization. Softw. Pract. Exper. **45**(1) (2015)
11. Mühlbauer, T., Rödiger, W., Seilbeck, R., Kemper, A., Neumann, T.: Heterogeneity-conscious parallel query execution: getting a better mileage while driving faster! In: DaMoN@SIGMOD (2014)
12. Ungethüm, A., et al.: Energy elasticity on heterogeneous hardware using adaptive resource reconfiguration LIVE. In: SIGMOD, pp. 2173–2176 (2016)
13. Ungethüm, A., Kissinger, T., Habich, D., Lehner, W.: Work-energy profiles: general approach and in-memory database application. In: Nambiar, R., Poess, M. (eds.) TPCTC 2016. LNCS, vol. 10080, pp. 142–158. Springer, Cham (2017). doi:10.1007/978-3-319-54334-5_10
14. Willhalm, T., et al.: Simd-scan: ultra fast in-memory table scan using on-chip vector processing units. PVLDB **2**(1), 385–394 (2009)
15. Xu, Z., et al.: Dynamic energy estimation of query plans in database systems. In: 2013 IEEE 33rd International Conference on Distributed Computing Systems (ICDCS), pp. 83–92. IEEE (2013)

Asynchronous Graph Pattern Matching on Multiprocessor Systems

Alexander Krause$^{(\boxtimes)}$, Annett Ungethüm, Thomas Kissinger, Dirk Habich,
and Wolfgang Lehner

Database Systems Group, Technische Universität Dresden, Dresden, Germany
{Alexander.Krause,Annett.Ungethuem,Thomas.Kissinger,Dirk.Habich,
Wolfgang.Lehner}@tu-dresden.de

Abstract. Pattern matching on large graphs is the foundation for a variety of application domains. Strict latency requirements and continuously increasing graph sizes demand the usage of highly parallel in-memory graph processing engines that need to consider non-uniform memory access (NUMA) and concurrency issues to scale up on modern multiprocessor systems. To tackle these aspects, graph partitioning becomes increasingly important. Hence, we present a technique to process graph pattern matching on NUMA systems in this paper. As a scalable pattern matching processing infrastructure, we leverage a data-oriented architecture that preserves data locality and minimizes concurrency-related bottlenecks on NUMA systems. We show in detail, how graph pattern matching can be asynchronously processed on a multiprocessor system.

1 Introduction

Recognizing comprehensive patterns on large graph-structured data is a prerequisite for a variety of application domains such as biomolecular engineering [11], scientific computing [17], or social network analytics [12]. Due to the ever-growing size and complexity of the patterns and underlying graphs, *pattern matching* algorithms need to leverage an increasing amount of available compute resources in parallel to deliver results with an acceptable latency. Since modern hardware systems feature main memory capacities of several terabytes, state-of-the-art graph processing systems (e.g., Ligra [16] or Galois [10]) store and process graphs entirely in main memory, which significantly improves scalability, because hardware threads are not limited by disk accesses anymore. To reach such high memory capacities and to provide enough bandwidth for the compute cores, modern servers contain an increasing number of memory domains resulting in a *non-uniform memory access (NUMA)*. To further scale up on those NUMA systems, pattern matching on graphs needs to carefully consider issues such as the increased latency and the decreased bandwidth when accessing remote memory domains, as well as the limited scalability of synchronization primitives such as atomic instructions [21].

The widely employed *bulk synchronous parallel (BSP)* processing model [19], which is often used for graph processing, does not naturally align with pattern

© Springer International Publishing AG 2017
M. Kirikova et al. (Eds.): ADBIS 2017, CCIS 767, pp. 45–53, 2017.
DOI: 10.1007/978-3-319-67162-8_6

matching algorithms [3]. That is because a high number of intermediate results is generated and need to materialized and transferred within the communication phase. Therefore we argue for an asynchronous processing model that neither requires a full materialization nor limits the communication to a distinct global phase. For efficient *pattern matching* on a single NUMA system, we employ a fine-grained *data-oriented architecture (DORA)* in this paper, which turned out to exhibit a superior scalability behavior on large-scale NUMA systems as shown by Pandis et al. [13] and Kissinger et al. [6]. This architecture is characterized by implicitly partitioning data into small partitions that are explicitly pinned to a NUMA node to preserve a local memory access.

Contributions. Following to a discussion of the foundations of graph pattern matching in Sect. 2, the contributions of the paper are as follows:

(1) We adapt the data-oriented architecture for scale-up graph pattern matching and identify the *partitioning strategy* as well as the design of the *routing table* as the most crucial components within such an infrastructure (Sect. 3).
(2) We describe an asynchronous query processing model for graph pattern matching and present the individual operators a query is composed of. Based on the operator characteristics, we identify *redundancy* in terms of partitioning as an additional critical issue for our approach (Sect. 4).
(3) We thoroughly evaluate our graph pattern matching approach on multiple graph datasets and queries with regard to scalability on NUMA systems. Within our evaluation, we focus on different options for the partitioning strategy, routing table, and redundancy as our key challenges (Sect. 5).

Finally, we discuss the related work in Sect. 6 and conclude the paper in Sect. 7 including promising directions for future work.

2   Foundations of Graph Pattern Matching

Within this paper, we focus on pattern matching for *edge-labeled multigraphs* as a general and widely employed graph data model [12,14]. An edge-labeled multigraph $G(V, E, \rho, \Sigma, \lambda)$ consists of a set of vertices V, a set of edges E, an incidence function $\rho : E \to V \times V$, and a labeling function $\lambda : E \to \Sigma$ that assigns a label to each edge, according to which edge-labeled multigraphs allow any number of labeled edges between a pair of vertices. A prominent example for edge-labeled multigraphs is RDF [2].

Pattern matching is a declarative topology-based querying mechanism where the query is given as a graph-shaped pattern and the result is a set of matching subgraphs [18]. For instance, the *query pattern* depicted in Fig. 1 searches for a vertex V_1, that has two outgoing edges targeting V_2 and V_3. Additionally, the query pattern seeks a fourth vertex V_4 which also has two outgoing edges to the same target vertices. The query pattern forms a rectangle with four vertices and four edges of which we search for all matching subgraphs in a graph. A well-studied mechanism for expressing such query patterns are *conjunctive queries*

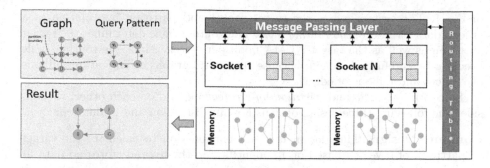

Fig. 1. Scalable graph pattern matching based on a data-oriented architecture [6,13].

(CQ) [20], which decompose the pattern into a set of *edge predicates* each consisting of a pair of vertices and an edge label. Assuming a wildcard label, the exemplary query pattern in Fig. 1 can be decomposed into the conjunctive query $\{(V_1, *, V_2), (V_1, *, V_3), (V_4, *, V_3), (V_4, *, V_2)\}$, where the bold vertices represent the source vertex of an edge. These four edge predicate requests form a sequence, that is processed by starting at each vertex in the data graph, because the query pattern does not specify a specific starting vertex.

3   Scalable Graph Pattern Matching Architecture

In this section, we briefly describe our target architecture and refer to the extended version of our paper [7] for more details. Figure 1 illustrates a NUMA system with N sockets which can run multiple worker threads concurrently, based on the underlying hardware. A graph of the form described in Sect. 2 can be distributed among the main memory regions, which are attached to one of the sockets. The distribution of the graph among these memory regions inherently demands graph partitioning and an appropriate partitioning strategy.

Partitioning Strategy. However, partitioning a graph will most likely lead to edges, which span over multiple partitions, like the edges $A \rightarrow B$ and $D \rightarrow B$ on the left hand side of Fig. 1. For instance, if vertex A is considered as a potential match for a query pattern, the system needs to lookup vertex B in another partition. Moving to another partition requires that the complete matching state needs to be transferred to another worker, which requires communicational efforts between the two responsible workers. Hence, the selection of the *partitioning strategy* is crucial when adapting the data-oriented architecture for graph pattern matching, because locality in terms of the graph topology is important [8].

Routing Table. Because one partition can not always contain all the necessary information for one query, it is inevitable to communicate intermediate results between workers. The communication is handled by a high-throughput message passing layer, which hides the latency of the communication network, as depicted in Fig. 1. The system stores the target socket and partition information in a crucial data structure, the *routing table*. The routing table determines the target partition as well as the target NUMA node per vertex. Thus, the routing table needs to be carefully designed, because real world graphs often feature millions of vertices and billions of edges.

Since *routing table* and *partitioning strategy* depend on each other, we consider the following three design options for our discussion and evaluation:

Compute Design. This design uses a hash function to calculate the target partition of a vertex based on its identifier on-the-fly and stores no data at all. Nevertheless, due to the simplicity of the routing table, the partitioning strategy can not take any topology-based locality information into account.

Lookup Design. The lookup design consists of a hash map, which stores a precomputed graph partitioning, i.e. one partition entry per vertex in the graph, thus this design doubles the memory footprint, since the graph is stored once as graph data and once in the routing table as topology data. As partitioning strategy, we use the well known *multilevel k-Way* partitioning to create a disjoint set of partitions. This heuristical approach creates partitions with high locality and tries to minimize the edge cut of the partitioning [5].

Hybrid Design. We created this design to combine the advantages of the two previous approaches, i.e. a small and locality preserving routing table. To enable this combination, we employ a dictionary as auxiliary data structure that maps virtual vertex ids to the original vertex ids of the locality aware graph partitioning. The dictionary is only used for converting the vertex ids of final query results. This range-based routing table maps dense ranges of virtual ids to the respective partition and has very low memory footprint such that the routing table easily fits into the cache of the multiprocessors.

4   Graph Pattern Matching Processing Model

The architecture introduced in Sect. 3 needs specific operators for pattern matching on NUMA systems. We identified three logical operators, which are necessary to model *conjunctive queries* as described in Sect. 2:

Unbound Operator. The unbound operator performs a parallel vertex scan over all partitions and returns edges matching the specified label. The unbound operator is always the first operator in the pattern matching process.

Vertex-Bound Operator. The vertex-bound operator takes an intermediate matching result as input and tries to match a new vertex in the query pattern.

Edge-Bound Operator. The edge-bound operator ensures the existence of additional edge predicates between vertices which are matching candidates for certain vertices of the query pattern. It performs a data lookup with a given source and target vertex as well as a given edge label. If the lookup fails, both vertices are eliminated from the matching candidates. Otherwise the matching state is passed to the next operator or is returned as final result.

To actually compose a *query execution plan (QEP)*, the query compiler sequentially iterates over the *edge predicates* of the *conjunctive query*. For each edge predicate, the query compiler determines whether source and/or target vertex are bound and selects the appropriate operator for the respective edge predicate. For the example query pattern in Fig. 1, the resulting operator assignments of the QEP are shown in Fig. 2(c).

Each operator is asynchronously processed in parallel and generates new messages that invoke the next operator in the QEP. Hence, different worker threads can process different operators of the same query at the same point in time. Based on the operator and its parametrization, we distinguish two ways of addressing a message that are related to the *routing table*:

Unicast. A unicast addresses a single graph partition and requires that the source vertex is known respectively bound by the operator. This case occurs for the *vertex-bound operator* if the source vertex is bound and for the *edge-bound* operator.

Broadcast. A broadcast targets all partitions of a graph, which increases the pressure on the message passing layer and requires the message to be processed on all graph partitions and thus, negatively affects the scalability. Additionally, *vertex-bound operators* that bound the target vertex require a broadcast.

Broadcasts generated by *vertex-bound operators* significantly hurt the scalability of our approach. The cause of this problem is inherently given by the data-oriented architecture, because a graph can either be partitioned by the source or the target vertex of the edges. Hence, we identify *redundancy* in terms of partitioning as an additional challenge for our approach. To reduce the need for broadcasts to the initial *unbound operator*, we need to redundantly store the graph partitioned by source vertex and partitioned by target vertex. However, the need for redundancy depends on the query pattern as well as on the graph itself as we will show within our evaluation.

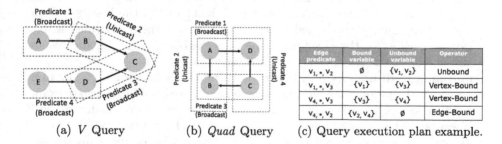

Edge predicate	Bound variable	Unbound variable	Operator
$v_{1,*,} v_2$	$\emptyset$	$\{v_1, v_2\}$	Unbound
$v_{1,*,} v_3$	$\{v_1\}$	$\{v_3\}$	Vertex-Bound
$v_{4,*,} v_3$	$\{v_3\}$	$\{v_4\}$	Vertex-Bound
$v_{4,*,} v_2$	$\{v_2, v_4\}$	$\emptyset$	Edge-Bound

(a) *V* Query   (b) *Quad* Query   (c) Query execution plan example.

Fig. 2. Query patterns and example operators for the query from Fig. 1.

5   Evaluation

In this section, we briefly describe our findings and refer to our extended version for an in depth explanation of the individual results [7]. We used a bibliographical like graph, which we call *biblio* for the remainder of this paper. The *biblio* graph was generated using the graph benchmark framework gMark [1] and has 546 k vertices, 780 k edges and an average out degree of 2.85 per vertex. Our NUMA system consists of four sockets equipped with an Intel Xeon E7-4830 CPU and a total of 128 GB of main memory. We defined two queries which are shaped like shown in Figs. 2(a) and (b) and ran them on the *biblio* graph.

Routing Table and Partitioning Strategy. Based on Fig. 3 we examine the infleunce of the routing table on the query performance. The figure shows the query runtime for the V query on the *biblio* graph, which we scaled up from factor 1 to factor 32. On the left hand side, we show the sole influence of the routing table and on the right hand side of the figure we show the query runtime per routing table design, if redundancy is used. In Fig. 3(a) we can see that our hybrid design marginally outperforms the memory intensive lookup design with a k-Way partitioning. The compute design and the lookup design which uses a hash function perform equally in terms of query performance. The advantage of our hybrid design and the k-Way based lookup design stems from the better graph partitioning algorithm, because neighborhood locality of adjacent vertices is considered. Our experiments showed, that the compute design results in the lowest time spent in the routing table per worker, which is not surprising. However, our hybrid design almost reaches the same routing table time due its the small size.

Avoiding Broadcasts with Redundancy. In Sect. 4 we mentioned that broadcasts hurt the scalability of a system. This issue is depicted in Fig. 4. The figure shows the scalability of our systems for both query types from Fig. 2. On the right hand side, we see that the *Quad* query suffers more from broadcasts. The reason is, that many tuples are matched for predicate 2 (c.f. Fig. 2(b)), which leads to a high number of broadcasts during the evaluation of predicate 3. For the V query on the left hand side of the Figure, we can see that the employment of redundancy still decreases the query runtime, but not as much as for the *Quad* query, because the broadcasting predicates are not dominant for this specific query instance.

Combining Redundancy and Routing Table Optimizations. Aside from testing both optimization techniques individually, we combined them to examine their synergy. In Fig. 3(b), we demonstrate the query performance of the V query on the *biblio* graph, again scaled up to factor 32. By adding redundancy to the query execution, all routing table designs greatly benefit in terms of query performance. However, we can now see a bigger advantage of our hybrid design, compared to the lookup k-Way design.

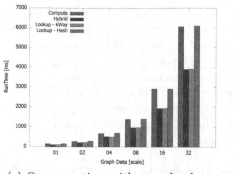
(a) Query runtime without redundancy

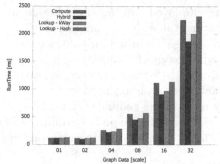
(b) Query runtime with redundancy

Fig. 3. *V* query on the *biblio* graph using different scale factors.

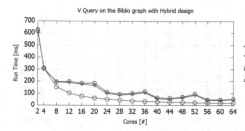

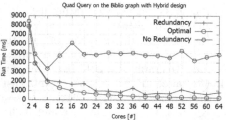

Fig. 4. Impact of redundancy, both queries on the *biblio* graph.

6 Related Work

Graph analytics is a widely studied field, as the survey from McCune et al. [9] shows. Many systems leverage the increased compute performance of scale-up or scale-out systems to compute graph metrics like PageRank and the counting of triangles [15], Single Source Shortest Path [15] or Connected Components [15]. Many of the available systems are inspired by the Bulk Synchrones Processing Model [19], which features local processing phases which are synchronized by a global superstep. A general implementation is the Gather-Apply-Scatter paradigm, as described in [4]. Despite working on NUMA systems, these processing engines are globally synchronized and lack the scalability of a lock-free architecture. We improve this issue by leveraging a high throughput message passing layer for asynchronous communication between the worker threads. However, in contrast to the systems mentioned above, we are calculating graph pattern matching and not graph metrics, like for instance GraphLab which is the only asynchronous graph processing engine according to [9].

7 Conclusions

In this paper, we showed that the performance of graph pattern matching on a multiprocessor system is determined by the communication behavior, the employed routing table design and the partitioning strategy. Our *Hybrid* routing table design implementation allows the system to leverage both the advantages from a *Compute* design and a *Lookup* design. Because of an intermediate step, the underlying graph partitioning algorithm is interchangable and can thus be adapted to specific partitioning requirements. Furthermore we could show that avoiding broadcasts is equally important. This issue was mitigated by introducing redundancy in the system. The added memory footprint can be mitigated with the positive influence of our *Hybrid* design, since it scales directly with the number of data partitions in the system.

Acknowledgments. This work is partly funded within the DFG-CRC 912 (HAEC).

References

1. Bagan, G., et al.: Generating flexible workloads for graph databases. PVLDB **9**, 1457–1460 (2016)
2. Decker, S., et al.: The semantic web: the roles of xml and rdf. IEEE **4**, 63–73 (2000)
3. Fard, A., et al.: A distributed vertex-centric approach for pattern matching in massive graphs. In: 2013 IEEE International Conference on Big Data (Oct 2013)
4. Gonzalez, J.E., et al.: Powergraph: Distributed graph-parallel computation on natural graphs. In: OSDI (2012)
5. Karypis, G., et al.: A fast and high quality multilevel scheme for partitioning irregular graphs. SIAM J. Sci. Comput. **20**(1), 359–392 (1998)
6. Kissinger, T., et al.: ERIS: A numa-aware in-memory storage engine for analytical workload. In: ADMS (2014)
7. Krause, A., et al.: Asynchronous graph pattern matching on multiprocessor systems (2017). https://arxiv.org/abs/1706.03968
8. Krause, A., et al.: Partitioning Strategy Selection for In-Memory Graph Pattern Matching on Multiprocessor Systems (2017). http://wwwdb.inf.tu-dresden.de/europar2017/. Accepted at Euro-Par 2017
9. McCune, R.R., et al.: Thinking like a vertex: A survey of vertex-centric frameworks for large-scale distributed graph processing. ACM Comput. Surv. **48**(2), 25:1–25:39 (2015)
10. Nguyen, D., et al.: A lightweight infrastructure for graph analytics. In: SIGOPS (2013)
11. Ogata, H., et al.: A heuristic graph comparison algorithm and its application to detect functionally related enzyme clusters. Nucleic Acids Res. **28**, 4021–4028 (2000)
12. Otte, E., et al.: Social network analysis: a powerful strategy, also for the information sciences. J. Inf. Sci. **28**, 441–453 (2002)
13. Pandis, I., et al.: Data-oriented transaction execution. PVLDB **2**, 928–939 (2010)
14. Pandit, S., et al.: Netprobe: A fast and scalable system for fraud detection in online auction networks. In: WWW (2007)
15. Seo, J., et al.: Distributed socialite: A datalog-based language for large-scale graph analysis. PVLDB **6**, 1906–1917 (2013)
16. Shun, J., et al.: Ligra: a lightweight graph processing framework for shared memory. IN: SIGPLAN (2013)
17. Tas, M.K., et al.: Greed is good: Optimistic algorithms for bipartite-graph partial coloring on multicore architectures. CoRR (2017)

18. Tran, T., et al.: Top-k exploration of query candidates for efficient keyword search on graph-shaped (RDF) data. In: ICDE (2009)
19. Valiant, L.G.: A bridging model for parallel computation. Commun. ACM **33**, 103–111 (1990)
20. Wood, P.T.: Query languages for graph databases. SIGMOD **41**, 50–60 (2012)
21. Yasui, Y., et al.: Numa-aware scalable graph traversal on SGI UV systems. IN: HPGP (2016)

Predicting Access to Persistent Objects
Through Static Code Analysis

Rizkallah Touma[1](✉), Anna Queralt[1], Toni Cortes[1,2], and María S. Pérez[3]

[1] Barcelona Supercomputing Center (BSC), Barcelona, Spain
{rizk.touma,anna.queralt,toni.cortes}@bsc.es
[2] Universitat Politècnica de Catalunya (UPC), Barcelona, Spain
[3] Ontology Engineering Group (OEG), Universidad Politécnica de Madrid,
Madrid, Spain
mperez@fi.upm.es

Abstract. In this paper, we present a fully-automatic, high-accuracy approach to predict access to persistent objects through static code analysis of object-oriented applications. The most widely-used previous technique uses a simple heuristic to make the predictions while approaches that offer higher accuracy are based on monitoring application execution. These approaches add a non-negligible overhead to the application's execution time and/or consume a considerable amount of memory. By contrast, we demonstrate in our experimental study that our proposed approach offers better accuracy than the most common technique used to predict access to persistent objects, and makes the predictions farther in advance, without performing any analysis during application execution.

1 Introduction

Persistent Object Stores (POSs), such as object-oriented databases and object-relational mapping systems (e.g. Hibernate, DataNucleus, dataClay [11]), are storage systems that expose persistent data in the form of objects and relations between these objects. This structure is rich in semantics ideal for predicting access to persistent data [8] and has invited a significant amount of research due to the importance of these predictions in areas such as prefetching, cache replacement and dynamic data placement.

In this paper, we present a fully-automatic approach that predicts access to persistent objects through static code analysis of object-oriented applications. Our approach takes advantage of the symmetry between application objects and POS objects to perform the prediction process before the application is executed

This work has been supported by the European Union's Horizon 2020 research and innovation program (grant H2020-MSCA-ITN-2014-642963), the Spanish Government (grant SEV2015-0493 of the Severo Ochoa Program), the Spanish Ministry of Science and Innovation (contract TIN2015-65316) and Generalitat de Catalunya (contract 2014-SGR-1051). The authors would also like to thank Alex Barceló for his feedback on the formalization included in this paper.

© Springer International Publishing AG 2017
M. Kirikova et al. (Eds.): ADBIS 2017, CCIS 767, pp. 54–62, 2017.
DOI: 10.1007/978-3-319-67162-8_7

and does not cause any overhead. In our experimental study, we demonstrate the viability of the proposed approach by answering the following research questions:

- **RQ1:** What is the accuracy of the proposed approach?
- **RQ2:** How much in advance can the approach make the predictions?

We also compare our approach with the *Referenced-Objects Predictor* (see Sect. 2) and the experimental results show that our approach offers better accuracy in all of the studied benchmarks, with reductions in false positives of as much as 30% in some cases. Moreover, our approach predicts accesses farther in advance giving additional time for the predictions to be utilized.

2   Related Work

The simplest technique to predict access to persistent objects is the *Referenced-Objects Predictor* (ROP), which is based on the following heuristic: each time an object is accessed, all the objects referenced from it are likely to be accessed as well [8]. In spite of its simplicity, this predictor is widely used in commercial POSs because it does not involve a complex and costly prediction process.

More complex approaches have been based on analysis done during application execution using various techniques such as Markov-Chains [9], traversal profiling [5,6] and the Lempel-Ziv compression algorithm [2]. The approach presented in [4] introduces type-level prediction based on the argument that patterns do not necessarily exist between individual objects but rather between object types. Type-level access prediction can capture patterns even when different objects of the same type are accessed and does not store information for each individual object, which reduces the amount of used memory. In general, the main drawbacks of these approaches are the overhead they add to application execution time and the fact that they are based on a most-common case scenario which might lead to erroneous predictions in some cases.

The approach brought forward in this paper combines the idea of application type graphs, presented in [6], with type-level prediction. The work in [6] proceeds by creating an object graph and generating object-level access hints based on profiling done during application execution. On the other hand, our approach generates type-level access hints based on static code analysis, thus benefiting from the advantages of type-level prediction while avoiding the issues stemming from performing the process during application execution.

Previous approaches that use static analysis to predict access to persistent data have targeted specific types of data structures such as linked data structures [1,3,7], recursive data structures [10,13] or matrices [12]. To the best of our knowledge, our work is the first that predicts access to persistent objects of any type prior to application execution.

3   Running Example

Figure 1 shows the partial implementation of a bank management system, all classes represent persistent types except the *BankManagement* class. The

```
 1  public class BankManagement {
 2    private List<Transaction> trans;
 3    private Customer manager;
 4
 5    public void setAllTransCustomers()
        {
 6      for (Transaction t : trans) {
 7        t.getAccount()
              .setCustomer(manager);
 8      }
 9    }
10  }
11
12  /* Persistent Classes */
13  public class Transaction {
14    private Account acc;
15    private Employee emp;
16    private TransactionType type;
17
18    public Account getAccount() {
19      if (type.typeID == 1) {
20        emp.doSomething();
21      } else {
22        emp.dept.doSomethingElse();
```

```
23    }
24    return acc;
25  }
26  }
27
28  public class Account {
29    private Customer cust;
30  }
31
32    public void setCustomer(Customer
        newCust) {
33      if (cust.comp == newCust.comp) {
34        cust = newCust;
35      }
36    }
37  }
38
39  public class Customer {
40    public Company comp;
41  }
42
43  public class Employee {
44    public Department dept;
45  }
```

Fig. 1. Example object-oriented application code

method *setAllTransCustomers()* updates the customers of all the transactions to a new customer restricting such updates to customers of the same company. In order to do so, it retrieves and iterates through all the *Transaction* objects and then navigates to the referenced *Account* and *Customer* until reaching the *Company* of each transaction and compares it with the new customer's company.

For this example, ROP would predict that each time a *Transaction* object is accessed, the referenced *Transaction Type*, *Account* and *Employee* objects will be accessed as well. However, the method *setAllTransCustomers()* does not access the predicted *Transaction Type* and *Employee* objects but needs the *Customer* and *Company* objects which are not predicted.

On the other hand, using static code analysis we can see that when *setAllTransCustomers()* is executed it accesses: (1) the object *BankManagement. manager*, (2) all the *Transaction* objects, and (3) the *Account, Customer* and *Company* objects of each transaction by calling *getAccount()* and *setCustomer()*. We can also see that *getAccount()* might access the *Department* of the *Employee* of a *Transaction*, depending on which branch of the conditional statement starting on line 19 is executed. Using this information, we can automatically generate method-specific access hints that predict which objects are going to be accessed.

4   Proposed Approach

Assuming we have an object-oriented application that uses a POS, we define T as the set of types of the application and $PT \subseteq T$ as its subset of persistent types. Furthermore, $\forall t \in T$ we define (1) F_t : the set of persistent member fields of t such that $\forall f \in F_t : type(f) \in PT$, (2) M_t : the set of member methods of t.

4.1   Type Graphs

Application Type Graph. The type graph of an application, as defined in [6], is a directed graph $G_T = (T, A)$ where:

- T is the set of types defined by the application.
- A is a function $T \times F \to PT \times \{single, collection\}$ representing a set of associations between types. Given types t and t' and field f, if $A(t, f) \to (t', c)$ then there is an association from t to t' represented by $f \in F_t$ where $type(f) = t'$ with cardinality c indicating whether the association is $single$ or $collection$.

Example. Figure 2 shows the type graph of the application from Fig. 1. Some of the associations of this type graph are: (1) A(Bank Management, trans) $\mapsto$ (Transaction, collection), (2) A(Transaction, acc) $\mapsto$ (Account, single).

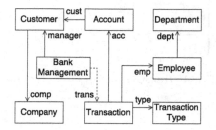

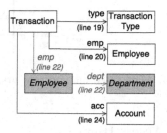

Fig. 2. Type graph G_T of the application from Fig. 1. Solid lines represent single associations and dashed lines represent collection associations.

Fig. 3. Method type graph G_m of the method $getAccount()$ from Fig. 1. Navigations highlighted in gray are branch-dependent.

Method Type Graph. We construct the type graph G_m of a method $m \in M_t$ from the associations that are navigated by the method's instructions. A navigation of an association $t \to^f t'$ is triggered when an instruction accesses a field f in an object of type t (navigation source) to navigate to an object of type t' (navigation target) such that $A(t, f) \to (t', c)$. A navigation of a collection association has multiple target objects corresponding to the collection's elements.

Example. Figure 3 shows G_m of method $getAccount()$ from Fig. 1.

Augmented Method Type Graph. We construct the augmented method type graph AG_m of a method $m \in M_t$ by adding association navigations that are caused by the invocation of another method $m' \in M_{t'}$ to G_m as follows:

- The type graph of the invoked method $G_{m'}$ is added to G_m through the navigation $t \to^f t'$ that caused the invocation of m'.

– The association navigations that are triggered by passing a persistent object as a parameter to m' are added directly to G_m.

Example. Figure 4 shows the augmented method type graph AG_m of method *setAllTransCustomers()*. Note that the navigations *BankManagement* $\rightarrow^{manager}$ *Customer* $\rightarrow^{comp}$ *Company* are triggered by passing the persistent object *Bank Management.manager* as a parameter to the method *setCustomer(newCust)*.

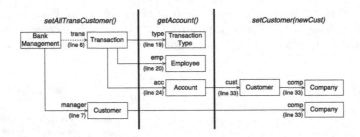

Fig. 4. Augmented method type graph AG_m of *setAllTransCustomers()* from Fig. 1.

4.2   Access Hints

We traverse a method's augmented graph and generate its set of access hints as:

$$AH_m = \{ ah \mid ah = f_1.f_2.\ldots.f_n \ where \ t_i \rightarrow^{f_i} t_{i+1} \in AG_M : 1 \leq i < n \}$$

Each access hint $ah \in AH_m$ corresponds to a sequence of association navigations in AG_m and indicates that the navigations' target object(s) is/are accessed.

Example. The augmented method type graph AG_m of Fig. 4 results in the following set of access hints for method *setAllTransCustomers()* (hints starting with the collection *trans* predict that *all* its elements will be accessed):

$$AH_m = \{ trans.type, \ trans.emp, \ trans.acc.cust.comp, \ manager.comp \}$$

4.3   Nondeterministic Application Behavior

Branch-Dependent Navigations. They are navigations that might not be triggered depending on a method's branching behavior and which of its branches are executed. Including branch-dependent navigations in G_m might result in false positives if the branch from which the navigation is triggered is not executed while excluding them might result in a miss if the branch is indeed executed (both strategies are evaluated in Sect. 5). We divide them in two types:

- Navigations *not* triggered inside all the branches of a conditional statement.
- Navigations of collection associations *not* triggered in all the iterations of a loop statement due to branching instructions (*continue, break, return*).

Example. In Fig. 3, the navigations $Transaction \rightarrow^{emp} Employee \rightarrow^{dept} Department$, highlighted in gray, are branch dependent (they are only triggered in one of the conditional statement's branches) while the navigation $Transaction \rightarrow^{emp} Employee$ is *not* (it is triggered inside both branches).

Overridden Methods. A method $m \in M_t$ might have overridden methods OM_m in the subtypes of its type ST_t. When an object is defined of type t but initialized to a subtype $t' \in ST_t$, the methods executed on the object are not known until runtime. Hence, using the access hints of m might lead to erroneous predictions. We propose to handle this case by adding one of the following sets of access hints to AH_m (both strategies are evaluated in Sect. 5):

- $\bigcap_{m' \in OM_m} AH_{m'}$: *intersection* of access hints of overridden versions of m.
- $\bigcup_{m' \in OM_m} AH_{m'}$: *union* of access hints of overridden versions of m.

5   Evaluation

We implemented a prototype of our approach in Java using **IBM Wala** and evaluated it on two benchmarks specifically designed for POSs and two benchmarks typically used for computation-intensive workloads:

- **OO7:** the *de facto* standard benchmark for POSs and OO databases.
- **JPAB:** measures the performance of ORMs compliant with Java Persistent API (JPA) using 4 types of workloads (persist, retrieve, query and update).
- **K-Means:** a clustering algorithm typically used as a big data benchmark.
- **Princeton Graph Algorithms (PGA):** a set of various graph algorithms with different types of graphs (undirected, directed, weighted).

We compared our approach with the *ROP* explained in Sect. 2 using the minimum possible depth of 1 as well as a depth of 3 to predict access to objects. In all the experiments, we used Hibernate 4.1.0 with PostgreSQL 9.3 as the persistent storage. In the following, we present our experimental results.

RQ1: What Is the Accuracy of the Proposed Approach? We answered this question by testing the different strategies proposed in Sect. 4.3 to deal with branch-dependent navigations and overridden methods. Figure 5 shows the True Positive Ratio (correctly predicted objects / accessed objects) and False Positive Ratio (incorrectly predicted objects / total predicted objects) of these strategies compared with *ROP*. Regardless of the used strategy, our approach results in fewer false positives in all of the studied benchmarks.

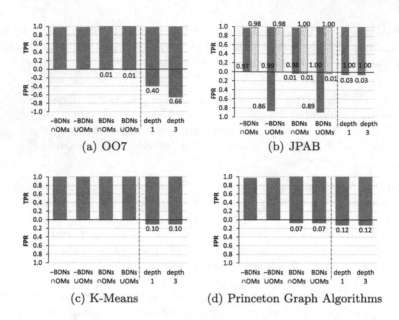

Fig. 5. True Positive Ratio (TPR) and False Positive Ratio (FPR) of our approach (left of the dashed line) compared with ROP (right of the dashed line). Columns represent:
- $\neg BDNs$ / $BDNs$: exclude / include branch-dependent navigations
- $\cup OMs$ / $\cap OMs$: intersection / union of overridden methods' access hints

The only exception is taking the union of overridden methods' access hints with *JPAB*, represented by the solid-colored set of columns in Fig. 5(b), which results in a sharp increase in false positives. This is due to the implementation of *JPAB* which includes five different tests each with its independent persistent classes, all of which are subclasses of a common abstract class. Hence, taking the union of overridden methods' access hints results in predicting access to many objects unrelated to the test being executed. We reran the analysis excluding the methods of the common abstract class and their overridden versions. The results are shown by the dotted set of columns in Fig. 5(b) and indicate that excluding this case, the behavior of *JPAB* is similar to that of the other benchmarks.

Based on the results of this experiment, we recommend excluding branch-dependent navigations when memory resources are scarce, since this strategy does not result in any false positives. By contrast, branch-dependent navigations should be included when we are willing to sacrifice some memory, which will be occupied with false positives, in return of a higher true positive ratio. Finally, we recommend to always take the intersection of overridden methods' access hints in order to avoid problems with special cases similar to *JPAB*.

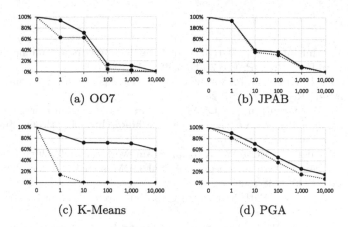

Fig. 6. The x-axis represents the number of persistent accesses between the prediction that a persistent object o will be accessed and the actual access to o. The y-axis represents the percentage of accesses that are predicted for each x-axis value. Solid lines represent our approach and dashed lines represent ROP.

RQ2: How Much in Advance Can the Approach Make the Predictions?
We measured how much in advance our approach can make the predictions and compared it with ROP by calculating the number of persistent accesses between the time that an object o is predicted to be accessed and the actual access to o. For example, Fig. 6 shows that with *OO7*, 95% of predictions made by our approach are done at least 1 persistent access in advance and 70% of predictions at least 10 persistent accesses in advance. The results shown in Fig. 6 indicate that in the case of *JPAB*, the improvement we obtain over ROP is very small because the benchmark's data model does not allow for predictions to be made far in advance. However, with the other benchmarks, most significantly with *K-Means*, our approach is able to predict accesses much farther in advance.

6   Conclusions

In this paper, we presented a novel approach to automatically predict access to persistent objects through static code analysis of object-oriented applications. The approach performs the analysis before application execution and hence does not add any overhead. The experimental results show that our approach achieves better accuracy than the most common prediction technique, the Referenced-Objects Predictor. Moreover, the true advantage of our approach comes from the fact that it can predict access to persistent objects farther in advance which indicates that the predictions can be exploited to apply smarter prefetching, cache replacement policies and/or dynamic data placement mechanisms.

References

1. Cahoon, B., McKinley, K.S.: Data flow analysis for software prefetching linked data structures in Java. Proc. PACT **2001**, 280–291 (2001)
2. Curewitz, K.M., Krishnan, P., Vitter, J.S.: Practical prefetching via data compression. SIGMOD Rec. **22**(2), 257–266 (1993)
3. Gornish, E.H., Granston, E.D., Veidenbaum, A.V.: Compiler-directed data prefetching in multiprocessors with memory hierarchies. In: Proceedings of ICS 1990, pp. 354–368. ACM (1990)
4. Han, W., Whang, K., Moon, Y.: A formal framework for prefetching based on the type-level access pattern in object-relational DBMSs. IEEE Trans. Knowl. Data Eng. **17**(10), 1436–1448 (2005)
5. He, Z., Marquez, A.: Path and cache conscious prefetching (PCCP). VLDB J. **16**(2), 235–249 (2007)
6. Ibrahim, A., Cook, W.R.: Automatic prefetching by traversal profiling in object persistence architectures. In: Thomas, D. (ed.) ECOOP 2006. LNCS, vol. 4067, pp. 50–73. Springer, Heidelberg (2006). doi:10.1007/11785477_4
7. Karlsson, M., Dahlgren, F., Stenström, P.: A prefetching technique for irregular accesses to linked data structures. In: Proceedings of HPCA, pp. 206–217 (2000)
8. Knafla, N.: A prefetching technique for object-oriented databases. In: Small, C., Douglas, P., Johnson, R., King, P., Martin, N. (eds.) BNCOD 1997. LNCS, vol. 1271, pp. 154–168. Springer, Heidelberg (1997). doi:10.1007/3-540-63263-8_19
9. Knafla, N.: Analysing object relationships to predict page access for prefetching. In: Proceedings of POS-8 and PJW-3, pp. 160–170. Morgan Kaufmann (1999)
10. Luk, C.-K., Mowry, T.C.: Compiler-based prefetching for recursive data structures. In: Proceedings of ASPLOS VII, pp. 222–233. ACM (1996)
11. Mart, J., Queralt, A., Gasull, D., et al.: Dataclay: a distributed data store for effective inter-player data sharing. J. Syst. Softw. **131**, 129–145 (2017)
12. Mowry, T.C., Lam, M.S., Gupta, A.: Design and evaluation of a compiler algorithm for prefetching. In: Proceedings of ASPLOS V, pp. 62–73. ACM (1992)
13. Stoutchinin, A., Amaral, J.N., Gao, G.R., Dehnert, J.C., Jain, S., Douillet, A.: Speculative prefetching of *induction pointers*. In: Wilhelm, R. (ed.) CC 2001. LNCS, vol. 2027, pp. 289–303. Springer, Heidelberg (2001). doi:10.1007/3-540-45306-7_20

Query-Driven Knowledge-Sharing for Data Integration and Collaborative Data Science

Andreas M. Wahl$^{(\boxtimes)}$, Gregor Endler, Peter K. Schwab, Sebastian Herbst, and Richard Lenz

Computer Science 6 (Data Management),
FAU Erlangen-Nürnberg, Erlangen, Germany
andreas.wahl@fau.de

Abstract. Writing effective analytical queries requires data scientists to have in-depth knowledge of the existence, semantics, and usage context of data sources. Once gathered, such knowledge is informally shared within a specific team of data scientists, but usually is neither formalized nor shared with other teams. Potential synergies remain unused. We introduce our novel approach of *Query-driven Knowledge-Sharing Systems (QKSS)*. A QKSS extends a data management system with knowledge-sharing capabilities to facilitate user collaboration without altering data analysis workflows. Collective knowledge from the query log is extracted to support data source discovery and data integration. Knowledge is formalized to enable its sharing across data scientist teams.

1 Introduction

Data scientists work according to their expert knowledge gained by solving previous data analysis challenges and maintain individual mental models of the available data sources. These models encompass knowledge about when certain data sources are useful, how they can be linked, how their content can be interpreted or what domain vocabulary is used. Within a team of data scientists, this knowledge is shared through personal interaction. In most cases however, it is not formally documented or shared between teams. Due to the complexity of analytical questions, it is common that multiple teams of data scientists work separately on data analysis challenges, especially in larger organizations. When different teams do not directly interact with each other, they miss out on opportunities to share their knowledge and to profit from experiences of others.

Contribution. To overcome the above deficiencies, we propose *Query-driven Knowledge-Sharing Systems (QKSS)*. A QKSS extends a data management system by adding services that formalize knowledge implicitly contained in a centralized query log and make it available to other users. Parts of the underlying mental model of each query are extracted. This model is mapped to actually available data sources by using previously generated mappings of related queries.

© Springer International Publishing AG 2017
M. Kirikova et al. (Eds.): ADBIS 2017, CCIS 767, pp. 63–72, 2017.
DOI: 10.1007/978-3-319-67162-8_8

Our contribution comprises multiple aspects: (**1**) We introduce *shards of knowledge* as an abstraction for the concepts behind query-driven knowledge-sharing. (**2**) We provide a formal model for building, evolving, and querying shards. (**3**) We explain the integration of a QKSS with existing data analysis tools and processes by suggesting a reference architecture.

2   Query-Driven Knowledge-Sharing Systems

We consider the term *knowledge* to denote domain knowledge about data sources required to query them. Such knowledge contains, among others, the following aspects:

(**1**) What data sources are available? (**2**) What parts of data sources can be used for specific analytical purposes? (**3**) Which vocabulary and semantics are used to describe data sources? (**4**) How can data sources be related to each other? (**5**) Who is using which data sources in which temporal context?

2.1   Services of a QKSS

To explain the services a QKSS offers to data scientists and the benefits it provides, we describe a user story from a clinical research scenario, including simplified example queries.

Consider three teams of data scientists accessing a QKSS (Fig. 1). The QKSS manages different data sources containing data from electronic health records. Alice and Bob from team 1 use medication plans from data source D1 (JSON format) and the relational database D2 for their main focus of drug dosage analysis. Team 2 (Carol and Dan) also conduct drug dosage analysis, but rely on data from the relational databases D2, D3 and data source D4 (CSV format). Erin constitutes team 3 and specializes in time series analysis of patient monitoring data. She uses two data sources D5 (Avro format) and D6 (Parquet format) from a distributed file system.

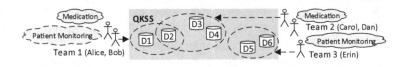

Fig. 1. Collaboration through a QKSS

Shared-Knowledge Support for Querying. Initially, none of the teams is aware of the others. The QKSS provides a SQL interface for data access. While Alice from team 1 is waiting for the completion of a query (Fig. 2), the QKSS detects that both teams rely on D2 by analyzing previous queries (Fig. 3).

The QKSS subsequently presents Alice with information and hints for future queries based on the collective knowledge of team 2. By analyzing queries of

```
SELECT D2.id, D2.Department
FROM D1 JOIN D2 ON D2.id = D1.PatNr
WHERE D1.Agent LIKE 'Dexametha%';
```

```
SELECT MIN(D2.Age)
FROM D4 JOIN D2 USING id
WHERE D4.ActiveAgent = 'Salbutamol';
```
```
SELECT D2.id, D3.Substance, D3.Dose
FROM D3 JOIN D2 ON D2.id = D3.Patient
```

Fig. 2. Exemplary query by Alice   **Fig. 3.** Exemplary queries by Team 2

team 2, the QKSS finds that data sources D3 and D4 have already been linked to D2. It therefore shows Alice a unified view of D2 and these sources. To help Alice with exploring the newly discovered sources, the QKSS ranks them according to frequency of use and temporal occurrence within query sessions of team 2.

Bob has recently received the order to investigate whether the ingestion of certain active agent combinations correlates with the occurrence of critical vital parameters. He is not aware of any data sources containing vital parameters yet, but has a notion of the kind of data he is looking for. Bob imagines vital parameter entries to have attributes such as `HeartRate` and `BloodPressure`. He uses this mental model to formulate a query referencing the hypothetical data source `VitalParameters` (Fig. 4).

```
SELECT HeartRate, BloodPressure
FROM VitalParemeters
WHERE HeartRate >= 130;
```

```
SELECT BloodPr AS BloodPressure
FROM D5
WHERE HeartRate < 90 AND Time > '16-01-10';
```
```
SELECT HRt AS HeartRate, BP AS
BloodPressure
FROM D6
WHERE PId = 'P41' AND TStmp = '1475693932';
```

Fig. 4. Exemplary query by Bob   **Fig. 5.** Exemplary queries by Erin

The QKSS utilizes his assumptions about the structure and semantics of a fictional data source named `VitalParameters` to suggest actual data sources. Using the knowledge extracted from Erin's queries (Fig. 5), the QKSS detects similarities between `VitalParameters` and the data sources D5 and D6. These have been queried before using similar structural assumptions and vocabulary. Thus, the QKSS can recommend D5 and D6 as replacement for the fictional data source in Bob's query. Erin's alias names for schema elements automatically become part of an ontology. Bob can use this ontology for easier comprehension of the vocabulary used by other teams. He can provide feedback about the suggestions through an interactive dialog. The QKSS remembers the mapping between Bob's expectations and the actually available data sources and offers to automatically generate a view that corresponds to his mental model. Bob can directly use this view in future queries.

Management of Shared Knowledge. Whenever synergies between teams are discovered, data scientists may decide to incorporate the shared knowledge of others in the representation of their own mental models. The QKSS provides mechanisms to subscribe to the knowledge contained in queries of others and therefore enables collaborative pay-as-you-go data integration. As long as Bob is involved in patient monitoring data analysis, he subscribes to Erin's queries and augments the formalized representation of his own mental model with hers. He can automatically see the same unified view of the patient monitoring data sources that she does. When he is no longer interested in this topic, he may unsubscribe from her queries and return to his prior medication-centric view.

2.2 Shards of Knowledge

To implement this subscription process and other QKSS services, we introduce *shards of knowledge* as an abstraction for knowledge from the query log. A shard of knowledge captures the mental model of data sources a group of data scientists forms over a period of time. We use the expression *shard* because single mental models may be incomplete, while the combination of all mental models yields the overall organizational knowledge inferable from the query log. As depicted in Fig. 6, shards encapsulate knowledge models constructed from specific portions of the query log. Subsequently, we formally describe the lifecycle of shards to illustrate their usage by data scientists.

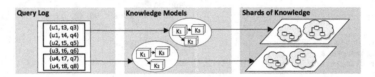

Fig. 6. From log entries to shards of knowledge

Instantiation. Shards are instantiated according to the preferences of data scientists. They determine which knowledge, in form of queries, is to be considered for each shard.

→ **Example:** An initial set of shards for a QKSS could encompass one shard for each team, incorporating all previous queries of the team members. In our example scenario, the QKSS manages an initial set of three shards.

Query Log. The query log L is a set of log entries of type $\mathcal{L}$. Each entry contains a user identifier u of type $\mathcal{U}$, a timestamp t of type $\mathcal{T}$ and a query q of type $\mathcal{Q}$ (Fig. 7(1)).

Only certain portions of the log L are relevant for the data scientists using the QKSS. These portions are extracted using functions $\texttt{extr}$ that apply a set of filter predicates of type $\mathcal{F}$ on the log (Fig. 7(2)). We provide exemplary definitions suitable for a QKSS (Fig. 7(3)/(4)): Each filter predicate denotes the user u

whose queries are to be considered, as well as the time period of relevant log entries using the timestamps t_{start} and t_{end}. Using a set of filter predicates F of type $\mathcal{F}$, we can extract log entries from a log L using the function `extr`.

$\rightarrow$ **Example:** For our scenario of three initial shards we would create the filter predicates F_1, F_2 and F_3 (Fig. 8). t_α and t_ω indicate that log entries from the whole lifespan of the QKSS are included and the QKSS considers future updates to the log. Fixed time spans could also be specified.

$$L : Set\ \mathcal{L}\quad \text{with } \mathcal{L} = (u : \mathcal{U},\ t : \mathcal{T},\ q : \mathcal{Q})\quad (1)$$
$$\mathtt{extr} : Set\ \mathcal{F} \times Set\ \mathcal{L} \rightarrow Set\ \mathcal{L}\quad (2)$$
$$\mathcal{F} = (u : \mathcal{U},\ t_{start} : \mathcal{T},\ t_{end} : \mathcal{T})\quad (3)$$
$$\mathtt{extr}(F, L) = \{l \in L | f \in F.((l.u = f.u)\quad (4)$$
$$\wedge(f.t_{start} \leq l.t \leq f.t_{end}))\}$$

$$F_1 = \{(Alice, t_\alpha, t_\omega), (Bob, t_\alpha, t_\omega)\}$$
$$F_2 = \{(Carol, t_\alpha, t_\omega), (Dan, t_\alpha, t_\omega)\}$$
$$F_3 = \{(Erin, t_\alpha, t_\omega)\}$$

Fig. 7. Extracting relevant queries   **Fig. 8.** Exemplary filter predicates

Knowledge Models. Query log extracts are used by log mining algorithms to create knowledge models (Fig. 6). The algorithms a_i create individual data structures K_i from a subset of the query log to represent the extracted knowledge (Fig. 9(5)). The indices i are elements of an index set I that can be used to label all algorithms provided by the QKSS.

Each knowledge model of type $\mathcal{K}$ consists of the product of the different knowledge aspects extracted by the log mining algorithms (Fig. 9(6)). The function `createModel`$_{L,\mathtt{extr},I}$ instantiates a model by applying the algorithms a_i on an extracted portion of the query log (Fig. 9(7)).

$\rightarrow$ **Example:** For our example scenario, the QKSS provides a variety of log mining algorithms. An algorithm a_{links} extracts join edges between data sources from the query log to form a graph of data sources to reason over. Another algorithm $a_{session}$ partitions the query log into explorative sessions. An algorithm a_{struct} tracks how referenced portions of data sources are structured. To detect synonyms between the vocabularies of different teams, a_{onto} maintains an ontology. By using the resulting index set I (Fig. 10(a)), the knowledge model for team 1 is created (Fig. 10(b)), for example.

Shards. A shard of type $\mathcal{S}$ formalizes the mental model shared by multiple data scientists (Fig. 11(8)). A set F of filter predicates is used to instantiate the shard with the relevant portions of the query log. These are used to create a knowledge model K of type $\mathcal{K}$ which contains all knowledge of the shard.

The function `createShard`$_{L,\mathtt{extr},I}$ takes a set of filter predicates F of type $\mathcal{F}$ to create a shard for a given Log L, a given extraction function `extr` and a given index set I of log mining algorithms (Fig. 11(9)). L, `extr` and I are identical for all shards of a specific QKSS instance.

$$a_i : Set\ \mathcal{L} \to K_i \quad \text{für } i \in I \qquad (5)$$

$$\mathcal{K} = \prod_{i \in I} K_i = (K_{i_1}, ..., K_{i_n}) \qquad (6)$$

$$\text{createModel}_{L,\text{extr},I} : Set\ \mathcal{F} \to \mathcal{K} \qquad (7)$$

$$\text{createModel}_{L,\text{extr},I}(F) = \prod_{i \in I} a_i\ (\text{extr}(F, L))$$

Fig. 9. Creating knowledge models

$$I = \{links, session, struct, onto\} \qquad (a)$$

$$\mathcal{K}_1 = \text{createModel}_{L,\text{extr},I}(F_1) = \qquad (b)$$

$$= (a_{links}\ (\text{extr}(F_1, L)), a_{session}\ (\text{extr}(F_1, L)),$$

$$a_{struct}\ (\text{extr}(F_1, L)), a_{onto}\ (\text{extr}(F_1, L))) =$$

$$= (K_{links}, K_{session}, K_{struct}, K_{onto})$$

Fig. 10. Exemplary knowledge model

$$\mathcal{S} = (F : Set\ \mathcal{F},\ K : \mathcal{K}) \qquad (8)$$

$$\text{createShard}_{L,\text{extr},I} : Set\ \mathcal{F} \to \mathcal{S} \qquad (9)$$

$$\text{createShard}_{L,\text{extr},I}(F) = (F, \text{createModel}_{L,\text{extr},I}(F))$$

Fig. 11. Creating shards

$$s_1 = \text{createShard}_{L,\text{extr},I}(F_1)$$

$$s_2 = \text{createShard}_{L,\text{extr},I}(F_2)$$

$$s_3 = \text{createShard}_{L,\text{extr},I}(F_3)$$

Fig. 12. Exemplary shards

$\to$ **Example:** In our scenario, each team might create a shard using the filter predicates F_1–F_3 (Fig. 12).

Evolution. Shards are dynamic and evolve over time. New log entries matching the filter predicates of a shard can become part of the underlying knowledge models. To provide data scientists with additional flexibility, the QKSS supports several operations to manage shards. While we consider shards to be immutable in our functional model, implementations may destroy or recycle shards.

Lifecycle Operations. Two shards can be merged into a single shard, to reflect in-depth collaboration between data scientist teams (Fig. 13(10)). Shards can also be expanded or narrowed (Fig. 13(11)/(12)). Thereby, specific parts of the query log can be added to or excluded from a shard.

$$\text{merge}_{L,\text{extr},I} : \mathcal{S} \times \mathcal{S} \to \mathcal{S} \quad \text{merge}_{L,\text{extr},I}(s1, s2) = \text{createShard}_{L,\text{extr},I}\ (s1.F \cup s2.F) \qquad (10)$$

$$\text{expand}_{L,\text{extr},I} : \mathcal{S} \times \mathcal{F} \to \mathcal{S} \quad \text{expand}_{L,\text{extr},I}(s, F) = \text{createShard}_{L,\text{extr},I}\ (s.F \cup F) \qquad (11)$$

$$\text{narrow}_{L,\text{extr},I} : \mathcal{S} \times \mathcal{F} \to \mathcal{S} \quad \text{narrow}_{L,\text{extr},I}(s, F) = \text{createShard}_{L,\text{extr},I}\ (s.F - F) \qquad (12)$$

Fig. 13. Basic operations on shards

Shard Comparison. Two shards are considered similar if their knowledge models are similar. The knowledge extracted by log mining algorithms a_i can be compared by using algorithm-specific functions $\text{sim}K_i$ (Fig. 14(13)). Two knowledge models of type $\mathcal{K}$ can be compared to each other using a function $\text{sim}\mathcal{K}$ (Fig. 14(14)). A weight function w allows to determine the influence of specific

$$\text{sim}K_i : K_i \times K_i \rightarrow [0;1] \tag{13}$$
$$\text{sim}\mathcal{K} : (I \rightarrow [0;1]) \times \mathcal{K} \times \mathcal{K} \rightarrow [0;1] \tag{14}$$
$$\text{sim}\mathcal{K}(w, \mathcal{K}_1, \mathcal{K}_2) = \sum_{i \in I}(w(i) \cdot \text{sim}K_i(\mathcal{K}_1.K_i, \mathcal{K}_2.K_i)) \text{ with } \sum_{i \in I}w(i)=1$$
$$\text{sim}\mathcal{S} : (I \rightarrow [0;1]) \times \mathcal{S} \times \mathcal{S} \rightarrow [0;1] \tag{15}$$
$$\text{sim}\mathcal{S}(w, s_1, s_2) = \text{sim}\mathcal{K}(w, s_1.K, s_2.K)$$

Fig. 14. Similarity functions

algorithms from the index set I on model similarity. The function $\text{sim}\mathcal{S}$ derives a numerical value for shard similarity by comparing the knowledge models of two shards (Fig. 14(15)). Data scientists can manually compare shards or rely on automatic comparisons by the QKSS. The system suggests evolution operations based on these comparisons, which can be reviewed by the data scientists.

$\rightarrow$ **Example:** Assume that all algorithms except a_{struct} are assigned a weight factor of 0. Because of the structural similarity of s_1 and s_2 determined by $\text{sim}K_{struct}$ (s_2 incorporates exactly half of the sources of s_1), the QKSS automatically suggests team 1 to merge s_2 into s_1.

Query Processing. The knowledge model of a shard represents the mental model of the data scientists. Queries can be written against this mental model and do not have to reference actually available data sources. Before they can be evaluated against these data sources, the QKSS reasons over the knowledge model (Fig. 15). After logging a query, the QKSS determines if the query is fully specified, which means it only contains schema elements that belong to actually available data sources. If this is the case, the query is processed normally. The knowledge model belonging to the shard of the querying user is evaluated in parallel to generate recommendations. These may include hints about similar data sources, similar users, or synonyms for schema elements.

We also allow queries to be underspecified, as data scientists should be able to express queries using their mental models of the data sources. Whenever the QKSS encounters an underspecified query, it evaluates the knowledge model in order to modify the query to be processable using the actually available data sources. If it cannot decide about relevant data sources, it collects feedback from the user through an interactive dialog. Otherwise, the knowledge model is used to rewrite the query to be consistent with the mental model of the user while using actual data sources.

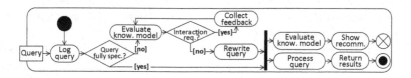

Fig. 15. QKSS query processing workflow

3   Reference Architecture

To demonstrate the feasibility of our approach, we are developing a reference
QKSS [7]. This implementation adheres to our QKSS architecture (Fig. 16).

Users pose queries via the *SQL API* which is a wrapper for the native query
interface of an existing data management system (DMS). Established analysis
workflows remain intact, as analysis tools simply connect to the QKSS instead
of a DMS. The *GUI* acts as a companion to present relevant knowledge and
suggests modifications to the queries and shard lifecycle operations.

Incoming queries are stored in the centralized *query log* by the *query inter-
ception* component. Intercepted queries can be rewritten or extended and subsequently forwarded to the DMS. Result retrieval is handled by the DMS. The
shard management oversees the lifecycle of all shards in the QKSS and generates
knowledge models from the query log to be stored in the *model repository*. It
also monitors the query log to update models if necessary. The *knowledge sharing*
component provides relevant knowledge to rewrite or extend intercepted queries
by evaluating the models from the model repository using the *inference engine*.
It adjusts models according to the queries of the data scientists. Additionally, it
monitors all shards of the QKSS to provide suggestions for lifecycle operations,
such as merging similar shards.

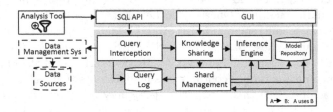

Fig. 16. QKSS reference architecture

4   Evaluation Methods

We assess the overall usefulness of our approach by analyzing how a QKSS
supports data scientists with their data analysis tasks. Additionally, we examine
if the performance of our reference implementation is sufficient for analytical
ad-hoc queries and interactive usage.

Usefulness: The usefulness of our approach is evaluated during a user study.
Knowledge models are created from queries of the participating users. The models are presented to the users to judge to what extent their intentions are correctly captured. By using logs of varying size, the number of queries required to
create meaningful models can be assessed. To assess the result quality of shard
comparisons, participating users validate if knowledge models that are marked
similar by the QKSS are actually built from similar log portions. Subsequently,
two groups of users are formed. Both groups get the task to answer specific

analytical questions using a given set of unfamiliar data sources. While both groups use the QKSS to access the data sources through a single interface, only one group receives recommendations based on prepared knowledge models from the QKSS. We analyze how a QKSS can support the users by comparing the working speed and the result quality of both groups.

Performance: We measure the computational effort required for initial knowledge model creation, model maintenance when new queries are added to the log, model evaluation during query processing, and model comparison. To simulate different application environments, query logs of varying size and complexity are processed by the log mining algorithms. We also measure the overall response time of the system during the processing of ad-hoc queries.

5 Related Work

Dataspace systems [4] rely on user feedback to incrementally adapt the managed data to the expectations of their users. However, existing implementations of dataspace systems do not sufficiently consider scenarios where different groups of users with heterogeneous expectations work with a common set of data sources.

Mental models of data scientists may differ from the actual schema and content of data sources. Some approaches allow queries with references to unknown schema elements [2,6]. However, they do not consider advanced temporal and social connections inferable from the query log.

Our approach differs from query recommendation [3] and completion [5], as we want to enable the users to specify complete queries using individual mental models. We aim to adjust the actually available data sources to the mental models of the users and not to force users to adjust to the data sources. Thus, we minimize bias caused by anchoring and adjustment, psychological phenomena that have been found to have adverse effects on query and result quality [1].

6 Summary

We introduce *Query-driven Knowledge-Sharing Systems (QKSS)* to support data scientists in integrating these data sources and querying them for data analysis tasks. Using a QKSS, data scientists can externalize tacit knowledge about data sources without manual documentation effort, explore how others interact with data sources, and discover relevant data sources. *Shards of knowledge* provide an intuitive abstraction for the user-facing concepts of a QKSS. They encapsulate knowledge models derived from relevant portions of the query log.

References

1. Allen, G., Parsons, J.: Is query reuse potentially harmful? Anchoring and adjustment in adapting existing database queries. ISR **21**(1), 56–77 (2010)
2. Eberius, J., Thiele, M., Braunschweig, K., Lehner, W.: DrillBeyond: processing multi-result open world SQL queries. In: SSDBM 2015 (2015)
3. Eirinaki, M., Abraham, S., Polyzotis, N., Shaikh, N.: QueRIE: collaborative database exploration. KDE **26**(7), 1778–1790 (2014)
4. Franklin, M., Halevy, A., Maier, D.: From databases to dataspaces: a new abstraction for information management. SIGMOD Rec. **34**(4), 27–33 (2005)
5. Khoussainova, N., Kwon, Y., Balazinska, M., Suciu, D.: SnipSuggest: context-aware autocompletion for SQL. PVLDB **4**(1), 22–33 (2010)
6. Li, F., Pan, T., Jagadish, H.V.: Schema-free SQL. In: SIGMOD 2014 (2014)
7. Wahl, A.M.: A minimally-intrusive approach for query-driven data integration systems. In: ICDEW 2016 (2016)

A Declarative Approach to Analyzing Schema Objects and Functional Dependencies

Christiane Engels[1], Andreas Behrend[1(✉)], and Stefan Brass[2]

[1] Institut für Informatik III, Rheinische Friedrich-Wilhelms-Universität Bonn,
Bonn, Germany
{engelsc,behrend}@cs.uni-bonn.de
[2] Institut für Informatik, Martin-Luther-Universität Halle-Wittenberg,
Halle, Germany
brass@informatik.uni-halle.de

Abstract. Database schema elements such as tables, views, triggers and functions are typically defined with many interrelationships. In order to support database users in understanding a given schema, a rule-based approach for analyzing the respective dependencies is proposed using Datalog expressions. We show that many interesting properties of schema elements can be systematically determined this way. The expressiveness of the proposed analysis is exemplarily shown with the problem of computing induced functional dependencies for derived relations.

Keywords: Schema analysis · Functional dependencies · Datalog

1 Introduction

The analysis of database schema elements such as tables, views, triggers, user-defined functions and constraints provide valuable information for database users for understanding, maintaining and managing a database application and its evolution. In the literature, schema analysis has been investigated for improving the quality of SQL/program code or detecting program errors [3] and for determining the consequences of schema changes [10], versioning [8], or matching [11]. In addition, the analysis of schema objects plays an important role for tuning resp. refactoring database applications [2]. All these approaches rely on exploring dependencies between schema objects and an in-depth analysis of their components and interactions. A comprehensive and flexible analysis of schema elements, however, is not provided as these approaches are typically restricted to some subparts of a given schema.

The same is true for analysis features provided by commercial systems where approaches such as integrity checking, executing referential actions or query change notification (as provided by Oracle) already use schema object dependencies but in an implicit and nontransparent way, only. That is, no access to the underlying meta-data is provided to the user nor can be freely analyzed by means of user-defined queries. Even the meta-data about tables and SQL views

© Springer International Publishing AG 2017
M. Kirikova et al. (Eds.): ADBIS 2017, CCIS 767, pp. 73–81, 2017.
DOI: 10.1007/978-3-319-67162-8_9

which are sometimes provided by system tables cover only certain information of the respective schema elements. In this paper, we propose a uniform approach for analyzing schema elements in a comprehensive way. To this end, the schema objects are compiled and their meta-data is stored into a Datalog program which employs queries for deriving interesting properties of the schema. This way, indirect dependencies between tables, views and user-defined functions (UDFs) can be determined which is important for understanding follow-up changes. In order to show the expressiveness of the proposed analysis, our rule-based approach is applied to the problem of deducing functional dependencies (FDs) for derived relations, i.e., views, based on FDs defined for base relations. This so-called *FD propagation* or *FD-FD implication* problem has been studied since the 80s [7,9,12] and has applications in data exchange [6], data integration [4], data cleaning [7], data transformations [5], and semantic query optimization. We show that our rule-based approach to schema analysis is well-suited for realizing all known techniques for FD propagation indicating the expressiveness of the proposed analysis. In particular, our contributions are as follows:

- We propose an approach for analyzing the properties of views, tables, trigger and functions in a uniform way.
- Our declarative approach can be easily extended for refining the analysis by user-defined queries.
- The employed Datalog solution can be simply transfered into SQL systems.
- In order to show the expressiveness of our approach, the implication problem for functional dependencies is investigated using our approach.

2   Rule-Based Schema Analysis

A database schema describes the structure of the data stored in a database system but also contains views, triggers, integrity constraints and user defined functions for data analysis. Functions and these different rule types, namely deductive, active and normative rules, are typically defined with various inter-dependencies. For example, views are defined with respect to base relations and/or some other views inducing a hierarchy of derived queries. In particular, the expression CREATE VIEW q AS SELECT ... FROM $p_1, p_2, ..., p_n$ leads to the set $\{p_1 \rightarrow q, ..., p_n \rightarrow q\}$ of direct dependencies between the derived relation q and derived or a base relations p_i. which are typically represented by means of a predicate dependency graph for analyzing indirect dependencies via the transitive closure, too. This allows for understanding the consequences of changes made to the instances of the given database schema (referred to as update propagation in the literature) or to its structure. This is important when a database user wants to know all view definitions potentially affected by these changes. Various dependencies can occur in a database schema such as table-to-table dependencies induced by triggers or view-to-table dependencies which can be induced by functions. The analysis of such dependencies can be further refined by structural details (e.g., negative vs. positive dependencies as needed for update propagation) as well as by considering the syntactical components of schema objects such as column names (attributes) or operator types (sum, avg, insert, delete, etc.). To this end, the definitions of schema objects need to be parsed and the obtained tokens stored as queryable facts. This kind of analysis is well-known from meta-programming in Prolog which led to the famous vanilla interpreter. For readability reasons we use Datalog with facts such as base(R,A) (base relation R with arity A), derived(V,A) (view V with arity A), dep(To,From) (dependency between relations), call(V,I,O,F) (input I and output O of function F in view V), attr(R,P,N) (position P of attribute named N in relation R) for representing meta-information about a given view or user-defined function. Based on these facts, schema analysis can be realized by queries like

```
attr_dups(N)      ← attr(R1,_,N),attr(R2,_,N),R1<>R2
idb_func_pred(V)  ← derived(V,_),call(V,_,_,_)
base_changes(B)   ← path(B,f₁),base(B,_),func(f₁,_)
tbl_dep(A,B)      ← base(A,_),base(B,_),path(A,F),path(F,B),func(F,_)
```

for determining reused attribute names, views calling a function, base tables possibly changed by function f_1, and cyclic dependencies between two base tables through a function. This way, many interesting properties of schema elements can be systematically determined which supports users in understanding the interrelationships of schema elements. Most database systems already allow for storing and querying meta-data about schema elements in a simple way but a comprehensive (and in particular user-driven) analysis like this is still missing.

3 Functional Dependency Propagation

In order to show the expressiveness of our approach, we investigate the possibility to compute induced FDs for derived relations using the deductive rules introduced above. FDs form special constraints which are assumed to hold for any possible valid database instance. The FD propagation problem is undecidable in the general setting for arbitrary relational expressions [9]. Even restricted to SC views, i.e., relational expressions allowing selection and cross product only, the propagation problem turns out to be coNP-complete (for an in-depth discussion on complexity see [7]). In favor of addressing the general setting, we drop the ambition of achieving completeness by considering a special case, only. Instead, we allow for arbitrary expressions over all relational operators, multiple propagation steps and possibly finite domains[1] in order to cover the majority of practical cases.

3.1 Preliminaries

A functional dependency $\alpha = \{A_1, \ldots, A_n\} \to B$ states that the attribute values of α determine those of B. The restriction to univariate right sides can be done

---

[1] Finite domains may introduce new FDs because of limited value combinations.

without loss of generality as well as the representation of FDs satisfying $B \notin \alpha$, only.[2] We allow $\alpha = \emptyset$ which means that the attribute values of B are constant. For our FD propagation rules, we employ a Datalog variant with special data types for finite, one-leveled sets and finite, possibly nested lists. In our approach we use the extended transitivity axiom

$$\alpha \to B, \ \gamma \to D, \ B \in \gamma, \ D \notin \alpha \ \Rightarrow \ \alpha \cup (\gamma - B) \to D \qquad (1)$$

to derive transitive FDs. Note that if $B \notin \alpha$ and $D \notin \gamma$, then the derived FD also satisfies $D \notin \alpha \cup (\gamma - B)$.

Rule Normalization. For our systematic FD propagation approach, we assume the Datalog rules defining views to be in a normal form, where each rule corresponds to exactly one of the relational operators π, π', σ, $\times$, $\cup$, $\cap$, $-$, or $\bowtie$.[3] Any set of Datalog rules can be transformed into an equivalent set of normalized rules while preserving important properties like being stratifiable [1].

3.2 Representation of FDs and Normalized Rules

We assume that functional dependencies for EDB predicates are given in a relation edb_fd(p, α, B, ID). Here α and B are (sets of) column numbers of the relation p. The fact represents the functional dependency $\alpha \to B$ for the relation p. The ID is of type list and used to identify the dependency in later steps, e.g., in case of union. The derived functional dependencies will be represented in the same way in an IDB predicate fd(p, α, B, ID'). Here ID' is related to the dependency's ID where the FD is derived from for propagated FDs or to a newly created ID for FDs that arise during the propagation process.

As in normal form every rule corresponds to exactly one operator, we can refine the above defined dependency relation dep/2 to rel/3 by adding the respective operator. A fact rel(p,q,op) indicates that a relation p depends (positively) on q via an operator op which is one of 'projection', 'extension', 'selection' 'product', 'join', 'negation', 'intersection', and 'union'. We further introduce an EDB predicate pos(head,body,pos_head,pos_body) for storing information on how the positions of non position preserving operators (cf. Table 1) transform from rule body to head (as FDs are represented via column numbers). Remembering that each relation is defined via one operator only and that we exclude self joins for simplicity (cf. Sect. 3.1), the above defined relation pos/4 is non-ambiguous. Finally, we have two additional EDB predicates eq(pred,pos1,pos2) and const(pred,pos,val) for information on equality conditions (e.g., X = Y or X = const) in extension and selection rules.

---

[2] Multivariate right sides and omitted FDs are retrievable via Armstrong's axioms.
[3] In order to simplify the FD propagation process we limit w.l.o.g. a union rule to two relations and do not allow self joins or cross products.

Table 1. Properties of FD propagation categorized by operator

Properties	π	π'	σ	$\times$	$\cup$	$\cap$	$-$	$\bowtie$
FDs are preserved	×[a]	×	×	×	−	×	×[b]	×
Positions are preserved	−	×	×	−[c]	×	×	×	−
Transitive FDs can appear	−	×	×	−	−	−	−	×
Additional FDs from equality conditions (variables and constants)	−	×	×	−	−	−	−	−
Additional FDs caused by instance reduction may appear	−	−	×	−	−	×	×	×

$\times \,\hat{=}\,$ yes, $- \,\hat{=}\,$ no
[a] Those where all contained variables are maintained
[b] Those of the minuend
[c] Positions of the first factor are preserved, positions of the second factor get an offset

3.3 Propagation Rules

In this section, we present three different types of propagation rules for (a) propagating FDs to the next step, (b) introducing additional FDs arising from equality constraints, and (c) calculating transitive FDs.

Example 1. Consider the following rule set given in normal form together with two FDs $\mathtt{fd(s,\{1\},2,ID_1)}$ and $\mathtt{fd(t,\{1,2\},3,ID_2)}$ for the base relations $\mathtt{s}$ and $\mathtt{t}$:

$$
\begin{aligned}
\mathtt{p(W,Z)} &\leftarrow \mathtt{q(W,X,Y,Z)} \\
\mathtt{q(W,X,Y,Z)} &\leftarrow \mathtt{r(W,X,Y,Z)\,,\ Y{=}2} \\
\mathtt{r(W,X,Y,Z)} &\leftarrow \mathtt{s(W,X)\,,\ t(X,Y,Z)}
\end{aligned}
$$

Omitting IDs, we obtain the following propagation process: First, both FDs are propagated to $\mathtt{r}$ resulting in $\mathtt{fd(r,\{1\},2,-)}$ and $\mathtt{fd(r,\{2,3\},4,-)}$ (with the appropriate column renaming for the latter FD). By transitivity we have $\mathtt{fd(r,\{1,3\},4,-)}$ as a combination of the two. All three FDs are propagated to $\mathtt{q}$ together with $\mathtt{fd(q,\emptyset,3,-)}$ resulting from the equality constraint $\mathtt{Y=2}$. Applying transitivity results in three more FDs for $\mathtt{q}$, but only $\mathtt{fd(q,\{1\},4,-)}$ is propagated further to $\mathtt{p}$ as $\mathtt{fd(p,\{1\},2,-)}$. The complete list of propagated FDs including IDs is given in Example 2.

Table 1 summarizes the properties of how FDs are propagated via the different relational operators which form the basis for the propagation rules. In most cases, the FDs are propagated as they are (with adjustments on the positions for π, $\times$, and $\bowtie$). If there is a single rule defining a derived relation, the source FDs transform to FDs for the new relation (restricted to the attributes in use). Union forms an exception where even common FDs are only propagated in special cases (cf. Sect. 3.4). For extensions π' and selections σ where additional FDs can occur due to equality conditions as well as for joins $\bowtie$ transitive FDs may appear so that taking the transitive closure becomes necessary. In cases where the number of tuples is reduced (i.e., σ, $\cap$, $\bowtie$, and $-$) it is possible that new FDs appear as

pos_pres	non_pos_pres	trans(R) ← base(R,_).
'selection'	'projection'	trans(R) ← rel(R,_,'join').
'extension'	'product'	trans(R) ← eq(R,_,_).
'negation'	'join'	trans(R) ← const(R,_,_).
'intersection'		

Fig. 1. Position preserving (left) and non position preserving (middle) operators, and relations where transitive FDs may occur (right).

there are less tuples for which the FD constraint must be satisfied. The different propagation rules for all relational operators except union are specified in the following.

(a) Induced FDs. For direct propagation of FDs from one level to the next, we distinguish between *position preserving* and *non position preserving* operators. In the first case FDs can be directly propagated (2), whereas in the latter adjustments on the column numbers are necessary (3). The EDB predicates pos_pres and non_pos_pres comprise the respective operators as listed in Fig. 1.

$$fd(P, \alpha, B, -) \leftarrow fd(Q, \alpha, B, -), rel(P, Q, op), pos\_pres(op). \tag{2}$$

$$fd(P, \{X_1, \ldots, X_n\}, Y, -) \leftarrow fd(Q, \{A_1, \ldots, A_n\}, B, -), \tag{3}$$

$$pos(P, Q, X_1, A_1), \ldots, pos(P, Q, X_n, A_n), pos(P, Q, Y, B),$$
$$rel(P, Q, op), non\_pos\_pres(op).$$

(b) Additional FDs. For any equality constraint $X = Y$ we can deduce the dependencies $X \rightarrow Y$ and $Y \rightarrow X$. Similar, a constant constraint $X = c$ induces the dependency $\emptyset \rightarrow X$. That is for any fact eq(R,pos1,pos2) and const(R,pos,val) respectively we derive the following FDs:

$$fd(R, pos1, pos2, -) \leftarrow eq(R, pos1, pos2). \tag{4}$$

$$fd(R, pos2, pos1, -) \leftarrow eq(R, pos1, pos2). \tag{5}$$

$$fd(R, \emptyset, pos, -) \leftarrow const(R, pos, val). \tag{6}$$

(c) Transitive FDs. Since transitive FDs can only occur for certain operators it is sufficient to deduce them for those cases, only (cf. Table 1):

$$fd(P, \varepsilon, D, -) \leftarrow fd(P, \alpha, B, -), fd(X, \gamma, D, -), \tag{7}$$
$$B \in \gamma, D \notin \alpha, \varepsilon = \alpha \cup (\gamma - \{B\}), trans(P).$$

$$fd(P, X, Y, -) \leftarrow fd(P, \alpha, X, ID), fd(P, \alpha, Y, ID), trans(P). \tag{8}$$

The first rule implements the extended transitivity axiom (1) and the second equates the right sides of two identical FDs (identified by matching IDs). The IDB predicate trans/1 comprises all relations where transitive FDs may occur.

3.4   Union

In case of union $p = p_1 \cup p_2$ even common FDs are only propagated in special cases. Consider the following example of post codes. In each country, the post code uniquely identify the city associated with it. But the same post code can be used in different countries for different cities. So although we have the FD post code → city in the relations german_post_codes and us_post_codes, it is not a valid FD in the union of both. A common FD of p_1 and p_2 is only propagated to p if the domains of the FD are disjoint, or if they match on common instances. The first case can only be handled safely on schema level if constants are involved. The latter is the case if the FDs have the same origin and are propagated in a similar way. Whether two FDs have the same origin can be easily checked (e.g. using path from Sect. 2). This criteria is not yet enough as the FDs might have been manipulated during the propagation process (e.g., changes in the ordering, equality constraints, etc.). We employ identifiers to track changes made to certain FDs using a list structure that adopts the tree structure of [9] who represents FDs as trees with source domains as leaves and the target domain as the tree's root. As the target is already handled in the FD itself, we keep track of the source domains and transitively composed FDs, only.

At the beginning, each base FD $\alpha \rightarrow B$ gets a unique identifier ID_i. The idea is to propagate this ID together with the FD and to keep track of the modifications made to the FD. For this purpose we attach an ordered tuple, a (possibly nested) list, to the ID, i.e., $ID_i[A_1, \ldots A_n]$ for $\alpha = \{A_1, \ldots A_n\}$. For the position preserving operators (that in particular do not change the FD's structure) the ID is identically propagated in (2). For the non position preserving operators the positions are updated (using a UDF) similarly to the position adjustments of the FD itself in (3). The difference is that the ID maintains an ordering and the cardinality stays invariant. For constant constraints, we set the constant value as ID in (6), equality constraints in (4), (5) and (8) get the (column number of the) left side as ID. In (7) we replace the occurrences of the column number B in the ID of fd(X, Y, D, -) by the ID of fd(X, α, B, -).

Example 2. For the FD propagation in Example 1 we have the following IDs:

```
fd(s,   {1},   2, ID₁[1]).            fd(q,   {1},   2, ID₁[1]).
fd(t,   {1,2}, 3, ID₂[1,2]).          fd(q,   {2,3}, 4, ID₂[2,3]).
                                      fd(q,   {1,3}, 4, ID₂[ID₁[1],3]).
fd(r,   {1},   2, ID₁[1]).            fd(q,   ∅,     3, '3').
fd(r,   {2,3}, 4, ID₂[2,3]).          fd(q,   {2},   4, ID₂[2,'3']).
fd(r,   {1,3}, 4, ID₂[ID₁[1],3]).     fd(q,   {1},   4, ID₂[ID₁[1],'3']).
                                      fd(p,   {1},   2, ID₂[ID₁[1],'3']).
```

A common ID implies that the same modifications have been made to a common base FD. This means that the FD is preserved in the case of union:

$$\texttt{fd(P,}\alpha\texttt{,B,ID)} \leftarrow \texttt{fd(P}_1\texttt{,}\alpha\texttt{,B,ID), fd(P}_2\texttt{,}\alpha\texttt{,B,ID),} \tag{9}$$

$$\texttt{rel(P,P}_1\texttt{,'union'), rel(P,P}_2\texttt{,'union'), path(P}_1\texttt{,X), path(P}_2\texttt{,X).}$$

4 Conclusion

In Sect. 3.3 we introduced our propagation rules for propagating functional dependencies. To compute the set of propagated FDs these rules are simultaneously applied to the input Datalog program in normal form. The rules are based on the observations in Table 1 which can be easily verified. The propagated functional dependencies of our approach are not complete as the problem is undecidable in general. Also limited to a less expressive subset of the relational operators (e.g., restricted operator order SPC views) one has to assume the absence of finite domains to achieve completeness. Nevertheless, we are able to deal with many cases appearing in real world applications. Our FD propagation approach can be flexible extended to allow for user-defined functions in the extension operator and even recursion can be covered with some modifications.

In [9] the FD implication problem was addressed first. We provided a full declarative approach covering most cases stated in this work. In addition, we are able to cover linear recursion in a similar way as proposed by [12]. Other related approaches like the work of [7] for conditional FDs can be incorporated into our approach, too. Besides these rule-based approaches, a detailed comparison with *the chase*, an established algorithm for FD implication, is object for future research.

References

1. Behrend, A., Manthey, M.: A transformation-based approach to view updating in stratifiable deductive databases. In: FOIKS 2008, pp. 253–271 (2008)
2. Boehm, A.M., Seipel, D., Sickmann, A., Wetzka, M.: Squash: a tool for analyzing, tuning and refactoring relational database applications. In: Seipel, D., Hanus, M., Wolf, A. (eds.) INAP/WLP -2007. LNCS (LNAI), vol. 5437, pp. 82–98. Springer, Heidelberg (2009). doi:10.1007/978-3-642-00675-3_6
3. Brass, S., Goldberg, C.: Proving the safety of SQL queries. In: QSIC 2005, pp. 197–204 (2005)
4. Calì, A., Calvanese, D., De Giacomo, G., Lenzerini, M.: Data integration under integrity constraints. Inf. Syst. **29**(2), 147–163 (2004)
5. Davidson, S.B., Fan, W., Hara, C.S., Qin, J.: Propagating XML constraints to relations. In: ICDE 2003, pp. 543–554 (2003)
6. Fagin, R., Kolaitis, P.G., Popa, L., Tan, W.C.: Reverse data exchange: Coping with nulls. In: PODS 2009, pp. 23–32 (2009)
7. Fan, W., Ma, S., Hu, Y., Liu, J., Wu, Y.: Propagating functional dependencies with conditions. PVLDB **1**(1), 391–407 (2008)
8. Herrmann, K., Voigt, H., Behrend, A., Rausch, J., Lehner, W.: Living in parallel realities co-existing schema versions with a bidirectional database evolution language. In: SIGMOD 2017 (2017)
9. Klug, A.C.: Calculating constraints on relational expressions. TODS **5**(3), 260–290 (1980)
10. Maule, A., Emmerich, W., Rosenblum, D.S.: Impact analysis of database schema changes. In: ICSE 2008, pp. 451–460 (2008)

11. Milo, T., Zohar, S.: Using schema matching to simplify heterogeneous data translation. In: VLDB 1998, p. 122 (1998)
12. Paramá, J.R., Brisaboa, N.R., Penabad, M.R., Places, Á.S.: Implication of functional dependencies for recursive queries. In: Ershov Memorial Conference 2003, pp. 509–519 (2003)

Dear Mobile Agent, Could You Please Find Me a Parking Space?

Oscar Urra$^{(\boxtimes)}$ and Sergio Ilarri

Department of Computer Science and Systems Engineering, I3A,
University of Zaragoza, Zaragoza, Spain
ourra@itainnova.es, silarri@unizar.es

Abstract. Vehicular ad hoc networks (VANETs) have attracted a great
interest in the last years due to their potential utility for drivers in applications that provide information about relevant events (accidents, emer-
gency brakings, etc.), traffic conditions or even available parking spaces.
To accomplish this, the vehicles exchange data among them using wireless communications that can be obtained from different sources, such as
sensors or alerts sent by other drivers. In this paper, we propose searching
of parking spaces by using a mobile agent that jumps from one vehicle to
another to reach the parking area and obtain the required data directly.
We perform an experimental evaluation with promising results that show
the feasibility of our proposal.

Keywords: Vehicular networks · Mobile agents · Distributed query
processing · Data management · Parking spaces

1 Introduction

Vehicular networks (VANETs) [7] are mobile ad hoc networks where the vehicles
can establish connections among them by using short-range wireless communications (such as IEEE 802.11p [11]) and exchange data that can be interesting for
the drivers, such as information about traffic jams, accidents, or scarce resources
such as available parking spaces. However, deploying applications that retrieve
and exploit those data is not an easy task [4], since the vehicles are continuously
moving and the interval of time for exchanging data is very short.

The data exchanged by the vehicles can be stored locally on them and a
driver that wants to retrieve information can submit a query over those data
by following different approaches: (1) push-based query processing assumes that
the relevant data are proactively pushed into the vehicles [1], so the query can
be processed locally by exploiting the previously received data; (2) pull-based
query processing implies disseminating the query in the network in such a way
that a number of vehicles are explicitly asked about data that could be relevant
for the query. Push-based approaches are easier to deploy and simpler, but the
potential queries are constrained by the data that are actively exchanged.

On the other hand, we believe that mobile agents could be suitable for distributed query processing in VANETs. Mobile agents [8] are programs that have

© Springer International Publishing AG 2017
M. Kirikova et al. (Eds.): ADBIS 2017, CCIS 767, pp. 82–90, 2017.
DOI: 10.1007/978-3-319-67162-8_10

the capability to pause their execution, move to another computer, and resume their execution. In this way, they can locate the relevant data sources and move there to process the data locally and filter out the irrelevant information, instead of sending the collected data to a central location, which may not be possible in a VANET due to the limitations and high costs of mobile networks.

In this paper, we study the potential use of mobile agents to retrieve data about available parking spaces in a city. The structure of the paper is as follows. In Sect. 2, we describe an approach to solve the problem by using mobile agents. In Sect. 3, we analyze a use case scenario to retrieve information about available parking spaces, and present an experimental evaluation. In Sect. 4, we present some related work and, finally, in Sect. 5 we show our conclusions and outline some prospective lines of future work.

2   Retrieving Parking Spaces Using Mobile Agents

We consider a scenario where vehicles exchange data about nearby available parking spaces by using a push-based data sharing approach. However, push-based approaches only disseminate popular data that are expected to be relevant in a nearby area, based on the evaluation of spatio-temporal criteria. Therefore, a pure push-based approach cannot handle situations where a driver is interested in retrieving information about available parking spaces located in further areas. Instead, we need to explicitly disseminate a query (pull-based approach) to retrieve data stored by vehicles located near the destination area. For that purpose, a mobile agent can autonomously jump from car to car to reach the relevant data sources that store data about those parking spaces (i.e., vehicles near the destination area) and query them, following the steps described in [10]:

1. A mobile agent is created that will reach the destination area by jumping from car to car, by using only ad hoc short-range wireless communications.
2. The agent retrieves data about available parking spaces by querying the local databases available in vehicles inside the destination area.
3. Once the agent has retrieved enough data about available parking spaces, it returns to the vehicle of the driver interested in those parking spaces (again, by hopping from vehicle to vehicle) and provides the results collected.

3   Experimental Evaluation

In this section, we first describe the use case scenario that we propose and then we perform a number of experiments to evaluate the performance of our proposal. We repeated each experiment 50 times (with different random starting positions for the vehicles and different trajectories) and we report the average results.

3.1   Experimental Setup

In the use case considered, a person traveling along a highway approaches a city, and when he/she is 15 km from the city he/she wants to obtain information about available parking spaces near the city center. Then, a mobile agent is launched, that travels to that area to collect information about the parking spaces by querying the vehicles present there, and will return to the driver information about a specific number of parking spaces. For the evaluation, we have used the MAVSIM simulator [9], that allows the simulation of both mobile agents and traffic in a realistic way (it uses road maps extracted from OpenStreetMaps and simulates the blocking of signals by buildings using the method described in [6]).

The map scenario chosen (Fig. 1) is a portion of the city of Zaragoza (Spain) and some fragments of interurban roads where the driver asking for parking spaces is located when the experiment starts. We consider an *urban area*, where the speed of vehicles is 50 km/h and buildings block the wireless signals, and a *interurban area*, where vehicles can travel up to 120 km/h, and there are no buildings blocking radio signals. The rectangle shown in Fig. 1 shows the destination area (the city center), whose size is 0.25 km². Some parameters of the simulations are shown in Table 1, and a few of them deserve further explanations:

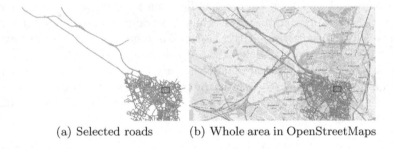

(a) Selected roads   (b) Whole area in OpenStreetMaps

Fig. 1. Scenario map used in the simulation

- The *hop strategy* refers to the algorithm used by the mobile agent when it travels by hopping among vehicles and it must choose, among several candidates, the most promising to reach the agent's final destination. We have chosen a greedy heuristic (called *map distance*) that uses the remaining distance (the shortest path) to the destination and has a good performance [10].
- The *mobile agent's hop delay* is the time needed by the mobile agent to hop from a vehicle to another one within the communication range. After performing experiments with real mobile devices [10], we decided not to assume a best-case scenario and establish the travel time, pessimistically, as one second.
- The default *percentage of vehicles with relevant data* is set to 50%, which means that only half of the vehicles have useful data for the query processing. This percentage allows to simulate cases where some vehicles do not participate in the data sharing of information about parking spaces.

Table 1. Simulation parameters

Parameter	Default value
Map dimensions	4 × 4 Km (urban area)
	20 Km (interurban area)
Map scenario	Zaragoza (Spain) and its surroundings
Size of the destination area	0.25 Km^2
Distance to the destination area	15000 m
Density of vehicles	50 vehicles/Km^2 (urban area)
	10 vehicles/Km (interurban area)
Speed of the vehicles	50 Km/h ± 10% (urban area)
	120 Km/h ± 15% (interurban area)
Hop strategy	MAP (map distance)
Percentage of vehicles with relevant data	50%
Total number of parking spaces in the destination area	220 *parking spaces*
Percentage of available parking spaces per vehicle	10%
Data processing delay	5 s
Number of available parking spaces to retrieve	10 *available parking spaces*
Parking occupancy	90%
Data collection and warning timeouts	4 and 3 min respectively
Communication range and mobile agents' hop delay	250 m and 1 s respectively

- The *percentage of available parking spaces per vehicle* is the percentage of available parking places within the area that are stored in a vehicle. We have chosen a default valor of 10% for this parameter, which is quite pessimistic. This means that, if there are 50 available parking spaces, each vehicle with data would have information about 5 of those available parking places. The smaller this value, the higher the number of vehicles that the mobile agent will need to visit to collect information about available parking places.
- The *data collecting timeout* is the maximum time that the agent will invest in collecting data. If this task takes too long and this timeout is exceeded, the agent will stop collecting data and try to return the results collected so far to the driver. Therefore, the amount of data collected could be smaller than the amount of available parking spaces required by the driver.
- The *warning timeout* represents an amount of time that, if exceeded, will lead the agent to enlarge the search area (100 m in each direction) to try to find more available spaces.

3.2   Influence of the Interurban Vehicle Density

In this experiment, we evaluate how the retrieval of information about available parking spaces is influenced by the density of vehicles in the interurban area. These areas are characterized by long roads (lengths of tens of kilometers), a small number of intersections, and the lack of buildings that block wireless signals. Due to the long length of the road, the most suitable traffic density measurement unit is the *number of vehicles per linear kilometer* (vehicles/km). We vary this value from 2 vehicles/km up to 20 vehicles/km.

Figure 2(a) shows how the time required to obtain the desired information about available parking spaces decreases as the density of vehicles increases. On the other hand, Fig. 2(b) shows the total number of hops performed by the agent. This value is the number of times that the mobile agent moves successfully from one vehicle to another using the wireless connection, and it can be used as a measure of the bandwidth required by the query processing. When the density of vehicles increases, the number of hops also increases, since the mobile agent is constantly looking for more promising vehicles to try to reach sooner the destination area. The growth in the number of hops, however, is not very steep and it stabilizes for about 8 to 10 vehicles/km.

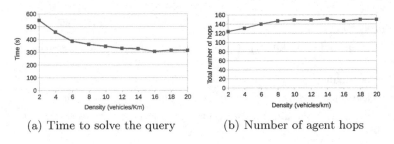

(a) Time to solve the query   (b) Number of agent hops

Fig. 2. Influence of the interurban vehicle density

We have also measured the number of vehicles whose data have been processed by the mobile agent looking for information about available parking spaces. This number refers to the number of vehicles that have data about parking places, since there also exist other irrelevant vehicles where the agent can move but that do not contain data about parking spots. Since the query is solved in the phase of data collection (in the urban area) the interurban density has no influence. The number of vehicles processed remains approximately constant, specifically between 6 and 7 vehicles. We omit the figure due to space constraints.

3.3   Influence of the Occupancy of the Parking Spaces

In this experiment, we vary the ratio of available parking places in the scenario, from 0% (all the parking places are available) to 100% (there is no available parking space) in which case the agent will always fail in its task.

Figure 3(a) shows the total time needed to solve the query. The rate of parking occupancy has little impact on the total time, since the mobile agent spends most of the time traveling to the destination area and returning to the originator car, whereas the actual data collection process is performed quite fast.

Figure 3(b) shows the number of vehicles whose data have been processed by the mobile agent. This number remains quite small until the parking occupancy increases considerably. When there is no available parking space (occupancy of 100%), the number of processed vehicles rises abruptly, due to the efforts performed by the agent to try to solve the query looking for non-existing data. In such a case, the timeout established for data collection is reached and the agent returns with an empty response, which might as well be an useful answer.

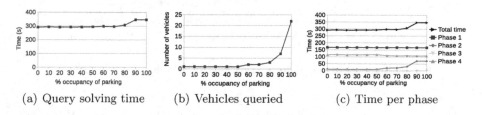

(a) Query solving time   (b) Vehicles queried   (c) Time per phase

Fig. 3. Influence of the occupancy of the parking spaces

The average time spent in the different phases of the query processing is shown in Fig. 3(c). The first phase (mobile agent traveling to the destination area) takes always the same time, since the ratio of available parking spaces does not have any influence on it. In the second phase (mobile agent collecting data), the time invested increases with the occupancy, since the agent will need to visit more vehicles to find information. Finally, the time invested in the fourth phase (agent traveling back to the origin vehicle) decreases slightly as the occupancy of parking spaces increases; this may seem surprising but, while the mobile agent is collecting data, the origin vehicle keeps traveling towards the target area, so the distance that the agent will need to traverse to reach the vehicle decreases.

We also measured the percentage of the number of parking spaces requested by the user that are collected and returned by the agent. For all the occupancy rates, the result obtained was 100% (i.e., information about all the spaces requested are found by the agent), with the exception of when there are no available parking spaces, where the result was 0%. As we increase the occupancy rate by 10% increments, the following worse case scenario evaluated corresponded to an occupancy of parking spaces of 90%, but with this rate there were still enough available parking spaces to satisfy the query (more specifically, 22).

3.4   Influence of the Number of Requested Available Parking Spaces

In this experiment, we vary the number of available parking spaces that the driver wants to retrieve. The parking occupancy ratio is set to 90%, as this is a quite challenging scenario. The number of available parking spots in the area is initially 22, so we vary the requested parking spaces from 2 to 22, and if the agent cannot find enough available parking spaces within the searching area, it will enlarge it according to the parameter *warning timeout*. Figure 4(a) shows the time needed to solve the query. As expected, the higher the number of parking spots to collect, the higher the time needed by the mobile agent to complete the process. However, the query processing times are not excessive in any case.

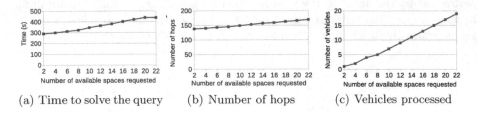

(a) Time to solve the query   (b) Number of hops   (c) Vehicles processed

Fig. 4. Influence of the number of requested available parking spaces

Figure 4(b) shows the number of hops performed by the agent. When the number of requested parking spaces increases, the number of hops also increases, since the agent needs to visit more vehicles. Consistently with this result, Fig. 4(c) shows the number of relevant vehicles that the mobile agent needs to visit, which grows, since the mobile agent needs to visit an increasing number of them to find the information they have about the existing available parking spots.

Finally, the percentage of collected data was 100% in all the cases, with the exception of the case of 22 places, where a 98% of the requested parking spaces (i.e., around 21.5 places, on average) were found despite the efforts performed by the mobile agent, due to the randomness of the information available in the vehicles and the high parking occupancy.

4   Related Work

Most research on query processing in vehicular networks has focused on push-based approaches (e.g., [1]) and only a few works consider pull-based approaches instead (e.g., [3]). The main reason is higher simplicity: push-based approaches avoid some challenges that appear when a query is disseminated in a VANET and the results need to be collected and communicated to the originating vehicle. Moreover, the use of mobile agents for query processing in vehicular networks has not been studied in depth, except for our previous work presented in [10].

Helping drivers to find parking spaces is a topic that has received considerable research attention. So, smart parking systems can be designed and deployed to provide information about parking spaces in specific areas (e.g., see [5]). However, these infrastructure-based solutions are quite expensive and not available globally. On the contrary, it would be interesting to have solutions that are flexible enough to obtain information about any available on-street parking. This motivated the development of proposals that exploit data dissemination in VANETs [4]. Information about the availability of parking spaces can be quite volatile, and some proposals try to take this into account. In [2], an allocation protocol is proposed for sharing information about available parking spaces using only ad hoc communications. This proposal guarantees that the information about a parking space is provided only to a single driver, which avoids competition problems if several drivers receive alerts about the same parking space.

Up to the authors' knowledge, existing proposals to provide information about available parking spaces are push-based, so they cannot be used in scenarios like the one studied in this paper. Moreover, the use of mobile agents to search available parking spaces is also new. Finally, the experimental evaluations focus either on urban scenarios or highways, whereas we have considered a mixed scenario that includes both an urban area and an inter-urban region.

5   Conclusions and Future Work

In this paper, we have presented an approach that uses the technology of mobile agents to find available parking places in a city area by looking *in situ* for that information among the data collected individually by the vehicles that circulate in that area. As opposed to the existing related work, the novelty of our study resides in the use of a pull-based approach to query about parking spaces in areas that are not located near the driver that submits the query (thus allowing new interesting use cases), the use of mobile agents in a parking space searching scenario, and the experimental evaluation with real maps combining both city and interurban areas. The experimental results presented show the performance of the proposal in different conditions and its feasibility.

As a future work, we plan to improve the behavior of the mobile agent to complete the query in less time and with a better use of the available bandwidth.

Acknowledgments. This work has been supported by the projects TIN2016-78011-C4-3-R (AEI/FEDER, UE), TIN2013-46238-C4-4-R, and DGA-FSE.

References

1. Cenerario, N., Delot, T., Ilarri, S.: A content-based dissemination protocol for VANETs: exploiting the encounter probability. IEEE Trans. Intell. Transp. Syst. **12**(3), 771–782 (2011)
2. Delot, T., Ilarri, S., Lecomte, S., Cenerario, N.: Sharing with caution: managing parking spaces in vehicular networks. Mob. Inf. Syst. **9**(1), 69–98 (2013)
3. Delot, T., Mitton, N., Ilarri, S., Hien, T.: GeoVanet: a routing protocol for query processing in vehicular networks. Mob. Inf. Syst. **7**(4), 329–359 (2011)
4. Ilarri, S., Delot, T., Trillo-Lado, R.: A data management perspective on vehicular networks. IEEE Commun. Surv. Tutor. **17**(4), 2420–2460 (2015)

5. Kotb, A.O., Shen, Y.C., Huang, Y.: Smart parking guidance, monitoring and reservations: a review. IEEE Intell. Transp. Syst. Mag. **9**(2), 6–16 (2017)
6. Martinez, F.J., Fogue, M., Toh, C.K., Cano, J.C., Calafate, C.T., Manzoni, P.: Computer simulations of VANETs using realistic city topologies. Wirel. Pers. Commun. **69**(2), 639–663 (2013)
7. Olariu, S., Weigle, M.C.: Vehicular Networks: From Theory to Practice, 1st edn. Chapman & Hall/CRC, Boca Raton (2009)
8. Trillo, R., Ilarri, S., Mena, E.: Comparison and performance evaluation of mobile agent platforms. In: The Third International Conference on Autonomic and Autonomous Systems (ICAS 2007), pp. 41–46. IEEE Computer Society (2007)
9. Urra, O., Ilarri, S.: MAVSIM: testing VANET applications based on mobile agents. In: Vegni, A.M., Agrawal, D.P. (eds.) Cognitive Vehicular Networks, pp. 199–224. CRC Taylor and Francis Group, Boca Raton (2016)
10. Urra, O., Ilarri, S., Trillo-Lado, R.: An approach driven by mobile agents for data management in vehicular networks. Inf. Sci. **381**, 55–77 (2017)
11. Uzcategui, R.A., Sucre, A.J.D., Acosta-Marum, G.: Wave: a tutorial. IEEE Commun. Mag. **47**(5), 126–133 (2009)

P2P Deductive Databases: Well Founded Semantics and Distributed Computation

L. Caroprese and E. Zumpano[✉]

DIMES, University of Calabria, 87036 Rende, Italy
{l.caroprese,e.zumpano}@dimes.unical.it

Abstract. This paper stems from previous works of the same authors in which a declarative semantics for *Peer-to-Peer* (P2P) systems, defined in terms of *Preferred Weak Models*, is proposed. Under this semantics only facts not making the local databases inconsistent can be imported. As in the general case a P2P system may admit many preferred weak models whose computational complexity is prohibitive, the paper looks for a more pragmatic solution. It assigns to a P2P system its *Well Founded Model*, a partial deterministic model that captures the intuition that if an atom is true in a preferred weak model, but it is false in another one, then it is undefined in the well founded model. The paper presents a distributed algorithm for the computation of the well founded model and a system prototype.

1 Introduction

Several proposals considering the issue of managing the coordination, the integration of information [4–6,9,13,18,20] as well as the computation of queries in a P2P system have been proposed in the literature [2,11]. This paper follows the proposal in [6–10] in which a different interpretation of mapping rules has led to the proposal of a semantics for a P2P system defined in terms of *Preferred Weak Models*. Under this semantics only facts not making the local databases inconsistent can be imported, and the preferred weak models are the consistent scenarios in which peers import maximal sets of facts not violating constraints.

Example 1. Consider a P2P in which the peer: (i) $\mathcal{P}_3$ contains two atoms: $r(a)$ and $r(b)$; (ii) $\mathcal{P}_2$ imports data from $\mathcal{P}_3$ using the (mapping) rule $q(X) \hookleftarrow r(X)$[1]. Moreover imported atoms must satisfy the constraint $\leftarrow q(X), q(Y), X \neq Y$ stating that the relation q may contain at most one tuple; (iii) $\mathcal{P}_1$ imports data from $\mathcal{P}_2$, using the (mapping) rule $p(X) \hookleftarrow q(X)$. $\mathcal{P}_1$ also contains the rules $s \leftarrow p(X)$ stating that s is *true* if the relation p contains at least one tuple, and $t \leftarrow p(X), p(Y), X \neq Y$, stating that t is *true* if the relation p contains at least two distinct tuples.

The intuition is that, with $r(a)$ and $r(b)$ *true* in $\mathcal{P}_3$, either $q(a)$ or $q(b)$ could be imported in $\mathcal{P}_2$ and, consequently, only one tuple is imported in the relation

---

[1] Please, note the special syntax we use for mapping rules.

© Springer International Publishing AG 2017
M. Kirikova et al. (Eds.): ADBIS 2017, CCIS 767, pp. 91–99, 2017.
DOI: 10.1007/978-3-319-67162-8_11

p of the peer $\mathcal{P}_1$. Note that whatever is the derivation in $\mathcal{P}_2$, s is derived in $\mathcal{P}_1$ while t is not derived. Therefore, s and t are, respectively, *true* and *false* in $\mathcal{P}_1$. □

A P2P system may admits many preferred weak models and the computational complexity is prohibitive. Therefore, a more pragmatic solution is needed. The paper first introduces a rewriting technique that allows modeling a P2P system $\mathcal{PS}$ as a unique logic program, $Rew_t(\mathcal{PS})$, that can be used as a computational vehicle to calculate the semantics of the P2P system; then presents the *Well Founded Model Semantics*, that allows obtaining a deterministic model whose computation is polynomial time.

Moreover, the paper presents a distributed algorithm for the computation of the well founded model and provides some details on the implementation of a system prototype for query answering in P2P network based on the proposed semantics.

2   P2P Systems: Syntax and Semantics

Familiarity is assumed with deductive database [1], logic programming, stable models, head cycle free (HCF) program, well founded model and computational complexity [3,12,16,17,19]. A *(peer) predicate symbol* is a pair $i : p$, where i is a *peer identifier* and p is a predicate symbol. A *(peer) atom* is of the form $i : A$, where i is a *peer identifier* and A is a standard atom. A *(peer) literal* is a peer atom $i : A$ or its negation *not* $i : A$. A conjunction $i : A_1, \ldots, i : A_m, not\ i : A_{m+1}, \ldots, not\ i : A_n, \phi$, where ϕ is a conjunction of built-in atoms, will be also denoted as $i : \mathcal{B}$, with $\mathcal{B}$ equals to $A_1, \ldots, A_m, not\ A_{m+1}, \ldots, not\ A_n, \phi$.

A *(peer) rule* can be of one of the following three types:

- STANDARD RULE. It is of the form $i : H \leftarrow i : \mathcal{B}$, where $i : H$ is an atom and $i : \mathcal{B}$ is a conjunction of atoms and built-in atoms.
- INTEGRITY CONSTRAINT. It is of the form $\leftarrow i : \mathcal{B}$, where $i : \mathcal{B}$ is a conjunction of literals and built-in atoms.
- MAPPING RULE. It is of the form $i : H \hookleftarrow j : \mathcal{B}$, where $i : H$ is an atom, $j : \mathcal{B}$ is a conjunction of atoms and built-in atoms and $i \neq j$.

$i : H$ is called *head* while $i : \mathcal{B}$ (resp. $j : \mathcal{B}$) is called *body*. Negation is allowed just in the body of integrity constraints. The definition of a predicate $i{:}p$ consists of the set of rules in whose head the predicate symbol $i{:}p$ occurs. A predicate can be of three different kinds: *base predicate*, *derived predicate* and *mapping predicate*. A base predicate is defined by a set of ground facts; a derived predicate is defined by a set of standard rules and a mapping predicate is defined by a set of mapping rules. An atom $i : p(X)$ is a *base atom* (resp. *derived atom*, *mapping atom*) if $i : p$ is a base predicate (resp. standard predicate, mapping predicate). Given an interpretation M, $M[\mathcal{D}]$ (resp. $M[\mathcal{LP}]$, $M[\mathcal{MP}]$) denotes the subset of base atoms (resp. derived atoms, mapping atoms) in M.

Definition 1. P2P SYSTEM. A *peer* $\mathcal{P}_i$ is a tuple $\langle \mathcal{D}_i, \mathcal{LP}_i, \mathcal{MP}_i, \mathcal{IC}_i \rangle$, where (i) $\mathcal{D}_i$ is a set of facts (*local database*); (ii) $\mathcal{LP}_i$ is a set of standard rules; (iii) $\mathcal{MP}_i$ is a set of mapping rules and (iv) $\mathcal{IC}_i$ is a set of constraints over predicates defined by $\mathcal{D}_i$, $\mathcal{LP}_i$ and $\mathcal{MP}_i$. A *P2P system* $\mathcal{PS}$ is a set of peers $\{\mathcal{P}_1, \ldots, \mathcal{P}_n\}$. □

Given a P2P system $\mathcal{PS} = \{\mathcal{P}_1, \ldots, \mathcal{P}_n\}$, where $\mathcal{P}_i = \langle \mathcal{D}_i, \mathcal{LP}_i, \mathcal{MP}_i, \mathcal{IC}_i \rangle$, $\mathcal{D}, \mathcal{LP}, \mathcal{MP}$ and $\mathcal{IC}$ denote, respectively, the global sets of ground facts, standard rules, mapping rules and integrity constraints, i.e. $\mathcal{D} = \bigcup_{i \in [1..n]} \mathcal{D}_i$, $\mathcal{LP} = \bigcup_{i \in [1..n]} \mathcal{LP}_i$, $\mathcal{MP} = \bigcup_{i \in [1..n]} \mathcal{MP}_i$ and $\mathcal{IC} = \bigcup_{i \in [1..n]} \mathcal{IC}_i$. In the rest of this paper, with a little abuse of notation, $\mathcal{PS}$ will be also denoted both with the tuple $\langle \mathcal{D}, \mathcal{LP}, \mathcal{MP}, \mathcal{IC} \rangle$ and the set $\mathcal{D} \cup \mathcal{LP} \cup \mathcal{MP} \cup \mathcal{IC}$.

We now review the *Preferred Weak Model* semantics in [6,7]. For each peer $\mathcal{P}_i = \langle \mathcal{D}_i, \mathcal{LP}_i, \mathcal{MP}_i, \mathcal{IC}_i \rangle$, the set $\mathcal{D}_i \cup \mathcal{LP}_i$ is a *positive normal program*, thus it admits just *one minimal model* that represents the *local knowledge* of $\mathcal{P}_i$. It is assumed that each peer is *locally consistent*, i.e. its local knowledge satisfies $\mathcal{IC}_i$ (i.e. $\mathcal{D}_i \cup \mathcal{LP}_i \models \mathcal{IC}_i$). Therefore, inconsistencies may be introduced just when the peer imports data. The intuitive meaning of a mapping rule $i : H \hookleftarrow j : \mathcal{B} \in \mathcal{MP}_i$ is that if the body conjunction $j : \mathcal{B}$ is *true* in the source peer $\mathcal{P}_j$ the atom $i : H$ can be imported in $\mathcal{P}_i$ only if it does not imply (directly or indirectly) the violation of some constraint in $\mathcal{IC}_i$.

Given a mapping rule $r = H \hookleftarrow \mathcal{B}$, the corresponding standard logic rule $H \leftarrow \mathcal{B}$ will be denoted as $St(r)$. Analogously, given a set of mapping rules $\mathcal{MP}$, $St(\mathcal{MP}) = \{St(r) \mid r \in \mathcal{MP}\}$ and given a P2P system $\mathcal{PS} = \mathcal{D} \cup \mathcal{LP} \cup \mathcal{MP} \cup \mathcal{IC}$, $St(\mathcal{PS}) = \mathcal{D} \cup \mathcal{LP} \cup St(\mathcal{MP}) \cup \mathcal{IC}$.

Given an interpretation M, an atom H and a conjunction of atoms $\mathcal{B}$:

- $val_M(H \leftarrow \mathcal{B}) = val_M(H) \geq val_M(\mathcal{B})$,
- $val_M(H \hookleftarrow \mathcal{B}) = val_M(H) \leq val_M(\mathcal{B})$.

Therefore, if the body is *true*, the head of a standard rule *must* be *true*, whereas the head of a mapping rule *could* be *true*. Intuitively, a *weak model* M is an interpretation that satisfies all standard rules, mapping rules and constraints of $\mathcal{PS}$ and such that each atom $H \in M[\mathcal{MP}]$ (i.e. each mapping atom) is *supported* from a mapping rule $H \hookleftarrow \mathcal{B}$ whose body $\mathcal{B}$ is satisfied by M. A *preferred weak model* is a weak model containing a maximal subset of mapping atoms.

Definition 2. (PREFERRED) WEAK MODEL. Given a P2P system $\mathcal{PS} = \mathcal{D} \cup \mathcal{LP} \cup \mathcal{MP} \cup \mathcal{IC}$, an interpretation M is a *weak model* for $\mathcal{PS}$ if $\{M\} = \mathcal{MM}(St(\mathcal{PS}^M))$, where $\mathcal{PS}^M$ is the program obtained from $ground(\mathcal{PS})$ by removing all mapping rules whose head is *false* w.r.t. M. Given two weak models M and N, M is said to *preferable* to N, and is denoted as $M \sqsupseteq N$, if $M[\mathcal{MP}] \supseteq N[\mathcal{MP}]$. Moreover, if $M \sqsupseteq N$ and $N \not\sqsupseteq M$, then $M \sqsupset N$. A weak model M is said to be *preferred* if there is no weak model N such that $N \sqsupset M$.

The set of weak models for a P2P system $\mathcal{PS}$ will be denoted by $\mathcal{WM}(\mathcal{PS})$, whereas the set of preferred weak models will be denoted by $\mathcal{PWM}(\mathcal{PS})$. □

Theorem 1. *For every consistent P2P system $\mathcal{PS}$, $\mathcal{PWM}(\mathcal{PS}) \neq \emptyset$.*

Example 2. Consider a P2P system $\mathcal{PS}$ in which the peer: $\mathcal{P}_2$ contains the facts $q(a)$ and $q(b)$, whereas $\mathcal{P}_1$ contains the mapping rule $p(X) \hookleftarrow q(X)$ and the constraint $\leftarrow p(X), p(Y), X \neq Y$. $\mathcal{WM}(\mathcal{PS})$ are: $M_0 = \{q(a), q(b)\}$, $M_1 = \{q(a), q(b), p(a)\}$ and $M_2 = \{q(a), q(b), p(b)\}$, whereas $\mathcal{PWM}(\mathcal{PS})$ are M_1 and M_2 as they import maximal sets of atoms from $\mathcal{P}_2$. □

3   Computing the Preferred Weak Model Semantics

This section presents an alternative characterization of the preferred weak model semantics, that allows to model a P2P system $\mathcal{PS}$ with a single logic program $Rew_t(\mathcal{PS})$. Let's firstly introduce some preliminaries. Given an atom $A = i : p(x)$, A^t denotes the atom $i : p^t(x)$ and A^v denotes the atom $i : p^v(x)$. A^t will be called the *testing atom*, whereas A^v will be called the *violating atom*.

Definition 3. Given a conjunction

$$\mathcal{B} = A_1, \ldots, A_h, not\ A_{h+1}, \ldots, not\ A_n, B_1, \ldots, B_k, not\ B_{k+1}, \ldots, not\ B_m, \phi \quad (1)$$

where A_i ($i \in [1.. \ n]$) is a mapping atom or a derived atom, B_i ($i \in [1.. \ m]$) is a base atom and ϕ is a conjunction of built in atoms, we define

$$\mathcal{B}^t = A_1^t, \ldots, A_h^t, not\ A_{h+1}^t, \ldots, not\ A_n^t, B_1, \ldots, B_k, not\ B_{k+1}, \ldots, not\ B_m, \phi \quad (2)$$

Therefore, given a negation free conjunction $\mathcal{B} = A_1, \ldots, A_h, B_1, \ldots, B_k, \ldots, \phi$, then $\mathcal{B}^t = A_1^t, \ldots, A_h^t, B_1, \ldots, B_k, \phi$. In the following, the rewriting of a P2P system is reported.

Definition 4. REWRITING OF AN INTEGRITY CONSTRAINT. Given an integrity constraint $i = \ \leftarrow \mathcal{B}$ (that is of the form (1)) its rewriting is defined as $Rew_t(i) = \{A_1^v \vee \cdots \vee A_h^v \leftarrow \mathcal{B}^t\}$. □

If $\mathcal{B}^t$ (of the form (2)), is *true*, at least one of the violating atoms $A_1^v, \ldots, A_h^v$ is *true*. Therefore, at least one of the atoms $A_1, \ldots, A_h$ cannot be inferred.

Definition 5. REWRITING OF A STANDARD RULE. Given a standard rule $s = H \leftarrow \mathcal{B}$, its rewriting is defined as $Rew_t(s) = \{H \ \leftarrow \mathcal{B}; \ H^t \leftarrow \mathcal{B}^t; \ A_1^v \vee \cdots \vee A_h^v \leftarrow \mathcal{B}^t, H^v \}$. □

In order to find the mapping atoms that, if imported, generate some inconsistencies (i.e. in order to find their corresponding violating atoms), all possible mapping testing atoms are imported and the derived testing atoms are inferred. In the previous definition, if $\mathcal{B}^t$ is *true* and the violating atom H^v is *true*, then the body of the disjunctive rule is *true* and therefore it can be deduced that at least one of the violating atoms $A_1^v, \ldots, A_h^v$ is *true* (i.e. to avoid such inconsistencies at least one of atoms $A_1, \ldots, A_h$ cannot be inferred).

Definition 6. REWRITING OF A MAPPING RULE. Given a mapping rule $m = H \hookleftarrow \mathcal{B}$, its rewriting is defined as $Rew_t(m) = \{H^t \leftarrow \mathcal{B}; \ H \leftarrow H^t, not \ H^v \ \}$. $\square$

Intuitively, to check whether a mapping atom H generates some inconsistencies, if imported in its target peer, a testing atom H^t is imported in the same peer. Rather than violating some integrity constraint, it (eventually) generates, by rules obtained from the rewriting of standard rules and integrity constraints, the atom H^v. In this case H, cannot be inferred and inconsistencies are prevented.

Definition 7. REWRITING OF A P2P SYSTEM. Given a P2P system $\mathcal{PS} = \mathcal{D} \cup \mathcal{LP} \cup \mathcal{MP} \cup \mathcal{IC}$, then: $Rew_t(\mathcal{MP}) = \bigcup_{m \in \mathcal{MP}} Rew_t(m)$, $Rew_t(\mathcal{LP}) = \bigcup_{s \in \mathcal{LP}} Rew_t(s)$, $Rew_t(\mathcal{IC}) = \bigcup_{i \in \mathcal{IC}} Rew_t(i)$ and $Rew_t(\mathcal{PS}) = \mathcal{D} \cup Rew_t(\mathcal{LP}) \cup Rew_t(\mathcal{MP}) \cup Rew_t(\mathcal{IC})$ $\square$

Definition 8. TOTAL STABLE MODEL. Given a P2P system $\mathcal{PS}$ and a stable model M for $Rew_t(\mathcal{PS})$, the interpretation obtained by deleting from M its violating and testing atoms, denoted as $\mathcal{T}(M)$, is a *total stable model* of $\mathcal{PS}$. The set of total stable models of $\mathcal{PS}$ is denoted as $\mathcal{TSM}(\mathcal{PS})$. $\square$

Example 3. Consider the P2P system $\mathcal{PS}$ in Example 2. From Definition (7) we obtain:
$$Rew_t(\mathcal{PS}) = \{q(a); \ q(b); \ p^t(X) \leftarrow q(X); \ p(X) \leftarrow p^t(X), not \ p^v(X);$$
$$p^v(X) \vee p^v(Y) \leftarrow p^t(X), p^t(Y), X \neq Y\}$$
The stable models of $Rew_t(\mathcal{PS})$ are: $M_1 = \{q(a), q(b), p^t(a), p^t(b), p^v(a), p(b)\}$, $M_2 = \{q(a), q(b), p^t(a), p^t(b), p(a), p^v(b)\}$. Then, the total stable models of $\mathcal{PS}$ are $\mathcal{TSM}(\mathcal{PS}) = \{\{q(a), q(b), p(b)\}, \{q(a), q(b), p(a)\}\}$. $\square$

Theorem 2. *For every P2P system* $\mathcal{PS}$, $\mathcal{TSM}(\mathcal{PS}) = \mathcal{PWM}(\mathcal{PS})$. $\square$

4 Well Founded Semantics and Distributed Computation

A P2P system may admit many preferred weak models whose computational complexity has been shown to be prohibitive [6,7]. Therefore, a deterministic model whose computation is guaranteed to be polynomial time is needed. In more details, the rewriting presented in Sect. 3 allows modeling a P2P system by a single disjunctive logic program. By assuming that this program is HCF, it can be rewritten into an equivalent normal program for which a *Well Founded Model Semantics* can be adopted. Such a semantics allows to compute in polynomial time a deterministic model describing the P2P system, by capturing the intuition that if an atom is *true* in a total stable (or preferred weak) *model* of $\mathcal{PS}$ and is *false* in another one, then it is *undefined* in the well founded model.

Theorem 3. *Let* $\mathcal{PS} = \mathcal{D} \cup \mathcal{LP} \cup \mathcal{MP} \cup \mathcal{IC}$ *be a P2P system, then* $Rew_t(\mathcal{PS})$ *is HCF iff there are not two distinct atoms occurring in the body of a rule in* $ground(\mathcal{LP} \cup \mathcal{IC})$ *mutually dependent by positive recursion.* $\square$

we assume each P2P system $\mathcal{PS}$ is s.t. $Rew_t(\mathcal{PS})$ is HCF. $\mathcal{PS}$ will be called as *HCF P2P system*. From previous hypothesis, it follows that $Rew_t(\mathcal{PS})$ can be normalized as $\mathcal{SM}(Rew_t(\mathcal{PS})) = \mathcal{SM}(Normalized(Rew_t \ (\mathcal{PS})))$.

Definition 9. REWRITING OF AN HCF P2P SYSTEM. Given an HCF P2P system $\mathcal{PS}$, $Rew_w(\mathcal{PS}) = Normalized(Rew_t(\mathcal{PS}))$. □

Therefore, the preferred weak models of an HCF P2P system $\mathcal{PS}$ corresponds to the stable models of the normal program $Rew_w(\mathcal{PS})$. Next step is to adopt for $Rew_w(\mathcal{PS})$ a three-valued semantics that allows computing *deterministic models* and in particular the *well founded model*.

Definition 10. WELL FOUNDED SEMANTICS. Given an HCF P2P system $\mathcal{PS}$ and the well founded model of $Rew_w(\mathcal{PS})$, say $\langle T, F \rangle$, the *well founded model semantics* of $\mathcal{PS}$ is given by $\langle \mathcal{T}(T), \mathcal{T}(F) \rangle$. □

Example 4. The rewriting of the HCF P2P system $\mathcal{PS}$ in Example 3 is:

$$Rew_w(\mathcal{PS}) = \{q(a); \; q(b); \; p^t(X) \leftarrow q(X); \; p(X) \leftarrow p^t(X), not \; p^v(X);$$
$$p^v(X) \leftarrow p^t(X), p^t(Y), X \neq Y, not \; p^v(Y);$$
$$p^v(Y) \leftarrow p^t(X), p^t(Y), X \neq Y, not \; p^v(X)\}$$

The well founded model of $Rew_w(\mathcal{PS})$ is $\langle \{q(a), q(b), p^t(a), p^t(b)\}, \emptyset \rangle$ and the well founded semantics of $\mathcal{PS}$ is given by $\langle \{q(a), q(b)\}, \emptyset \rangle$. The atoms $q(a)$ and $q(b)$ are *true*, while the atoms $p(a)$ and $p(b)$ are *undefined*. □

Theorem 4. *Let $\mathcal{PS}$ be a HCF P2P system, then deciding: (i) whether an interpretation M is a preferred weak model of $\mathcal{PS}$ is P-time; (ii) whether an atom A is* true *in some preferred weak model of $\mathcal{PS}$ is $\mathcal{NP}$-complete; (iii) whether an atom A is* true *in every preferred weak model of $\mathcal{PS}$ is $co\mathcal{NP}$-complete; (iv) whether an atom A is* true *in the well founded model of $\mathcal{PS}$ is P-time.* □

$Rew_w(\mathcal{PS})$ allows to compute the well founded semantics of $\mathcal{PS}$ in polynomial time. In this section we present a technique allowing to compute the well founded model in a distributed way. The basic idea is that each peer computes its own portion of the "unique logic program", sending to the other peers the result.

Definition 11. Let $\mathcal{PS}=$ be an HCF P2P system, where $\mathcal{P}_i = \langle \mathcal{D}_i, \mathcal{LP}_i, \mathcal{MP}_i, \mathcal{IC}_i \rangle$, for $i \in [1..n]$. Then, $Rew_w(\mathcal{P}_i) = Rew_w(\mathcal{D}_i \cup \mathcal{LP}_i \cup \mathcal{MP}_i \cup \mathcal{IC}_i)$.

Previous definition allows to derive a single normal logic program for each peer. Observe that, $Rew_w(\mathcal{PS}) = \bigcup_{i \in [1..n]} Rew_w(\mathcal{P}_i)$.

Example 5. The rewritings located on peers $\mathcal{P}_1$ and $\mathcal{P}_2$ of Example 2 are:

$$Rew_w(\mathcal{P}_1) = \{p^t(X) \leftarrow q(X); \; p(X) \leftarrow p^t(X), not \; p^v(X);$$
$$p^v(X) \leftarrow p^t(X), p^t(Y), not \; p^v(Y), X \neq Y;$$
$$p^v(Y) \leftarrow p^t(X), p^t(Y), not \; p^v(X), X \neq Y\}.$$
$$Rew_w(\mathcal{P}_2) = \{q(a); \; q(b)\}.$$

□

The idea is the following: if a peer receives a query, then it will recursively query the peer to which it is connected through mapping rules, before being able to calculate its answer. Formally, a local query submitted by a user to a peer does not differ from a remote query submitted by another peer. Once retrieved the necessary data from neighbor peers, the peer computes its well founded model and evaluates the query (either local or remote) on that model; then if the query is a remote query, the answer is sent to the requesting peer. For the sake of presentation, a partial model reports, using the syntax $[T, U]$, the sets of true and undefined atoms, instead of the sets of true and false atoms.

Example 6. Consider the P2P system in Example 2. If $\mathcal{P}_1$ receives the query $p(X)$, it submits the query $p(X)$ to the peer $\mathcal{P}_2$. Once $\mathcal{P}_2$ receives the query $q(X)$, it computes its well founded model $W_2 = [\{q(a), q(b)\}, \emptyset]$. $\mathcal{P}_1$ receives the data, populates its local database and computes its well founded model $W_1 = [\{p^t(a), p^t(b)\}, \{p^v(a), p^v(b), p(a), p(b)\}]$ and finally, evaluates the query $p(X)$ over W_1. The answer will be $[\emptyset, \{p(a), p(b)\}]$. Observe that, the peer replies providing two undefined atoms: $p(a)$ and $p(b)$. □

Algorithm: $ComputeAnswer_i$
input: 1) $i : q(X)$ - a query
       2) s - a sender which is a peer $\mathcal{P}_k$ (remote query) or *null* (local query)
output: $[T, U]$ where T (resp. U) is the set of *true* (resp. *undefined*) atoms
begin
   while true
      wait for an input $i : q(X)$ and s;
      $P = Rew_w(\mathcal{P}_i)$;
      for each $(i : h(X) \hookleftarrow j : b(X)) \in \mathcal{MP}_i$
        $[T, U] = ComputeAnswer_j(j : b(X), \mathcal{P}_i)$;
         $P = P \cup T \cup \{a \leftarrow not\ a \mid a \in U\}$;
      end for
      $[T_w, U_w] = ComputeWellFoundedModel(P)$;
      $answer = [\{i : q(x) \mid i : q(x) \in T_w\}, \{i : q(x) \mid i : q(x) \in U_w\}]$;
      if $isNull(s)$ **then** $show(answer)$;
      else $send(answer, \mathcal{P}_k)$;
      end if
   end while
end

We associate to a P2P system, $\mathcal{PS}$, a graph $g(\mathcal{PS})$ whose nodes are the peers of $\mathcal{PS}$ and whose edges are the pairs $(\mathcal{P}_i, \mathcal{P}_j)$ s.t. $(i : H \hookleftarrow j : B) \in \mathcal{PS}$. We say that $\mathcal{PS}$ is acyclic if $g(\mathcal{PC})$ is acyclic. In order to guarantee the termination of the computation, from now on we assume that our P2P systems are acyclic. Without loss of generality, we assume that each mapping rule is of the form $i : p(X) \hookleftarrow j : q(X)$. The behavior of the peer $\mathcal{P}_i = \langle \mathcal{D}_i, \mathcal{LP}_i, \mathcal{MP}_i, \mathcal{IC}_i \rangle$ within the P2P system can be modeled by the algorithm $ComputeAnswer_i$, running on the peer $\mathcal{P}_i$. It receives in input a query $i : q(X)$ submitted by a sender s which can be another peer or a user. For each mapping rule $i : h(X) \hookleftarrow j : b(X)$, the peer $\mathcal{P}_i$ queries the peer $\mathcal{P}_j$ asking for $j : b(X)$. $\mathcal{P}_i$ will receive true and undefined tuples that will enrich the local knowledge. Observe that, true atoms will be simply inserted into P while for each undefined atom a, a rule $a \leftarrow not\ a$ allowing to infer the correct truth value for a (undefined) in the well founded model, is inserted. Once the local knowledge has been updated, the local well founded model is computed by means of the function *ComputeWellFoundedModel* and the answer for the query is extracted. If the sender is a user then the answer is simply shown, otherwise (if it is a peer) it is sent back to it.

A system prototype for query answering in a P2P network based on the proposed semantics has benn implemented. The system has been developed using Java.

The communication among peers is performed by using JXTA libraries [15]. A peer is a node in the JXTA network. JXTA defines a set of protocols providing implementation for basic and complex P2P functionalities allowing each device in the network to communicate and interact as a peer and guarantees some advantages in the development of P2P applications (e.g. it can be run on different digital devices (PCs, PDAs)). *XSB* [21] is a logic programming and deductive database system used for computing the answer to a given query using the well founded semantics. InterProlog [14] is an open source front-end that provides Java with the ability to interact with XSB engine.

References

1. Abiteboul, S., Hull, R., Vianu, V.: Foundations of Databases. Addison-Wesley, Boston (1994)
2. Bertossi, L., Bravo, L.: Query answering in peer-to-peer data exchange systems. In: Lindner, W., Mesiti, M., Türker, C., Tzitzikas, Y., Vakali, A.I. (eds.) EDBT 2004. LNCS, vol. 3268, pp. 476–485. Springer, Heidelberg (2004). doi:10.1007/978-3-540-30192-9_47
3. Ben-Eliyahu, R., Dechter, R.: Propositional Semantics for Disjunctive Logic Programs. In: JICSLP, pp. 813–827 (1992)
4. Calì, A., Calvanese, D., De Giacomo, G., Lenzerini, M.: On the decidability and complexity of query answering over inconsistent and incomplete databases. In: PODS, pp. 260–271 (2003)
5. Calvanese, D., De Giacomo, G., Lenzerini, M., Rosati, R.: Logical foundations of peer-to-peer data integration. In: PODS, pp. 241–251 (2004)
6. Caroprese, L., Greco, S., Zumpano, E.: A logic programming approach to querying and integrating P2P deductive databases. In: FLAIRS, pp. 31–36 (2006)
7. Caroprese, L., Molinaro, C., Zumpano, E.: Integrating and querying P2P deductive databases. In: IDEAS, pp. 285–290 (2006)
8. Caroprese, L., Zumpano, E.: Consistent data integration in P2P deductive databases. In: Prade, H., Subrahmanian, V.S. (eds.) SUM 2007. LNCS (LNAI), vol. 4772, pp. 230–243. Springer, Heidelberg (2007). doi:10.1007/978-3-540-75410-7_17
9. Caroprese, L., Zumpano, E.: Modeling cooperation in P2P data management systems. In: An, A., Matwin, S., Raś, Z.W., Ślęzak, D. (eds.) ISMIS 2008. LNCS (LNAI), vol. 4994, pp. 225–235. Springer, Heidelberg (2008). doi:10.1007/978-3-540-68123-6_25
10. Caroprese, L., Zumpano, E.: Handling preferences in P2P systems. In: Lukasiewicz, T., Sali, A. (eds.) FoIKS 2012. LNCS, vol. 7153, pp. 91–106. Springer, Heidelberg (2012). doi:10.1007/978-3-642-28472-4_6
11. Franconi, E., Kuper, G., Lopatenko, A., Serafini, L.: A robust logical and computational characterisation of peer-to-peer database systems. In: Aberer, K., Koubarakis, M., Kalogeraki, V. (eds.) DBISP2P 2003. LNCS, vol. 2944, pp. 64–76. Springer, Heidelberg (2004). doi:10.1007/978-3-540-24629-9_6

12. Gelfond, M., Lifschitz, V.: The stable model semantics for logic programming. In: Proceedings of Fifth Conference on Logic Programming, pp. 1070–1080 (1998)
13. Halevy, A., Ives, Z., Suciu, D., Tatarinov, I.: Schema mediation in peer data management systems. In: Interenational Conference on Database Theory, pp. 505–516 (2003)
14. InterProlog. http://www.declarativa.com/interprolog/
15. Project JXTA. http://www.jxta.org/
16. Lone, Z., Truszczyński, M.: On the problem of computing the well-founded semantics. In: Lloyd, J., Dahl, V., Furbach, U., Kerber, M., Lau, K.-K., Palamidessi, C., Pereira, L.M., Sagiv, Y., Stuckey, P.J. (eds.) CL 2000. LNCS, vol. 1861, pp. 673–687. Springer, Heidelberg (2000). doi:10.1007/3-540-44957-4_45
17. Lloyd, J.W.: Foundations of Logic Programming. Springer-Verlag, Heidelberg (1987)
18. Madhavan, J., Halevy, A.Y.: Composing mappings among data sources. In: VLDB, pp. 572–583 (2003)
19. Papadimitriou, C.H.: Computational Complexity. Addison-Wesley, Boston (1994)
20. Tatarinov, I., Halevy., A.: Efficient Query reformulation in peer data management systems. In: SIGMOD, pp. 539–550 (2004)
21. XSB Project. http://xsb.sourceforge.net

Is Distributed Database Evaluation Cloud-Ready?

Daniel Seybold$^{(\boxtimes)}$ and Jörg Domaschka

Institute of Information Resource Management, Ulm University, Ulm, Germany
{daniel.seybold,joerg.domaschka}@uni-ulm.de

Abstract. The database landscape has significantly evolved over the last decade as cloud computing enables to run distributed databases on virtually unlimited cloud resources. Hence, the already non-trivial task of selecting and deploying a distributed database system becomes more challenging. Database evaluation frameworks aim at easing this task by guiding the database selection and deployment decision. The evaluation of databases has evolved as well by moving the evaluation focus from performance to distribution aspects such as scalability and elasticity. This paper presents a cloud-centric analysis of distributed database evaluation frameworks based on evaluation tiers and framework requirements. It analysis eight well adopted evaluation frameworks. The results point out that the evaluation tiers performance, scalability, elasticity and consistency are well supported, in contrast to resource selection and availability. Further, the analysed frameworks do not support cloud-centric requirements but support classic evaluation requirements.

Keywords: NoSQL · Distributed database · Database evaluation · Cloud

1 Introduction

Relational database management systems (RDBMS) have been the common choice for persisting data for many decades. Yet, the database landscape has changed over the last decade and a plethora of new database management systems (DBMS) have evolved, namely NoSQL [20] and NewSQL [15]. These are promising persistence solutions not only for Web applications, but also for new domains such as "BigData" and "IoT". While NewSQL database systems are inspired by the relational storage model, the storage models of NoSQL database system can be further classified into key-value stores, document-oriented stores, column-oriented stores and graph-oriented stores [20]. NoSQL and NewSQL DBMS are designed to satisfy requirements such as high performance or scalability by running on commodity hardware as a distributed database management system (DDBMS), providing a single DBMS, which is spread over multiple nodes. An element of the overall DDBMS is termed *database node*.

An enabler of the DBMS evolvement is cloud computing by providing fast access to commodity hardware via elastically, on-demand, self-service resource

© Springer International Publishing AG 2017
M. Kirikova et al. (Eds.): ADBIS 2017, CCIS 767, pp. 100–108, 2017.
DOI: 10.1007/978-3-319-67162-8_12

provisioning [17]. Infrastructure as a Service (IaaS) is the preferable way to deploy a DDBMS, requiring a high degree of flexibility in compute, storage and network resources [17].

With the number of available NoSQL and NewSQL systems and cloud resource offerings, the database selection and deployment on the cloud is a challenging task. Hence, DBMS evaluation is a common approach to guide these decisions. With the evolvement of the DBMSs, the landscape of database evaluation frameworks (DB-EFs) has evolved as well: from single node evaluation, *e.g.* TPC-E [1] of the Transaction Processing Performance Council (TPC), to DDBMS evaluation, *e.g.* the Yahoo Cloud Serving Benchmark (YCSB) [7], adding new evaluation tiers such as scalability, elasticity or consistency. Yet, these DB-EFs aim at different evaluation tiers and differ towards common DB-EF requirements [6,14], especially with respect to cloud computing.

In order to facilitate the selection and deployment of DDBMS in the cloud, we present an analysis of DB-EF with the focus on exploiting cloud computing. Our contribution is threefold by (1) defining relevant evaluation tiers for DDBMS deployed in the cloud; (2) extending existing requirements towards DDBMS evaluation with cloud specific requirements; (3) analyse existing evaluation frameworks based on the evaluation tiers and evaluation requirements.

The remainder is structured as follows: Sect. 2 introduces the background on DBMS evaluation. Section 3 defines the evaluation tiers while Sect. 4 defines the requirements towards evaluation frameworks. Section 5 analysis and discusses existing frameworks. Section 6 concludes.

2   Background

Evaluating DBMS imposes challenges for the evaluation frameworks itself, which have been discussed over decades. A first, but still valid guideline for evaluating RDBMS defines the requirements such as relevance to an application domain, portability to allow benchmarking of different systems, scalability to support benchmarking large systems, and simplicity to ensure that the results are easy to understand [14]. A more recent guideline adds the DDBMS and the resulting challenges for supporting different deployment topologies and coordination of distributed experiments [6]. By adopting these challenges, several DB-EFs have been established over the years, which are analysed in Sect. 5.

An overview of existing DB-EF focuses on the tiers availability and consistency. Yet, general requirements for DB-EF are not introduced and cloud specific characteristics are not considered [11]. An overview of DB-EFs for NoSQL database systems is provided by [19]. Yet, the focus lies on evaluating the data model capabilities, without taking explicitly into accout DDBMS aspects and the usage of cloud resources. Evaluating the dimensions of consistency in DDBMS, is also analysed by [5], introducing client-centric and data-centric consistency metrics. Related DB-EF for consistency are presented and missing features for fine-grained consistency evaluation are outlined. An overview of DB-EF is included

---

[1] http://www.tpc.org/tpce/.

in a recommendation compendiums [16] for distributed database selection based on functional and non-functional requirement Yet, the compendium considers only the performance evaluation.

3   Distributed Database Evaluation Tiers

With the evolving heterogeneity in DDBMSs their evaluation becomes even more challenging. DBMS evaluation is driven by **workload domains (WD)**. On a high level WDs can classified into *transactional* (TW) [9], *web-oriented* (WOW) [9], *Big Data* (BDW) [12] and *synthetic* workloads (SW) [7]. These WDs drive the need for considering various evaluation tiers, which are distilled out of database and cloud research.

Resource selection (RS) determines the best matching resources to run a DBMS. For traditional RDBMSs the focus lies on single node resources, (CPU, memory, storage). For DDBMSs network, locality and number of nodes became important factors. By using cloud resources for a DBMS, the cloud providers tend to offer more *heterogeneous resources* such as VMs with dedicated storage architectures[2], container based resources[3], or dedicated resource locations from data center to physical host level.

Performance (P) evaluates the behaviour of a DBMS against a specific kind of workload. Performance metrics are *throughput* and *latency*, which are measured by the evaluation framework.

Scalability (S) defines the capability to process arbitrary workload sizes by adapting the DBMS by scaling *vertically* (scale-up/down) or *horizontally* (scale-in/out) [1]. Scaling vertically changes the computing resources of a single node. Horizontal scaling adds nodes to a DDBMS cluster (scale-out) or removes nodes (scale-in) In the following the term scalability implies horizontal scalability Measuring scalability is performed by correlating throughput and latency for growing cluster sizes and workloads. A high scalability rating is represented by constant latency and proportionally growing throughput with respect to the number of nodes and the workload size [7].

Elasticity (E) defines the ability to cope with sudden workload fluctuations without service disruption [1]. Elasticity metrics are *speedup* and *scaleup* [7]. Speedup refers to the required time for a scaling action, *i.e.* adapting the cluster size, redistributing data and stabilising the cluster. Scaleup refers to the benefit of this action, *i.e.* the throughput/latency development with respect to the workload fluctuation.

Availability (A) represents the degree to which a DBMS is operational and accessible when required for use. The availability of a DBMS can be affected by *overload* (issuing more requests in parallel than theDBMS can handle or *failures on the resource layer* (a node failure). With respect to failures, DDBMSs apply replication of data to multiple database nodes. A common

---

[2] https://aws.amazon.com/de/ec2/instance-types/.
[3] https://wiki.openstack.org/wiki/Magnum.

metric to measure availability with respect to node failures are the takeover time, and the performance impact.

Consistency (C) Distributed databases offer different consistency guarantees as there is trade-off between consistency, availability and partitioning, *i.e.* the CAP theorem [13]. Consistency can be evaluated *client-centric* (*i.e.* from the application developer perspective) and *data-centric* (*i.e.* from the database administrator perspective) [5] . Here, we only consider client-centric consistency that can be classified into *staleness* and *ordering* [5]. Staleness defines how much a replica lags behind its master. It is measured either in time or versions. Ordering defines all requests must be executed on all replicas in the same chronological order.

4   Evaluation Frameworks Requirements

Besides the evaluation tiers, DBMS evaluation imposes requirements towards the DB-EF itself. We briefly present established requirements [6,14] as well as novel cloud-centric requirements.

Usability (U) eases the framework configuration, execution and extension by providing sufficient documentation and tools to run the evaluation. Hence, the evaluation process has to be transparent to provide objective results [14].

Distribution/Scalability (D/S) is provided by distributed workload generation, *i.e.* the framwork clients can be distributed across multiple nodes in order to increasing the workload by utilising an arbitrary amount of clients [6].

Measurements Processing (MP) defines that measurements are gathered not only in an aggregated but also in a fine-grained manner for further processing [6]. As the amount of measurements can grow rapidly for multiple or long running evaluation runs, file-based persistence might not be sufficient. Hence, advanced persistence options such as time series databases (TSDBS), will ease the dedicated processing and visualisation.

Monitoring (MO) data improves the significance of evaluation results. Hence, monitoring of the involved resources, clients ,and DBMSs should be supported by the evaluation framework to provide the basis of a thorough analysis. Again an advanced persistence solution is beneficial.

Database Abstraction (DA) enables the support of multiple DBMSs by abstracting database driver implementations. Yet, the abstraction degree needs to be carefully chosen as a too high abstraction might limit specific DBMS features and distort results. Therefore, the abstraction interface should be aligned with the specified workload scenarios [6].

Client Orchestration (CO) enables automated evaluation runs. Therefore, the framework should provide tools that orchestrate evaluations, *i.e.* provision (cloud) resources, create, execute and collect the results and clean-up the clients. Hence, CO eases the creation of arbitrary load patterns and the simulation of multi-tenant workload patterns.

Database Orchestration (DO) enables the management of the DDBMSs to facilitate repetitive evaluation for different resources, configurations and the adaptation of the DDBMS based on predefined conditions. Hence, the evaluation framework should provide tools to automatically orchestrate DDBMSs, *i.e.* provision resources, setup, configure and adapt generic DDBMSs.

Multi-phase Workloads (MpW) define the support of multiple workloads that run in parallel This is crucial to execute advanced evaluation scenarios. Further, the specification of the load development over a certain time frame per workload is required to simulate real world scenarios.

Extensibility (E) defines the need to provide an architecture, which eases the extension of the framework capabilities, *e.g.* by adding support for additional DBMSs or workload types.

5   Analysis of Evaluation Frameworks

In this section we analyse DB-EFs, which focus on DDBMSs. Hereby, we consider only DB-EFs, which have been published within the evolvement of DDBMSs, *i.e.* from 2007 on. In addition, we only consider the original evaluation frameworks and no minor extensions or evaluations based on these frameworks.

First, we analyse each framework based on the workload domain and supported evaluation tiers. The results are shown in Table 1. Second, we analyse each frameworks capabilities against the presented DB-EF requirements from Sect. 4. The results are shown in Table 2. The analysis applies ✗ = *not supported*, (✓) = *partially supported*, ✓ = *supported*. A detailed analysis can be found in an accompanying technical report [21].

Table 1. Distributed database evaluation tiers

Evaluation framework	Evaluation tier						
	WD	RS	P	S	E	A	C
TPC-E	TW	(✓)	✓	✓	✗	✗	✗
YCSB [7]	SW	✗	✓	✓	✓	✗	✗
YCSB++ [18]	TW, BDW, SW	✗	✓	✓	✓	✗	✓
BG [3]	WOW	✗	✓	✓	✗	✗	✓
BigBench [12]	TW, BDW	✗	✓	✗	✗	✗	✗
OLTP-bench [9]	WOW, SW	✗	✓	✓	✗	✗	✗
YCSB-T [8]	TW, SW	✗	✓	✓	✓	✗	✓
LinkBench [2]	WOW	✗	✓	✗	✗	✗	✗

The first insight of our analysis is that performance, scalability, elasticity and consistency tiers are well covered, but the resource selection and availability tier lack support (*cf.* Sect. 5.2). The second insight points out that the traditional DB-EF requirements [14] such as usability, distribution and extensibility are well supported, while monitoring and cloud orchestration are not (*cf.* Sect. 5.2).

Table 2. Distributed database evaluation tiers

Evaluation framework	Evaluation framework requirement								
	U	D/S	MP	MO	DA	CO	DO	MpW	E
TPC-E	✓	✓	(✓)	✗	(✓)	✗	✗	✓	✓
YCSB [7]	✓	✓	(✓)	✗	✓	✗	✗	✗	✓
YCSB++ [18]	(✓)	✓	✓	✓	(✓)	✓	✗	✓	✗
BG [3]	✓	✓	✓	✓	✓	✗	✗	✓	✓
BigBench [12]	✓	✓	✗	✗	(✓)	✗	✗	✗	✓
OLTP-bench [9]	(✓)	✓	(✓)	✓	(✓)	✓	✗	✓	✓
YCSB+T [8]	(✓)	✓	(✓)	✗	✓	✗	✗	✗	✓
LinkBench [2]	✓	(✓)	(✓)	✓	(✓)	✗	✗	✓	✓

5.1 Results for Evaluation Tiers

The resulting table (*cf.* Table 1) shows that the early DB-EFs focus on performance, scalability and elasticity, while newer frameworks focus as well on consistency. Hereby, only the performance tier has established common rating indices such as throughput, latency or SLA based rating [3]. While multiple frameworks target the scalability and elasticity tier, a common methodology and rating index has not yet been established. Yet, the need for a common evaluation methodology [22] and rating index [10] is already carved out.

Currently not supported evaluation tiers are resource selection and availability. While resource selection is partially supported by TPC-E, which considers physical hardware configurations, cloud-centric resource selection is not in the scope of any of the frameworks. With the increasing heterogeneity of cloud resources from diverse virtual machines offering to even container based resources, the consideration of cloud-centric resource selection needs to move into the focus of novel DB-EFs. Yet, existing DB-EFs can be applied to evaluate DDBMS running on heterogeneous cloud resources, but the DB-EFs do not offer an explicit integration with cloud resource offerings. Hence, manual resource management, monitoring and client/DDBMS orchestration hinders cloud-centric evaluations.

As availability is a major feature of DDBMS it is surprising that it is not considered by the analysed DB-EFs. Especially, as cloud resources do fail, availability concepts for applications running on cloud resources is widely discussed topic. Again, the support of DDBMSs orchestration can enable database specific availability evaluations.

5.2 Results for Evaluation Framework Requirements

The analysis of the evaluation framework requirements (*cf.* Table 2) shows that usability, scalability, database abstraction and extensibility are covered by all frameworks. Measurement processing is covered as well but only a few frameworks support advanced features such as visualisation and none of the frameworks supports advanced storage solutions such as TSDBs. Multi-phase workloads are partially covered by the frameworks, especially by the frameworks from the TW and WOW domains. The monitoring of client resources is partially covered, but only OLTP-bench considers resource monitoring. While all frameworks support the distributed execution of evaluations, only two support the orchestration of clients, which complicates the distributed evaluation runs. Further, none of the frameworks supports DBMS orchestration. This fact leads to high complexity only for setting up the evaluation environment, especially when it comes to heterogeneous cloud resources. Further, dynamic DDBMS transitions for evaluating tiers such as elasticity or availability, always require custom implementations, which impedes the comparability and validity of the results.

6   Conclusion and Future Work

In the last decade the landscape of distributed database systems has evolved and NoSQL and NewSQL database systems appeared. In parallel, cloud computing enabled novel deployment option for database systems. Yet, these evolvements raise the complexity in selecting and deploying an appropriate database system.

In order to ease such decisions, several evaluation frameworks for distributed databases have been developed. In this paper, we presented an analysis of distributed database evaluation frameworks based on evaluation tiers and requirements towards the frameworks itself. The analysis is applied to eight evaluation frameworks and provides a thorough analysis of their evaluation tiers and capabilities. The results of this analysis shows that the performance, scalability, elasticity, and consistency tiers are well covered, while resource selection and availability are not considered by existing evaluation frameworks. With respect to the framework requirements, traditional requirements are covered [14], while cloud-centric requirements such as orchestration are only partially supported.

The analysis shows, that existing frameworks can be applied to evaluate distributed databases in the cloud, but there are still unresolved issues on the evaluation tier side, *i.e.* the support for resource selection and availability evaluation, and on the framework requirement side, *i.e.* the orchestration of clients and databases and exploitation of advanced storage solutions. This hinders repeatability [14] of evaluations on heterogeneous cloud resources as well as dynamic transition in the cluster. Yet, cloud computing research already offers approaches to enable automated resource provisioning and application orchestration in the cloud based on *Cloud Orchestration Tools (COTs)* [4]. Integrating COT into evaluation frameworks can be an option to ease the distributed execution of evaluation runs as well as orchestrating database clusters across different cloud resources. As COTs provide monitoring and adaptation capabilities, they can ease the evaluation of dynamic cluster transitions by defining advanced evaluation scenarios with dynamic database cluster adaptations.

Future work will comprise the analysis of COTs with respect to their exploitation in database evaluation frameworks. In addition, the design and implementation of a cloud-centric database evaluation framework is ongoing.

Acknowledgements. The research leading to these results has received funding from the EC's Framework Programme HORIZON 2020 under grant agreement number 644690 (CloudSocket) and 731664 (MELODIC). We thank Moritz Keppler and the Daimler TSS for their valuable and constructive discussions.

References

1. Agrawal, D., Abbadi, A., Das, S., Elmore, A.J.: Database scalability, elasticity, and autonomy in the cloud. In: Yu, J.X., Kim, M.H., Unland, R. (eds.) DASFAA 2011. LNCS, vol. 6587, pp. 2–15. Springer, Heidelberg (2011). doi:10.1007/978-3-642-20149-3_2
2. Armstrong, T.G., Ponnekanti, V., Borthakur, D., Callaghan, M.: Linkbench: a database benchmark based on the facebook social graph. In: SIGMOD (2013)
3. Barahmand, S., Ghandeharizadeh, S.: Bg: A benchmark to evaluate interactive social networking actions. In: CIDR (2013)
4. Baur, D., Seybold, D., Griesinger, F., Tsitsipas, A., Hauser, C.B., Domaschka, J.: Cloud orchestration features: Are tools fit for purpose? In: UCC (2015)
5. Bermbach, D., Kuhlenkamp, J.: Consistency in distributed storage systems. In: Gramoli, V., Guerraoui, R. (eds.) NETYS 2013. LNCS, vol. 7853, pp. 175–189. Springer, Heidelberg (2013). doi:10.1007/978-3-642-40148-0_13
6. Bermbach, D., Kuhlenkamp, J., Dey, A., Sakr, S., Nambiar, R.: Towards an extensible middleware for database benchmarking. In: Nambiar, R., Poess, M. (eds.) Performance Characterization and Benchmarking: Traditional to Big Data. LNCS, pp. 82–96. Springer, Cham (2015). doi:10.1007/978-3-319-15350-6_6
7. Cooper, B.F., Silberstein, A., Tam, E., Ramakrishnan, R., Sears, R.: Benchmarking cloud serving systems with ycsb. In: SoCC (2010)
8. Dey, A., Fekete, A., Nambiar, R., Rohm, U.: Ycsb+t: Benchmarking web-scale transactional databases. In: ICDEW (2014)
9. Difallah, D.E., Pavlo, A., Curino, C., Cudre-Mauroux, P.: Oltp-bench: An extensible testbed for benchmarking relational databases. VLDB **7**, 277–288 (2013)
10. Dory, T., Mejias, B., Roy, P., Tran, N.L.: Measuring elasticity for cloud databases. In: Cloud Computing (2011)
11. Friedrich, S., Wingerath, W., Gessert, F., Ritter, N., Pldereder, E., Grunske, L., Schneider, E., Ull, D.: Nosql oltp benchmarking: A survey. In: GI-Jahrestagung (2014)
12. Ghazal, A., Rabl, T., Hu, M., Raab, F., Poess, M., Crolotte, A., Jacobsen, H.A.: Bigbench: towards an industry standard benchmark for big data analytics. In: SIGMOD (2013)
13. Gilbert, S., Lynch, N.: Brewer's conjecture and the feasibility of consistent, available, partition-tolerant web services. ACM Sigact News **33**, 51–59 (2002)
14. Gray, J.: Benchmark Handbook: For Database and Transaction Processing Systems. Morgan Kaufmann Publishers Inc, San Francisco (1993)
15. Grolinger, K., Higashino, W.A., Tiwari, A., Capretz, M.A.: Data management in cloud environments: Nosql and newsql data stores. JoCCASA **2**, 22 (2013)

16. Khazaei, H., Fokaefs, M., Zareian, S., Beigi-Mohammadi, N., Ramprasad, B., Shtern, M., Gaikwad, P., Litoiu, M.: How do i choose the right NoSQL solution? a comprehensive theoretical and experimental survey. BDIA **2**, 1 (2016)
17. Mell, P., Grance, T.: The nist definition of cloud computing. Technical report, National Institute of Standards & Technology (2011)
18. Patil, S., Polte, M., Ren, K., Tantisiriroj, W., Xiao, L., López, J., Gibson, G., Fuchs, A., Rinaldi, B.: Ycsb++: benchmarking and performance debugging advanced features in scalable table stores. In: SoCC (2011)
19. Reniers, V., Van Landuyt, D., Rafique, A., Joosen, W.: On the state of nosql benchmarks. In: ICPE (2017)
20. Sadalage, P.J., Fowler, M.: NoSQL Distilled: A Brief Guide to the Emerging World of Polyglot Persistence. Pearson Education, London (2012)
21. Seybold, D., Domaschka, J.: A cloud-centric survey on distributed database evaluation. Technical report. Ulm University (2017)
22. Seybold, D., Wagner, N., Erb, B., Domaschka, J.: Is elasticity of scalable databases a myth? In: IEEE Big Data (2016)

ADBIS 2017 – Workshops

New Trends in Databases and Information Systems: Contributions from ADBIS 2017 Workshops

Andreas Behrend[1], Diego Calvanese[2], Tania Cerquitelli[3]([envelope]), Silvia Chiusano[3],
Christiane Engels[1], Stéphane Jean[4], Natalija Kozmina[5], Béatrice Markhoff[6],
Oscar Romero[7], and Sahar Vahdati[1]

[1] University of Bonn, Bonn, Germany
[2] Free University of Bozen-Bolzano, Bolzano, Italy
[3] Politecnico di Torino, Turin, Italy
`tania.cerquitelli@polito.it`
[4] ENSMA Poitiers, Poitiers, France
[5] University of Latvia, Riga, Latvia
[6] University of Tours, Tours, France
[7] Universitat Politècnica de Catalunya, Barcelona, Spain

Abstract. In the last few years, research on database and information system technologies has been rapidly evolving thanks to the new paradigms of software and hardware adopted by modern scientific and more invasive applications. A huge and heterogeneous amount of data should be efficiently stored, managed, and analyzed exploiting proper technologies for such novel and more interesting data-driven applications. New and cutting-edge research challenges arise that have been attracting great attention from both academia and industry. The 21st European Conference on Advances in Databases and Information Systems (ADBIS 2017), held on September 24–27, 2017 in Nicosia, Cyprus includes four thematic workshops covering some emerging issues concerning such new trends in database and information system research. The aim of this paper is to present such events, their motivations and topics of interest, as well as briefly outline their programs including interesting keynotes, invited papers and a wide range of research, application, and industrial contributions selected for presentations. The selected papers have been included in this volume.

1 Introduction

The ADBIS conferences aim at providing a forum for the dissemination of research accomplishments and promoting interaction and collaboration between the database and information system research communities from European countries and the rest of the world. The ADBIS conferences provide an international platform for the presentation of research on database theory, development of advanced DBMS technologies, and their applications. The 21st edition of ADBIS, held on September 24–27, 2017 in Nicosia, Cyprus includes four thematic workshops covering emerging and cutting-edge research topics concerning new trends

M. Kirikova et al. (Eds.): ADBIS 2017, CCIS 767, pp. 111–122, 2017.
DOI: 10.1007/978-3-319-67162-8_13

in database and information system research. Papers accepted at the ADBIS main conference span a wide range of topics in the field of databases and information systems ranging from innovative platforms for data handling to emerging hardware and software technologies for database and information systems, from data management on novel architectures (cloud and MapReduce environments) to interesting and novel machine learning algorithms for effectively supporting decision-making process. Meanwhile, the general idea behind each workshop event was to collect contributions from various domains representing new trends in the broad research areas of databases and information systems. Specifically, the following workshop events have been organized:

- The 1st Workshop on Novel Techniques for Integrating Big Data (BigNovelTI 2017)
- The 1st International Workshop on Data Science: Methodologies and Use-Cases (DaS 2017)
- The 2nd International Workshop on Semantic Web for Cultural Heritage (SW4CH 2017)
- The 1st Workshop on Data-Driven Approaches for Analyzing and Managing Scholarly Data (AMSD 2017)

Each workshop had its own international program committee, whose members served as the reviewers of papers included in the rest of this volume. In the following, for each workshop, a brief introduction of its main motivations and topics of interest is presented, as well as its program including interesting keynotes, invited papers and the selected papers for presentations. These papers have been included in this volume. Finally, some acknowledgements from the workshop organizers are provided.

2 The 1st Workshop on Novel Techniques for Integrating Big Data (BigNovelTI 2017)

Introduction. The 1st Workshop on Novel Techniques for Integrating Big Data (BigNovelTI 2017) has been organised by Diego Calvanese (Free University of Bozen-Bolzano, Italy) and Oscar Romero (Universitat Politècnica de Catalunya, Spain).

The challenges posed by Big Data require novel data integration techniques beyond Data Warehousing (DW), which has been the in-company de-facto standard for data integration. On the one hand, bottom-up data integration approaches, performing a virtual integration, are a good fit for integrating disparate autonomous and heterogeneous data sources in large-scale distributed decision support systems, a current trend to contextualise the in-house data. Virtual data integration however tends to suffer from poor performance, which hinders the right-time analysis approach, and needs to be adapted and combined with materialisation to meet the new requirements brought by Big Data. On the other hand, variety and the need to deal with external non-controlled sources in an automatic way require to look at this problem from a broader perspective than the one of traditional data management, and semantics need to be included in the data integration processes. In such setting, domain knowledge is represented in an ontology, and inference over such knowledge is used to support data integration. However, this approach poses significant computational challenges that still need to be overcome, and hence its potential has not been fully exploited yet in real world applications. For these reasons, (i) providing different degrees of coupling of the integrated data sources based on their heterogeneity and autonomy, and (ii) dealing with and integrating semantics as a first-class citizen are open questions for novel scenarios.

Program. The workshop included a panel session and four selected research papers. The panel session sought to promote the interaction and collaboration between the data management and knowledge representation communities, which tackle data integration from different perspectives. As main outcome, promising research lines in this research area were presented and discussed.

The four selected papers address different interesting research issues related to data integration for Big Data. Brandt et al. presented a framework for temporal ontology-based data access (OBDA). In particular, compared to the standard OBDA, the framework envisions the addition of static and temporal rules, where the latter are based on $datalog_{nr}MTL$. Ibáñez et al. introduced a collaborative approach that stores queries in a multidimensional-aware (based on an extension of the QB4OLAP RDF vocabulary) knowledge base that is then analysed to help users assessing semantic consistency of aggregation queries in self-service BI contexts. Theodorou et al. presented an adaptive Big Data reference architecture, named *Theta*, thought to introduce flexibility in the data analytics part of the system enabling optimization at runtime to meet the level of data quality required by the data analyst. Finally, Trajcevski et al. introduced the concept of knowledge-evolution trajectory as a model to formalise the spatial, temporal, and content-based aspects of scientific publications and enable novel querying techniques based on these three aspects.

Acknowledgements. The BigNovelTI chairs would like to acknowledge the help of all PC members. The alphabetically ordered list of PC members contais: Alberto Abelló (Universitat Politècnica de Catalunya, Spain), Andrea Calì (Birkbeck College London, U.K.), Theodoros Chondrogiannis (Free University of Bozen-Bolzano, Italy), Martin Giese (University of Oslo, Norway), Gianluigi Greco (University of Calabria, Italy), Katja Hose (Aalborg University, Denmark), Petar Jovanovic (Universitat Politècnica de Catalunya, Spain), Maria C. Keet (University of Cape Town, South Africa), Roman Kontchakov (Birkbeck College London, U.K.), Antonella Poggi (Sapienza University of Rome, Italy), Mantas Simkus (TU Vienna, Austria), Sergio Tessaris (Free University of Bozen-Bolzano, Italy), Christian Thomsen (Aalborg University, Denmark), Stefano Rizzi (University of Bologna, Italy), Alejandro Vaisman (Instituto Tecnológico de Buenos Aires, Argentina), Stijn Vansummeren (Université Libre de

Bruxelles, Belgium), Panos Vassiliadis (University of Ioannina, Greece), Robert Wrembel (Poznan University of Technology, Poland), and Esteban Zimanyi (Universit Libre de Bruxelles, Belgium).

3   The 1st International Workshop on Data Science: Methodologies and Use-Cases (DaS 2017)

Introduction. The 1st International Workshop on Data Science: Methodologies and Use-Cases (DaS 2017) has been organized by Tania Cerquitelli (Politecnico di Torino, Italy), Silvia Chiusano (Politecnico di Torino, Italy), and Natalija Kozmina (University of Latvia, Latvia).

Data science is an interdisciplinary field about scientific processes, methodologies, and systems to extract useful knowledge or insights from data in various forms. Data can be analyzed using data mining, machine learning, data analysis and statistics, optimizing processes and maximizing the knowledge exploitation in real-life settings. DaS 2017 is a workshop aimed at fostering and sharing research and innovation on data science. The workshop allows researchers and practitioners to exchange their ideas and experiences on designing and developing data science applications, discuss the main open issues, and share novel solutions for data management and analytics.

The DaS topics of interest include methodologies, models, algorithms, and architectures for data science, scalable and/or descriptive data mining and machine learning techniques for knowledge discovery, data warehouses and large-scale databases, and experiences with data-driven project development and deployment in various application scenarios.

Keynote Presentation. Rossano Schifanella (computer scientist at the University of Turin) gave a talk entitled *Combined Effect of Content Quality and Social Ties on User Engagement*. He is a visiting scientist at Nokia Bell Labs, Cambridge, UK, and a former visiting scientist at Yahoo Labs and at Indiana University School of Informatics and Computing where he was applying computational methods to model social behavior in online platforms. His research mainly focuses on a data-driven analysis of the behavior of (groups of) individuals and their interactions on social media platforms.

In his talk, Rossano presented a large scale study on the complex intertwinement between quality, popularity, and social ties in an online photo sharing platform, proposing a methodology to democratize exposure and foster long term users engagement.

Invited Paper. Genoveva Vargas-Solar contributed to DaS with an invited paper entitled *Efficient Data Management for Putting forward Data-centric Sciences*. She is a senior scientist of the French Council of Scientific Research (CNRS) and since 2008, Genoveva is deputy director the Franco-Mexican Laboratory of Informatics and Automatic Control (LAFMIA) an international research unit established at CINVESTAV. Her research interests concern distributed and heterogeneous databases, reflexive systems and service based database systems. She conducts fundamental and applied research activities for addressing these challenges on different architectures ARM, raspberry, cluster, cloud, and HPC.

The novel and multidisciplinary data-centric and scientific movement promises new and not yet imagined applications that rely on massive amounts of evolving data that need to be cleaned, integrated, and analysed for modelling, prediction, and critical decision making purposes. In her paper, Genoveva explains the key challenges and opportunities for data management in this new scientific context, and discusses how data management can best contribute to data-centric sciences applications through clever data science strategies.

Selected Papers. The DaS workshop is composed of 11 papers discussing a variety of interesting research issues, application domains, and experience reports in the area of data science. Specifically, the workshop is composed of three sections covering different aspects in the area of *data analytics for data science* (4 papers authored by Datta et al., Cagliero et al., Saleh and El-Tazi, and Podapati et al.), *data management for data analytics* (3 papers authored by Galkin et al., Bobrov et al., and Borzovs et al.), and *data science use cases* (4 papers authored by Haq and Wilk, Quemy, Gončarovs and Grabis, and Hernandez-Mendez et al.). In the following we present the research papers for each of the three areas above.

In the *Data analytics for data science* section, the paper entitled *Parallel Subspace Clustering using Multi-core and Many-core Architectures* (Datta et al.) presents a multi-core approach exploiting CPU computation to parallelize the SUBSCALE algorithm. The authors report results over performances and discuss the concurrency problems that effect the parallelization of the SUBSCALE algorithm. A novel pattern, named Generalized High-Utility Itemset (GHUI), has been proposed in the paper entitled *Discovering High-utility Itemsets at Multiple Abstraction Levels* (Cagliero et al.) to represent sets of item groups, each one characterized by a high total profit. The GHUI pattern exploits the available data taxonomies to model high-utility itemsets at different abstraction levels. The significance of the proposed pattern and the performance of the GHUI mining algorithm have been evaluated on retail data with the aim of planning advertising campaigns of retail products. Aimed at grouping tags associated with the StackOverflow posts in semantically coherent clusters, a technique based on the Latent Dirichlet Allocation (LDA) has been proposed in the paper entitled *Automatic Organization of Semantically Related Tags using Topic Modelling* (Saleh and El-Tazi). First, by analyzing post texts through LDA, topics are discovered and associated with the posts. Then, groups of tags associated with the topics are built. A measure to evaluate the quality of the extracted groups is presented. The approach has been experimentally validated on a real dataset extracted from StackOverflow. The paper entitled *Fuzzy Recommendations in Marketing Campaigns* (Podapati et al.) proposes a technique based on fuzzy logic to put in relationship the proneness of the customers of telecommunication companies to be infrastructure-stressing (i.e., operating in zones of high demand that might produce network service failures) with geo-demographic customer segments. The aim is to shape the marketing campaigns understanding whether there are segments of users that are too infrastructure-stressing and should be avoided.

In the *Data management for data analytics* section, the paper entitled *Towards a Multi-way Similarity Join Operator* (Galkin et al.) puts forward an approach to overcome limitations of most of the existing query engines that rely on binary join-based query planners and execution methods with complexity depending on the number of data sources involved. The authors propose a multi-way similarity join operator (MSimJoin) that accepts more than two inputs and is able to identify duplicates corresponding to similar entities. In the paper titled *Workload-Independent Data-driven Vertical Partitioning*, Bobrov et al. introduce a new class of vertical partitioning algorithms that don't exploit workload information and are data-driven. The proposed algorithm relies on the database logical schema to extract functional dependencies from tables and perform partitioning accordingly. The authors experimentally compared their algorithm with two existing workload-dependent algorithms. Borzovs et al. in the paper entitled *Can SQ and EQ Values and their Difference Indicate Programming Aptitude to Reduce Dropout Rate?* discuss the inadequacy of metrics and models available in the reviewed literature when predicting the potential dropouts among students of Computer Science study programs. Gained results in their empirical evaluation disprove the hypothesis that students' programming aptitude would correlate with the systemizing quotient (SQ), empathy quotient (EQ), and its difference.

Four interesting experience reports respectively in health care, law and justice, financial, and business domains are discussed in the *Data science use cases* section. A critical issue in health care domain is the ability to predict the proper treatment for patients affected by a certain pathology. The paper entitled *Fusion of Clinical Data: A Case Study to Predict the Type of Treatment of Bone Fractures* (Haq and Wilk) presents the application of data fusion techniques, such as combination of data (COD) and interpretation (COI) approaches, in building a decision model for patients with bone fractures to distinguish between those who have to undergo a surgery and those who should be treated non-surgically. An extensive and critical review of the current state-of-the-art research in data science techniques for law and justice is presented in the paper *Data Science Techniques for Law and Justice: Current State of Research and Open Problems* (Quemy). The author describes the difficulties of analyzing legal environments, identifies four fundamental problems, and discusses how they are covered by the current top approaches available in the literature. In the paper entitled *Using Data Analytics for Continuous Improvement of CRM Processes: Case of Financial Institution*, Gončarovs and Grabis present a case study where data mining strategies have been exploited to identify the Next Best Offer (NBO) for selling financial products to bank's customers. Based on the experience gained on this case study, the authors eventually point out that the interaction between business owners and data analysts is necessary to improve the models. Aimed at supporting the business ecosystems modeling (BEM) management process, the Business Ecosystem Explorer (BEEx) tool is introduced in the paper entitled *Towards a Data Science Environment for Modeling Business Ecosystems: The Connected Mobility Case* (Hernandez-Mendez et al.). BEEx empowers end-users to adapt not only the BEM but also the visualizations. The authors describe their experience on modeling the Connected Mobility Ecosystem in the context of the project TUM Living Lab Connected Mobility (TUM LLCM).

Acknowledgements. The DaS co-chairs would like to acknowledge the help of all PC members who provided comprehensive, critical, and constructive comments. The alphabetically ordered list of PC members contains: Julien Aligon (University of Toulouse - IRIT, France), Antonio Attanasio (Istituto Superiore Mario Boella, Italy), Agnese Chiatti (Pennsylvania State University, USA), Evelina Di Corso (Politecnico di Torino, Italy), Javier A. Espinosa-Oviedo (Barcelona Supercomputing Center, Spain), Göran Falkman (University of Skövde, Sweden), Fabio Fassetti (Università della Calabria, Italy), Patrick Marcel (University François Rabelais of Tours, France), Kjetil Nørvåg (Norwegian University of Science and Technology, Norway), Erik Perjons (Stockholm University, Sweden), Emanuele Rabosio (Politecnico di Milano, Italy), Francesco Ventura (Politecnico di Torino, Italy), Gatis Vītols (Latvia University of Agriculture, Latvia), Xin Xiao (Huawei Technologies Co., Ltd., China), José Luis Zechinelli Martini (Universidad de las Americas - Puebla, Mexico).

4   The 2nd International Workshop on Semantic Web for Cultural Heritage (SW4CH 2017)

Introduction. The 2nd International Workshop on Semantic Web for Cultural Heritage (SW4CH 2017) has been organized by Béatrice Markhoff (University of Tours, France) and Stéphane Jean (ENSMA, Poitiers, France). The aim of this workshop is to bring together interdisciplinary research teams involved in Semantic Web solutions for Cultural Heritage. Interdisciplinarity is a key point in this field, as Software Engineers, Data Scientists, and various Humanity and Social Scientists have to work together for inventing, devising and implementing novel ways of digital engagement with heritage based on the semantic web principles, data, and services. It is crucial to have a place to exchange experiences, present states of the art, and discuss challenges. The University of Tours, ISAE-ENSMA Engineering School, and University of Poitiers are involved in such interdisciplinary projects, within the scope of the Intelligence des Patrimoines (I-Pat) programme, funded by the Région Centre Val de Loire, and led by the CESR (Centre d'tudes Supérieur de la Renaissance), a DARIAH laboratory and one of the principal European laboratories in the field of the Renaissance.

Invited Paper. An invited talk has been given by Carlo Meghini, prime researcher at CNR-ISTI and the head of the Digital Libraries group in the NeMIS Lab of ISTI. His presentation was about *Narratives in Digital Libraries*. One of the main problems of the current Digital Libraries (DLs) is the limitation of the discovery services offered to the users, which typically boil down to returning a ranked list of objects in response to a natural language query. No semantic relation among the returned objects is usually reported, which could help the user in obtaining a more complete knowledge on the subject of the search. The introduction of the Semantic Web, and in particular of the Linked Data, has the potential of improving the search functionalities of DLs. In order to address this problem, Carlo Meghini introduced narratives as new first-class objects in DLs, and presented preliminary results on this endeavor.

Selected Papers. The workshop selected 6 papers addressing interesting research issues in the context of Semantic Web for Cultural Heritage. In this field, research teams work in knowledge engineering for building, extending and using ontologies, for representing their data. Ontologies are clearly identified as an integration means, a base for interconnecting projects and datasets. In this context, the CIDOC-CRM plays a central role, as shown in the number of projects that use it.

The first presented paper, *Introducing Narratives in Europeana: Preliminary Steps* (C. Meghini et al.), reports on a particular case of the work introduced in the invited talk, namely using a tool developed to enrich the Europeana content. A narrative is here defined as a semantic network that consists of linked events, their spatio-temporal properties, participating entities, and objects included in the digital collection, which can be represented using the CIDOC-CRM.

The second paper, *Evaluation of Semantic Web Ontologies for Modelling Art Collections* (D. Liu et al.), presents an evaluation of three existing ontologies used in the Cultural Heritage field, CIDOC-CRM, the Europeana Data Model, and VRA (Core), with respect to the representation of Art Collections. The comparison is based on the concrete modelling of four different artworks.

The third paper, *The CrossCult Knowledge Base: a co-inhabitant of cultural heritage ontology and vocabulary classification* (A. Vlachidis et al.), discusses design rationales of the CrossCult Knowledge Base, devised to support a cross-cultural public access to European History, fully compliant with the CIDOC-CRM ontology and equipped with specific thesauri. Several choices made during the design process which involved inter-disciplinary teams are discussed.

The fourth paper, *The Port History Ontology* (B. Rohou et al.), describes the first results of a multidisciplinary research project on an extension of the CIDOC-CRM dedicated to the history of ports. The Port History Ontology is mainly based on a pre-defined model representing the spatio-temporal evolution of ports.

The fifth paper, *A WordNet ontology in advancing search digital dialect dictionary* (M. Mladenović et al.), introduces an ontology-based method for connecting a standard language (Serbian) and a dialect in Serbian to improve the search over the dialect dictionary using keywords entered in the standard language. The proposal defines SWRL rules to infer new synonyms.

The last presented paper, *When it comes to Querying Semantic Cultural Heritage Data* (B. Markhoff et al.), provides an overview of projects and approaches for querying semantic web data, including ontology-based mediation, federated query systems and full web querying, with a focus on solutions already used for semantic Cultural Heritage data.

All in all, on the one hand this second edition of SW4CH clearly reflects the increasing number of CIDOC-CRM uses, for modeling a variety of CH domains, and, on the other hand, it highlights the development of ontology-based methods for connecting, integrating and querying several Digital Cultural Heritage resources.

Acknowledgements. The SW4CH co-chairs would like to acknowledge the help of all PC members, who carefully evaluated each submission and provided useful comments to all authors. The alphabetically ordered list of PC members contains: Trond Aalberg (NTNU, Norway), Carmen Brando (EHESS-CRH, France), Benjamin Cogrel (Free University of Bozen-Bolzano, Italy), Donatello Conte (University of Tours, France), Peter Haase (Metaphacts, Germany), Mirian Halfeld Ferrari Alves (University of Orléans, France), Katja Hose (Aalborg University, Denmark), Efstratios Kontopoulos (University of Thessaloniki, Greece), Cvetana Krstev (University of Belgrade, Serbia), Nikolaos Lagos (Xerox, France), Denis Maurel (University of Tours, France), Carlo Meghini (CNR/ISTI Pisa, Italy), Isabelle Mirbel (University of Nice Sophia Antipolis, France), Alessandro Mosca (SIRIS Academic Barcelona, Spain), Dmitry Muromtsev (ITMO University, Russia), Cheikh Niang (AUF Paris, France), Xavier Rodier (University of Tours, France), Maria Theodoridou (FORTH ICS, Heraklion, Crete, Greece), Genoveva Vargas-Solar (University of Grenoble, France), Dusko Vitas (University of Belgrade, Serbia).

5   The 1st Workshop on Data-Driven Approaches for Analyzing and Managing Scholarly Data (AMSD 2017)

Introduction. The 1st Workshop on Data-Driven Approaches for Analyzing and Managing Scholarly Data (AMSD 2017) was organized by Andreas Behrend (University of Bonn, Germany) and Christiane Engels (University of Bonn, Germany).

Supporting new forms of scholarly data publication and analysis has become a crucial issue for scientific institutions worldwide. Even though there are already several search engines, bibliography websites and digital libraries (e.g., Google Scholar, CiteSeerX, Microsoft Academic Search, DBLP) supporting the search and analysis of publication data, the offered functionality is still rather limited. For example, the provided keyword-based search does not employ the full opportunities an advanced semantic search engine could offer. In addition, a comprehensive content analysis beyond title and abstract as well as the capability to take the quality and relevance of the publications into account is not provided so far.

International community forums such as FORCE11 and Research Data Alliance (RDA) already engage for open and semantic publishing which enable various opportunities for new analytics on scholarly data. The global vision is a uniform system combining all forms of scholarly data offering a comprehensive semantic search and advanced analytics including:

- Knowledge extraction and reasoning about scientific publications and events,
- Quality assessment of scientific publications and events, e.g. via rankings or quality metrics,
- Recommender systems for conferences and workshops, and
- Identification of research schools and key papers or key researcher within a field using citation analysis.

In this first edition of the AMSD workshop, we want to encourage especially researchers from the database community (but also from the fields information retrieval, digital libraries, graph databases, machine learning, recommender systems, visualization, etc.) to combine their resources in order to exploit the possibilities of scholarly data analysis. The main goals of the AMSD workshop are to facilitate the access of scholarly data, to enhance the ability to assess the quality of scholarly data and to foster the development of analysis tools for scholarly data. The following topics will be addressed:

- Managing scholarly data, i.e. representation, categorization, connection and integration of scholarly data in order to foster reusability, knowledge sharing, and analysis,
- Analyzing scholarly data, i.e. designing and implementing novel and scalable algorithms for knowledge extraction and reasoning about and assessing the quality of scientific publications and events with the aim of forecasting research trends, establishing recommender systems, and fostering connections between groups of researchers,
- Applications on scholarly data, i.e. providing novel user interfaces and applications for navigating and making sense of scholarly data and highlighting their patterns and peculiarities.

Program. The workshop consists of 3 accepted papers addressing interesting research challenges to support new forms of scholarly data publication and analysis. In the paper entitled *Publication Data Integration As a Tool for Excellence-Based Research Analysis at the University of Latvia* a publication data integration tool is presented designed and used at the University of Latvia for excellence-based research analysis. Besides the usual bibliographic data about authors (name, affiliation) and publications, quantity, quality and structural bibliometric indicators are included for a excellence-based analysis. In the paper *Evaluating Reference String Extraction Using Line-Based Conditional Random*

Fields: A Case Study with German Language Publications the authors address the problem of extracting individual reference strings from the reference section of scientific publications. They propose an approach named RefExt that apply a line-based conditional random fields rather than constructing the graphical model based on the individual words, dependencies and patterns that are typical in reference sections provide strong features while the overall complexity of the model is reduced.

Finally, in the paper *CEUR Make GUI A usable web frontend supporting the workflow of publishing proceedings of scientific workshops* a new GUI web front end for facilitating the generation of proceedings in CEUR-WS is described. A previously developed command line tool has proven to be not feasible for this purpose. The performed user evaluation shows a 40% less execution time of the GUI variant compared to the command line version and rather positive usability scores.

Keynote Presentations. This year, two keynote presentations were offered at AMSD. The first talk was provided by Johann Gamper from the University of Bozen-Bolzano who is an expert in temporal data management. In his talk he discussed key requirements for managing temporal data in current database systems. In particular, he showed how the requirements can be mapped to simple and powerful primitives for the relational model, and identifies a range of open problems when dealing with so-called long data.

The second talk was given by Andreas Behrend from the University of Bonn who works on a cloud-based tool for managing and analyzing publication data. In his talk he presented the OpenResearch project in which a semantic wiki is developed for storing and querying information about scientific events, research projects, publishers, journals, and articles.

Acknowledgements. The AMSD co-chairs would like to express their gratitude to all PC members, who carefully evaluated each submission and provided useful comments to the authors. This year, our PC consists of the following members: Sören Auer (TIB University of Hannover, Germany), Andreas Behrend (University of Bonn, Germany), Christiane Engels (University of Bonn, Germany), Said Fathalla (University of Bonn), Christoph Lange (Fraunhofer IAIS, Germany), Heba A. Mohamed (University of Bonn, Germany), Sara Soliman (University of Alexandria), Sahar Vahdati (Fraunhofer IAIS, Germany), Hannes Voigt (University of Dresden, Germany).

6   Conclusions

ADBIS 2017 workshop organizers would like to express their thanks to everyone who contributed to the success of their events and to this volume content. We thank the authors, who submitted papers to the workshops in the context of ADBIS 2017. Special thanks go to the Program Committee members as well as to the external reviewers of the ADBIS 2017 workshops, for their support in evaluating the submitted papers, providing comprehensive, critical, and constructive comments and ensuring the quality of the scientific program and of this volume.

Finally a special thank to ADBIS 2017 workshop chairs Johann Gamper (Free University of Bozen-Bolzano, Italy) and Robert Wrembel (Poznan University of Technology, Poland) to continuously support us during the ADBIS 2017 workshop organization time frame with useful suggestions and fruitful discussion.

We all hope you will find the volume content an useful contribution to promote novel ideas for further research and developments in the areas of databases and information systems. Enjoy the reading!

The 1st Workshop on Data-Driven Approaches for Analyzing and Managing Scholarly Data (AMSD 2017)

Publication Data Integration as a Tool for Excellence-Based Research Analysis at the University of Latvia

Laila Niedrite[✉], Darja Solodovnikova[✉], and Aivars Niedritis

Faculty of Computing, University of Latvia, Riga, Latvia
{laila.niedrite,darja.solodovnikova,
aivars.niedritis}@lu.lv

Abstract. The evaluation of research results can be carried out with different purposes aligned with strategic goals of an institution, for example, to decide upon distribution of research funding or to recruit or promote employees of an institution involved in research. Whereas quantitative measures such as number of scientific papers or number of scientific staff are commonly used for such evaluation, the strategy of the institution can be set to achieve ambitious scientific goals. Therefore, a question arises as to how more quality oriented aspects of the research outcomes should be measured. To supply an appropriate dataset for evaluation of both types of metrics, a suitable framework should be provided, that ensures that neither incomplete, nor faulty data are used, that metric computation formulas are valid and the computed metrics are interpreted correctly. To provide such a framework with the best possible features, data from various available sources should be integrated to achieve an overall view on the scientific activity of an institution along with solving data quality issues. The paper presents a publication data integration system for excellence-based research analysis at the University of Latvia. The system integrates data available at the existing information systems at the university with data obtained from external sources. The paper discusses data integration flows and data integration problems including data quality issues. A data model of the integrated dataset is also presented. Based on this data model and integrated data, examples of quality oriented metrics and analysis results of them are provided.

Keywords: Research evaluation · Research metrics · Data integration · Data quality · Information system · Data model

1 Introduction

Evaluation in science nowadays is turning into a routine work based on metrics [1]. In circumstances of a wide diversity of metrics, it is significant that they are chosen carefully, e.g. in compliance with research institution's strategic goals.

Research indicators can be used for different purposes [2]: science policy making at the state level, distribution of research funding, organization and management activities, e.g. in Human Resource Management for recruiting or promoting employees involved in research, content management and decisions at individual researchers'

© Springer International Publishing AG 2017
M. Kirikova et al. (Eds.): ADBIS 2017, CCIS 767, pp. 125–136, 2017.
DOI: 10.1007/978-3-319-67162-8_14

level, e.g. where and what to publish, and providing consumer information, e.g. university rankings that include science indicators. These usage examples correspond to the group of research performance indicators and do not include input indicators, such as number of researchers.

At the institutions level, the research evaluation can be performed to support the achievement of strategic goals. Whereas quantitative measures such as number of scientific papers, amount of funding, number of scientific staff and many others are commonly used for such evaluation, the strategy of the institution can be set to achieve ambitious scientific goals. Therefore, a question arises as to how more quality oriented aspects of the research outcomes can be measured. For example, at the state level to allocate funds to excellent institutions, measurement methods from three categories are used: peer review-based models, publication count-based models and citation-based models [3].

We will propose a data integration architecture for bibliometric information analysis, which is one of the research performance indicators. Bibliometric indicators can be also classified in more detail as quantity, quality and structural indicators [4]. A quantity indicator is, for example, number of publications. An example of a quality indicator is h-index. Structural indicators allow to evaluate connections, for example, co-authors from different fields, institutions or countries.

To supply an appropriate dataset for evaluation of both types of metrics for measuring quantitative and qualitative aspects, a suitable framework should be provided, that ensures that neither incomplete, nor faulty data are used, that metric computation formulas are discussed and are valid and the computed metrics are interpreted correctly. To provide such a framework with the best possible features, data from various available sources should be integrated to achieve an overall view on the scientific activity of an institution along with solving data quality issues.

The principles characterizing the best practice in metrics-based research assessment are given in the "Leiden manifesto" [1]. Among these principles, some of them should be considered when building a data collection and integration system for effective science evaluation, for example, data collection should be transparent, the institutions and persons that are evaluated can verify the data provided for evaluation, the indicator values should be updated regularly.

Research information management has many properties that are typical for data integration scenarios: many data sources with inconsistent data models, heterogeneity, and many involved stakeholders with diverse goals [5]. Knowledge from data integration field can be used to simplify the data collection and integration in research information field. For example, a uniform data model or standard can be used [5].

One such standard is CERIF (Common European Research Information Format), which includes information about projects, persons, publications, organizations, and other entities. Many research information systems in Europe are built interrelated with this standard [6]. However, there are other standards and models that are significant to describe research information, among them DOI (Digital Object Identifier) to identify publications and ORCID (Open Researcher and Contributor ID) to identify authors of publications.

Today there are many efforts trying to implement information systems to support research evaluation activities. Institutions develop their own or use commercial or non-commercial products to maintain data about research results. Many information sources have been used at the University of Latvia (LU) for a while to get the insight into the actual situation with research activities and their outcomes, but this process needed improvements by providing an integrated information oriented to scientific excellence. The requirements for research evaluation in Latvia are declared in the regulations issued by the government and prescribe how the funding for scientific institutions is calculated [7, 8]. As stated in these regulations, the productivity of scientific work is evaluated according to the number of publications indexed in Scopus or Web of Science (WoS). These quantitative data must be extended with data necessary for computation of qualitative indicators.

The paper presents a publication data management system for excellence-based research analysis at LU. The system integrates data available at the university information system with data from the library information system as well as with data obtained via API from Scopus and WoS databases. The paper discusses data integration flows and data integration problems including data quality issues. A data model of the integrated dataset is also presented. Based on this data model and integrated data, examples of quality oriented metrics and analysis results of them are provided.

2   Related Work

A research information system in Scandinavia [9] is an example of such system that is implemented and used in Denmark, Finland, Norway, and Sweden and mostly contains integrated, high quality bibliometric data. The system is used for performance-based funding. It is remarkable, that this system also has its own publication indicator that by weighting the results from different fields allows to compare them.

In the field of data integration, the recent approaches are focused on mappings between models and integration processes. Also in research information systems, mappings between different systems should be considered as important parts, they provide specifications for data integration processes [5].

The Polish Performance-research funding system allows to evaluate 65 parameters. 962 research units provided data about more than a million research outcomes for the 4-year period. The data collection process was performed through submission of the questionnaire through the Information System on Higher Education in Poland. The study [10] was performed to find out the most important metrics to facilitate the transition to more targeted system to meet the excellence requirements, where only the most important metrics are reported. The research showed that many of existing metrics are not significant for the evaluation.

Italian experience [11] shows the implementation of a research information system in Italy, where 66 Italian institutions introduced IRIS, that is a system based on DSpace [12] and customized for Italian environment. ORCID also was used at a national level. Entities, attributes and relations in this system are compliant with the CERIF ontology [13]. The collected huge amount of data allowed to understand the whole situation in research and, for example, to develop new publication strategies. The authors of the study mention also the problems with the data quality, when not all institutions control the data collection process and do not implement data validation processes of data provided by researchers.

3   Data Model

To support analysis and evaluation of the scientific activity of LU members involved in research, both employees and students, we propose a system architecture that integrates information about publications from all accessible data sources that include the library information system ALEPH, LU management information system LUIS, SCOPUS and WoS databases. To maintain publication information and support reporting and analysis, the publication data from multiple sources are linked and stored in a repository in LUIS. The data model of the repository is depicted in Fig. 1.

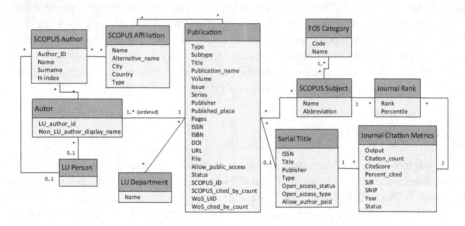

Fig. 1. Publication repository data model

The central class to store bibliographical data about a publication as well as number of citations in SCOPUS and WoS databases is *Publication*. The bibliographical data in this class are entered by the author of the publication or the faculty staff or they are populated during the data loading process from ALEPH or SCOPUS databases. For each publication indexed by SCOPUS or WoS, we also include the corresponding number of citations as well as publication identifiers used in both databases to maintain a link with SCOPUS and WoS as data sources.

The information about authors of the publication is reflected by the classes *Author* and *SCOPUS Author*. The class *Author* represents ordered authors of the publication, which are affiliated with LU as well as with other institutions.

Authors recognized as affiliated with LU are also linked to *LU Person* class, which stores personal and other information used for different functionality of LUIS. For foreign authors, we store only their name as it appears on the paper. If a publication is indexed by SCOPUS, we also collect author information from it, which includes name, surname, H-index and author ID assigned by SCOPUS, which is used in the author matching process.

We also store the information about the affiliation of the publication with LU department or faculty, which is represented as the class *LU Department*. This information is obtained automatically from data about the work place of the author and may be corrected by the responsible faculty or library staff. If the publication is indexed by SCOPUS, we also store the information about the institutions the publication authors are affiliated with in the class *SCOPUS Affiliation*. This information is necessary to analyze connections with co-authors from different institutions or countries.

For the analysis of the quality of publications, we use not only the citation number, but also other citation metrics provided by SCOPUS. The absolute values of such metrics are calculated for journals and serial conference proceedings yearly and their values are represented by the class *Journal Citation Metrics*. We also collect information about the open access status of the journal or conference proceedings and include it in the class *Serial Title*. In SCOPUS database, journals are also ranked among other journals that belong to the same subject areas according to their values for CiteScore metric [14], an alternative to WoS Journal Impact Factor. Journal rank information is represented by the class *Journal Rank* and it is connected with the corresponding subject area (class *SCOPUS Subject*). For reporting on publications of different OECD categories, we store the correspondence of SCOPUS subject areas and Field of Science and Technology (FOS) categories.

4   Scenarios of Obtaining Publication Data

To accumulate the most complete list of publications authored by LU staff and students in the central repository in LUIS, we gather publication data from different sources, link publications to the correspondent members of LU staff, correct errors and duplicates and provide the consolidated information using a set of reports used by the management of LU. In the following section the 4 scenarios of obtaining publication data as well as data flows related to each scenario are discussed.

4.1   Publication Data Added by Authors

LU employees, PhD and Master's degree students are able to add information about their co-authored publications to the repository themselves. The process when publication data are entered by authors is depicted in Fig. 2. LUIS system maintains user profiles for all LU members. Among various other information about a user, a profile includes a section devoted to research, which in turn contains a list of author's publications obtained from the repository. An author can supplement this list by adding newly published articles or articles, which were not loaded automatically. Before adding them, an author is automatically requested to search for his/her publications in the LUIS repository and library information system ALEPH, which are not linked with the author's profile, to avoid creation of duplicates. If a desired article is found, it is

Fig. 2. Publication data added by authors

possible to add it to the profile. In this case, the author does not need to supply any additional information about the article.

If, however, the article is not present in either of the systems, the author has to specify the type and subtype of the publication (for example, journal article, book chapter, book, etc.) and supply bibliographical information. Besides, an author must indicate the status of the publication: published, submitted for publication, developed or under development, attach publication file (full text or book cover) and indicate whether it can be accessed publicly at the e-resource repository of LU. After the author has finished entering publication data, they are transferred to the library information system ALEPH. To ensure the best possible data quality, members of the library staff validate publication data, correct any errors if present and approve a publication. Finally, publication data are synchronized back with LUIS and become available for evaluation and reports.

Besides entering new data, LUIS users can unlink mistakenly attached publications from their profiles, which were automatically loaded to the repository from other sources or confirm that author matching was performed correctly.

4.2 Publication Data Added by Faculty Staff

Data about publications authored by the faculty members can also be entered into the repository by specially designated faculty staff (Fig. 3). The procedure for adding data is similar to the one that is performed by authors. The differences are that faculty staff can record data about publications authored by other faculty members, correct erroneous links between publications and authors and adjust the list of affiliated LU departments.

Fig. 3. Publication data added by faculty staff

4.3 Publication Data Obtained from SCOPUS and Web of Science

There are two external sources of publication data that are used in the data loading process: SCOPUS and WoS citation database systems. SCOPUS offers API, which allows to search for publications authored by LU members and extract bibliographical data of such publications as well as various citation metrics. The extraction and loading process (Fig. 4) is run daily. Articles that were published during the last 2 years are inserted into the repository or updated daily, but all other publications are updated on a weekly basis. Bibliographical information is extracted from SCOPUS and loaded into the repository table which corresponds to the class *Publication* of the repository data model. Data about authors (unique author identifier, name, surname, H-index) and publication and author's affiliation are loaded into tables which correspond to the classes *Author, SCOPUS Author* and *SCOPUS Affiliation* of the repository data model.

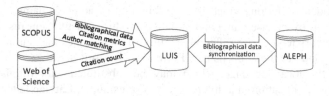

Fig. 4. Publication data obtained from SCOPUS and Web of Science

Affiliations are associated with authors as well as with publications directly. In addition to bibliographical and author information, citation metrics are also obtained that include current number of citations of individual publications as well as citation metrics obtained for the particular journal or conference proceedings: Source Normalized Impact per Paper (SNIP) [15], the SCImago Journal Rank (SJR) [16], CiteScore [14]. Previously, it was possible to obtain Impact per Publication (IPP) metric [17], which is not available from SCOPS anymore, so this number is retained for previously loaded publications and is not loaded for the new ones.

The first step of the SCOPUS data loading process that is executed on any new publication is a recognition phase. The main goal of this phase is to identify publications that are already registered in the repository, but that are newly indexed by SCOPUS, to avoid creation of duplicates. The first criterion used for the recognition is Document Object Identifier (DOI) which is unique for every publication. If the matching publication with the same DOI is not found in the repository, the search based on the similar title and publication year is performed. To determine the existing publication with the most similar title in the repository, Jaro-Winkler similarity [18] is used because there may be different alternatives of title spelling as well as data quality issues are sometimes present. Different thresholds for Jaro-Winkler similarity were tested and experimental evaluation of matching results revealed that the most suitable threshold is 0,93 and currently this coefficient is used to consider titles of publications similar.

If the recognition process detects an existing publication in the repository or if a publication has already been previously updated with SCOPUS data, we update the number of citations for the publication as well as journal citation metrics and establish a link with a corresponding Scopus record for newly indexed publications by means of filling SCOPUS ID attribute of the *Publication* class (Fig. 1).

If a processed publication is new to the system, a new instance of the *Publication* class is created with all the bibliographical information obtained from SCOPUS database, publication authors are also represented as instances of *Author* and *SCOPUS Author* classes and citation metrics as well as journal rank data are created or updated if information about a journal has been previously loaded.

In case of a new publication, the author matching process is performed, when for each author affiliated with LU, a corresponding instance of *LU Person* class is searched for and, if found, it is associated with a corresponding instance of the *Author* class. The primary criterion for author matching process is the SCOPUS author identifier, which allows to uniquely identify authors, whose publications have been previously loaded into the repository. If author search by identifier is unsuccessful, the matching process uses the secondary criterion, which is a combination of author's name and surname. Author matching by precise name and surname produces insufficient results, because publication authors tend to use different spelling of their names and surnames that not always corresponds to their full names. Furthermore, a full name can contain special characters of the local language, which may be substituted with English language characters in the publication in a different way. For example, Latvian letter 'Š' may be substituted with English letter 'S' or with two symbols 'SH'. To solve this data quality issue, the author matching based on names and surnames is performed using Jaro-Winkler similarity between the full name as it appears in the publication and official author's full name, i.e. *LU Person* instance with the highest Jaro-Winkler similarity coefficient that exceeds a threshold is linked to the publication. We use the same threshold for similarity coefficient 0.93, which was selected based on the experimental matching results. After a match is found, we also establish an association between the instance of the *SCOPUS Author* class and the corresponding instance of the *LU Person* class to use SCOPUS author identifier as the primary criterion for matching future publications.

When a new publication is loaded into the repository, the second process phase – publication data synchronization with the library information system ALEPH is executed. During this phase, publication data are exported to ALEPH, bibliographical information is supplemented and any errors are manually corrected by the library staff to maintain the best possible data quality and finally the updated data are imported back to the repository.

Another data source used in the integration process is WoS web services. The version of web services available at the University of Latvia does not include journal citation metrics and provides only limited bibliographical information, number of citations, full names and sometimes Researcher identifiers of publication authors if they have registered such identifier in WoS database. Since the affiliation of authors is not available, we have discovered that author matching process for WoS data produces too many incorrectly identified authors, therefore, a decision has been made to add new Web of Science publications to the repository manually. Besides, the integration process also matches publication data obtained from WoS with publications already available at the repository. Just as for publications loaded from SCOPUS, the primary criterion used for matching is DOI and the secondary criteria are title and publication year. The integration process regularly updates WoS number of citations for recognized publications.

4.4   Publication Data Added by Library Staff

There is a considerable number of publications that are not indexed by SCOPUS and are authored by LU members, especially in the humanities. The information about such publications is necessary to perform accurate evaluation of the scientific activity of the institution. Therefore, library staff members manually add bibliographical information about publications to the library system ALEPH (Fig. 5). This is done when a librarian comes across a new publication authored by any LU member in some journal or conference proceedings or when the information about a new publication is obtained from the list of recently indexed publications in Web of Science database, which is

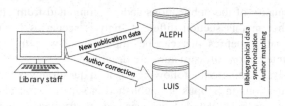

Fig. 5. Publication data added by library staff

monthly distributed by Web of Science. When new publication data appear in ALEPH, a synchronization process is conducted that integrates bibliographical information into the repository to ensure that it always represents an overall view on publication data.

The synchronization process includes author matching phase. During this phase, for each author of a publication, a corresponding LU person is searched for, using the same algorithm based on the full name similarity as in other matching phases. If the corresponding LU person is found, the publication is attached to his/her LUIS profile. We use the same minimal Jaro-Winkler similarity coefficient 0,93 to consider names similar.

In addition to entering new publications to ALEPH, library staff members are also responsible for correcting and supplementing bibliographical data for publications added by authors and by faculty staff and for publications data imported from Scopus database.

5 Data Analysis: Case Study

The context of the data collection and integration can be described with the total number of publications of LU researchers for the last 30 years that equals to 42417. 6967 publications out of them are indexed by Scopus and 7764 publications are indexed by WoS. The case study is performed using data that corresponds to the LU Faculty of Computing and the time frame that was chosen for research results evaluation was 2012–2016. We have already performed an initial evaluation of research performance by means of quantitative metrics [19]. To give a context for the further data analysis some numbers, e.g. publication count and Scopus publications, should be mentioned. The total number of publications is decreasing at the faculty from 99 in year 2012 to 79 in year 2015, but the number of Scopus publications is growing and in year 2015 was 55.

The goal of the case study was to define metrics based on the data attributes provided by the data model of the integrated publication information system, and with the goal to evaluate the quality of the publications, to find out the positive trends as well as the problems with the quality.

The quality aspects of a publication can be indirectly described with the source quality characteristics, e.g. journal quartiles that are computed from the citation count of all journal publications, because we can presume that the journal with the highest quartile Q1 will accept the best publications. Another group of quality indicators directly describe quality of the publications and are computed from citation counts of publications. Further in this section different analysis scenarios for research output evaluation with quality metrics that can be implemented with the new publication module and data integration infrastructure are described.

For the 1st analysis scenario the following research question was formulated: "How many faculty publications in Scopus are published in sources with and without computed quartiles?" Later more detailed analysis was performed to find out, how the publication count is divided among quartiles.

Results are shown in Fig. 6(a) and (b). The results show an unsatisfactory trend for the faculty, that in the year 2015 the number of publications in sources without quartiles was increasing. The detailed analysis showed that in all years, except the last year 2016, the biggest number of publications belongs to quartile Q3. Regarding the excellence, the number of publications with Q1 is increasing in the last 3 years.

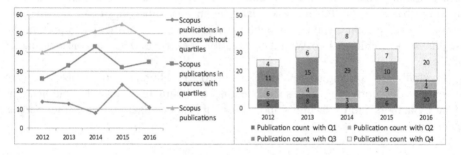

Fig. 6. Publication count in Scopus sources (a) with and without quartiles (b) detailed count by quartiles

For the 2nd analysis scenario, the following research question was formulated: "How many faculty publications in Scopus are not cited comparing with all publications and the publications in sources with computed quartiles?" Figure 7 shows the trend that the proportion of uncited publications remains unchanged in sources with computed quartiles, but is growing among all Scopus publications.

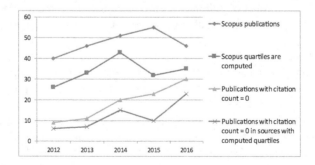

Fig. 7. Publications that are not cited

For the 3rd analysis scenario the research question was formulated: "How many are there citation counts in Scopus sources with quartiles and without computed quartiles?" Figure 8(a) shows the trend that the citation count in Scopus sources without computed quartiles is decreasing, but at the same time the citation count in the years 2012–2015 is remaining stable. More detailed analysis (see Fig. 8(b)) shows a significant citation count of Q3 publications, however this can be explained with bigger amount of Q3 publications among all others.

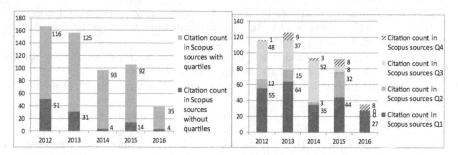

Fig. 8. Citation count in Scopus sources (a) with and without quartiles (b) detailed citation count by quartiles

These results can help to decide at each individual researcher's level to try publishing their works in sources with one quartile higher, but for the faculty, the shift from sources with Q3 to Q2 may be the most promising and realistic.

6   Conclusions

The main contribution of this paper is the architecture, that implements different flows of data and integrates them into one consistent system for research output evaluation. The architecture is based on the idea to ensure data quality, control, and also the integration process transparency, involving publication authors in different roles – as information providers or approvers.

The model of integrated dataset was determined mostly by existing information systems at the university and external interfaces (WoS and Scopus). The model is integrated with the global data model of the management information system of the University of Latvia, so the research evaluation can be extended with other types of metrics not only the bibliometric ones.

The proposed architecture ensures calculation of metrics for the publications evaluation both quantitative and qualitative ones. However, this paper concentrates mostly on the qualitative ones to support the scientific excellence. The quantitative metrics can help to set further goals, based on findings of performed analysis.

References

1. Hicks, D., Wouters, P., Waltman, L., De Rijcke, S., Rafols, I.: The Leiden Manifesto for research metrics. Nature **520**(7548), 429–431 (2015). doi:10.1038/520429a
2. Kosten, J.: A classification of the use of research indicators. Scientometrics **108**(1), 457–464 (2016). doi:10.1007/s11192-016-1904-7
3. Aagaard, K., Bloch, C., Schneider, J.W.: Impacts of performance-based research funding systems: the case of the Norwegian Publication Indicator. Res. Eval. **24**(2), 106–117 (2015). doi:10.1093/reseval/rvv003
4. Nikolić, S., Penca, V., Ivanović, D., Surla, D., Konjović, Z.: Storing of bibliometric indicators in CERIF data model. In: International Conference on Internet Society Technology (2013). doi:10.13140/2.1.2196.5121
5. Quix, C., Matthias, J.: Information integration in research information systems. Procedia Comput. Sci. **33**, 18–24 (2014). doi:10.1016/j.procs.2014.06.004
6. Jörg, B.: CERIF: the common European research information format model. Data Sci. J. **9**, 24–31 (2010). doi:10.2481/dsj.CRIS4
7. Rampāne, I., Rozenberga, G.: Latvijas Universitātes publikāciju citējamība datubāzēs (2012–2015). Alma Mater (vasara), pp. 26–28 (2016)
8. Cabinet Regulation No. 1316: Regulations regarding calculation and assignment of grant-based funding for research institutions. https://likumi.lv/doc.php?id=262508
9. Sivertsen, G.: Data integration in Scandinavia. Scientometrics **106**(2), 849–855 (2016). doi:10.1007/s11192-015-1817-x
10. Kulczycki, E., Korzeń, M., Korytkowski, P.: Toward an excellence-based research funding system: evidence from Poland. J. Informetr. **11**(1), 282–298 (2017). doi:10.1016/j.joi.2017. 01.001
11. Galimberti, P., Mornati, S.: The Italian model of distributed research information management systems: a case study. Procedia Comput. Sci. **106**, 183–195 (2017). doi:10. 1016/j.procs.2017.03.015
12. DSpace-CRIS Home. https://wiki.duraspace.org/display/DSPACECRIS/DSpace-CRIS+Home
13. The International Organisation for Research Information. http://eurocris.org/cerif/main-features-cerif
14. Teixeira da Silva, J.A., Memon, A.R.: CiteScore: a cite for sore eyes, or a valuable, transparent metric? Scientometrics **111**(1), 553–556 (2017). doi:10.1007/s11192-017-2250-0
15. Moed, H.F.: Measuring contextual citation impact of scientific journals. J. Informetr. **4**(3), 265–277 (2010). doi:10.1016/j.joi.2010.01.002
16. Gonzalez-Pereira, B., Guerrero-Bote, V.P., Moya-Anegon, F.: A new approach to the metric of journals' scientific prestige: the SJR indicator. J. Informetr. **4**(3), 379–391 (2010). doi:10. 1016/j.joi.2010.03.002
17. Hardcastle, J.: New journal citation metric – impact per publication (2014). http:// editorresources.taylorandfrancisgroup.com/new-journal-citation-metric-impact-per-publication/
18. Winkler, W.: The state of record linkage and current research problems. Technical report, Statistics of Income Division, US Census Bureau (1999)
19. Niedrite, L., Solodovnikova, D.: University IS architecture for the research evaluation support. In: 11th International Scientific and Practical Conference "Environment.Technology. Resources", pp. 112–117. Rezekne Academy of Technologies, Rezekne (2017). doi:10. 17770/etr2017vol2.2528

Evaluating Reference String Extraction Using Line-Based Conditional Random Fields: A Case Study with German Language Publications

Martin Körner[1]($\boxtimes$) (iD), Behnam Ghavimi[2] (iD), Philipp Mayr[2] (iD),
Heinrich Hartmann[3] (iD), and Steffen Staab[1] (iD)

[1] Institute for Web Science and Technologies,
University of Koblenz-Landau, Koblenz, Germany
{mkoerner,staab}@uni-koblenz.de
[2] GESIS – Leibniz Institute for the Social Sciences, Cologne, Germany
{behnam.ghavimi,philipp.mayr}@gesis.org
[3] Independent, Munich, Germany
heinrich@heinrichhartmann.com

Abstract. The extraction of individual reference strings from the reference section of scientific publications is an important step in the citation extraction pipeline. Current approaches divide this task into two steps by first detecting the reference section areas and then grouping the text lines in such areas into reference strings. We propose a classification model that considers every line in a publication as a potential part of a reference string. By applying line-based conditional random fields rather than constructing the graphical model based on individual words, dependencies and patterns that are typical in reference sections provide strong features while the overall complexity of the model is reduced. We evaluated our novel approach RefExt against various state-of-the-art tools (CERMINE, GROBID, and ParsCit) and a gold standard which consists of 100 German language full text publications from the social sciences. The evaluation demonstrates that we are able to outperform state-of-the-art tools which rely on the identification of reference section areas.

Keywords: Reference extraction · Citations · Conditional random fields · German language papers

1 Introduction

Citation data shows the link between efforts of individual researchers, topics, and research fields. Despite the widely acknowledged benefits, the open availability of citation data is unsatisfactory. Some commercial companies such as Elsevier and Google do have access to citation data and utilize them to supply their users with effective information retrieval features, recommendation systems, and other knowledge discovery processes. Yet, the majority of smaller information retrieval

© Springer International Publishing AG 2017
M. Kirikova et al. (Eds.): ADBIS 2017, CCIS 767, pp. 137–145, 2017.
DOI: 10.1007/978-3-319-67162-8_15

systems, such as Sowiport [1] for the social sciences, lack comprehensive citation data.

Recent activities like the "OpenCitations Project" or the "Initiative for Open Citations" aim to open up this field and improve the current situation. The "Extraction of Citations from PDF Documents" (EXCITE) project[1] at GESIS and University of Koblenz-Landau is in line with these initiatives and aims to make more citation data available to researchers with a particular focus on the German language social sciences. The shortage of citation data for the international and German social sciences is well known to researchers in the field and has itself often been subject to academic studies [2]. In order to open up citation data in the social sciences, the EXCITE project develops a set of algorithms for the extraction of citation and reference information from PDF documents and the matching of reference strings against bibliographic databases.

In this paper, we will consider the earlier steps in the extraction process that result in individual reference strings. There are several factors that result in the difficulty of the reference extraction task. One such factor is the high number of possible reference styles. According to Zotero[2], there exist more than four hundred difference citation styles in the social sciences alone. Further, there exists a large variety of layouts for publications including different section headings, headers, footers, and varying numbers of text columns. Figure 1 shows three challenging examples where the reference section does not contain a heading, where the reference strings contain a line break after the author names, and where reference strings strongly differ in their length, respectively.

Current solutions that perform reference string extraction have in common that they first identify the reference section and then, in a separate step, segment the reference section into individual reference strings. Thereby, errors that are made during the classification of reference sections directly impact the accuracy of the reference string extraction. For example, if a paragraph that contains reference strings was not recognized as part of the reference section, its reference strings will not be considered in the following step. To prevent this, our approach does not extract reference strings from an area that is first identified as the reference section. Instead, this indicator of a possible reference section is considered as only one of many features in a machine learning model that directly classifies the text lines as reference strings given the full text of the research paper. Other features are based on the text layout and the content of a given text line. A key observation here is that a text line usually does not contain information of more than one reference string. This allows the model to operate not on a word level but on a text line level.

The performance of our approach was evaluated using a novel gold standard for publications in the German language social sciences. To allow for a fair comparison, existing methods were retrained on the same data set that is used in our approach. As a result, the evaluation also provides insights into how well existing methods adapt to publications in the German language social sciences.

---

[1] https://west.uni-koblenz.de/en/research/excite.

[2] https://www.zotero.org/styles/.

haben wir in einem weiteren Analyseschritt die Einstellungsmuster der Bevölkerung in den einzelnen Ländern nach den oben unterschiedenen Konstellationen untersucht (vgl. Tabelle 3). Die Verteilung auf die vier Einstellungscluster bestätigt im Wesentlichen die bei der Analyse der nationalen Durchschnittswerte bereits erkennbaren Tendenzen. Die Unterschiede in den nationalen Einstellungsprofilen treten dabei oft noch deutlicher hervor.

mit eher etatistischer Tradition.

Als Fazit bleibt festzuhalten, dass das *Konzept* des Wohlfahrtsstaates, d. h. die dem Staat zugeschriebene sozialpolitische Verantwortung und die damit implizierten Zielvorstellungen nach wie vor in allen untersuchten EU-Ländern – entgegen der verbreiteten Wohlfahrtsstaatskritik – auf eine hohe Akzeptanz stößt und dass die Erwartungshaltungen und Ansprüche der

sozialen Investitionen die sozialpolitische Relevanz der Bildungspolitik verstärkt thematisiert (vgl. Allmendinger/Leibfried 2003).

Allmendinger, Jutta, Leibfried, Stefan, 2003: Education and the welfare state: the four worlds of competence production. In Journal of European Social Policy, 13. Jg., S. 63–81.

Esping-Andersen, Gösta, 1990: The Three

Literatur

Antonil:

Cultura e Opulencia do Brasil por suas drogas e minas (Kultur und Reichtum Brasiliens anhand seiner Pflanzungen und Bergwerke, port.), Paris 1968

Associação Brasileira de Ensino de Engenharia:

Formação do engenheiro industrial (Die Ausbildung des Industrieingenieurs, port.), São Paulo 1982

---

möglichen.

Die Zusammenarbeit mit der Europäischen Gemeinschaft – dem wichtigsten Handelspartner *(Tab. 4)* – erfolgt seit 1992 entsprechend dem früheren Handels- und Kooperationabkommen zwischen der EG und Jugoslawien: fast freier Zutritt zum EG-Markt, abgesehen von einigen Beschränkungen bei landwirtschaftlichen, Stahl- und Textilprodukten, Export der

Literatur:

Belec, B. (1992): Denationalisierung und ihre Konsequenzen für die Agrarwirtschaft in Slowenien. In: Arbeitsmaterialien zur Raumordnung und Raumplanung der Universität Bayreuth, Heft 108, S. 19-23.
Der Fischer Weltalmanach 1992.
Die Zwölfergemeinschaft (1991): Schlüsselzahlen. Luxemburg.
Gams, I. (1991): The Republic of Slovenia – Geographical Constants of the New Central-

Informationsmaterial. Ljubljana.
Zentrum für Internationale Zusammenarbeit und Entwicklung (1992): Slowenien Ihr Wirtschaftspartner. Ljubljana.

Fig. 1. Examples for difficult reference sections. The publications are part of the evaluation dataset and have the SSOAR-IDs 35306, 43525, and 48511, respectively. Documents available from http://www.ssoar.info/ssoar/handle/document/<ID> by replacing <ID> with the corresponding SSOARID.

The remainder of this paper is structured as follows. In Sect. 2, we present the related work in the area of reference string extraction. Section 3 introduces a novel approach[3] to reference string extraction that does not rely on the detection of reference zones. This approach is evaluated in Sect. 4 using a new gold standard for reference string extraction in the area of German language social sciences. Section 5 contains a summary and possible future work.

2   Related Work

There exists a considerable amount of literature about the extraction of bibliographic information from the reference section of scientific publications [4–10].

Reviewing this literature shows that there are two categories of approaches. One group concentrates on the reference string segmentation task by assuming the reference strings to be given [4–6]. The other group considers the reference string extraction from an article in the PDF or text format [7–10]. Further, all reference string extraction approaches follow two common steps.

---

[3] This approach was described in a preprint by Körner [3].

The first step identifies the text areas of the publication that contain the reference strings. Councill, Giles, and Kan [8] as well as Wu et al. [9] use a set of regular expressions to locate the beginning and end of reference sections. Tkaczyk et al. [10] apply a layout analysis on publications given as PDF files which results in textual areas that are grouped into zones. These zones are then classified as "metadata", "body", "references", or "other" using a trained Support Vector Machines (SVMs) model [10]. Lopez [7] trains a conditional random field (CRF) [11] model that performs a segmentation of textual areas into zones similar to Tkaczyk et al. [10].

In a second step, the lines in the identified areas are grouped into individual reference strings. Councill, Giles, and Kan [8] as well as Wu et al. [9] apply regular expressions to detect possible markers of reference strings such as numbers or identifiers surrounded by brackets. If such markers are found, the lines are grouped accordingly. If no markers are found, the lines are grouped based on the line length, ending punctuation, and strings that appear to be author name lists [8]. Tkaczyk et al. [10] use the k-means learning algorithm to perform a clustering into two groups: The first lines of reference strings and all other lines. The features for this clustering include layout information such as the distance to the previous line and textual information such as a line ending with a period [10]. As with the reference area detection, Lopez [7] learn a CRF model for this task. This model uses an input format that is different from the one that is used for their first CRF model. Tokens are split at white spaces and for each token, a list of features is created. Such features include layout information such as the font size and font weight as well as textual features such as the capitalization of the token and whether the token resembles a year, location, or name.

3   Approach

As previously discussed, a typical problem of existing reference string extraction approaches is a wrong classification of textual areas during the first step (see Sect. 2). Another key insight is that reference strings commonly start in a new line. This is used in the previously described k-means clustering algorithm by Tkaczyk et al. [10] and provides potential advantages over a word-based approach. For example, a line-based model drastically reduces the number of assigned target variables while still allowing the expression of relevant features. Further, it can capture patterns that repeat every few lines more naturally than a word-based model which focuses on a more local context.

These two insights are leveraged by applying a line-based classification model on the whole publication. For this, a possible set of labels consists of B-REF, I-REF, and O where B-REF denotes the first line of a reference string, I-REF a line of a reference string which is not the first line, and O any other line. This is based on the Beginning-Intermediate-Other (BIO) notation [12]. For our evaluation, we assigned one of the three labels to every text line in a publication. Having such a labeling, it is then possible to automatically extract the reference strings by concatenating a line labeled with B-REF together with the following lines labeled with I-REF until reaching a line that is labeled with B-REF or O.

Our model RefExt[4] uses both textual and layout features. In our evaluation we used textual features that signalize whether a line only consists of a number, starts with a capitalized letter, ends with a number, ends with a period, ends with a comma, or contains a year, a year surrounded by braces, a page range, an ampersand character, a quotations mark, a colon, a slash, or opening and closing braces. Another type of textual features counts the occurrences of numbers, words, periods, commas, and words that only consist of one capitalized letter. Further, we used four layout features. One signalizes whether the current line is indented when compared to the previous line. Another detects a gap between the current and previous line that is larger than a predefined value. The third layout feature is assigned to a line that contains less characters than the previous one. The last layout feature signalizes the position of a given line in the whole document. For this, the current line number is divided by the total number of lines. A more detailed description of the used features, together with all evaluation results, is provided on GitHub[5].

One advantage of using CRFs is that features do not have to be independent from each other due to the modeled conditional probability distribution [13]. Another advantage is the possibility to include contextual information. To do so, a CRF model with a high Markov order can be applied. This is feasible due to the line-based approach and the resulting lower number of random variables in the model when compared to a word-based approach.

4   Evaluation

The gold standard that is used in the following evaluation is based on 100 German publications from the SSOAR repository[6] in the PDF format. Since this evaluation focuses on the reference string extraction, documents that consist of scanned pages or that do not contain a reference section were excluded beforehand from the otherwise random selection. The resulting papers contain an average of 54 reference strings with a total number of 5,355 reference strings. Figure 2 gives an overview of the publication types and publication years of the gold standard documents. Resulting from the fact that existing reference string extraction tools use different input formats for their training procedures and also show differences in text-encoding of the resulting reference strings, a number of annotation file formats were created and manually inspected. This resulting gold standard is available on GitHub[7].

Since most existing citation information extraction tools focus on English language publications, the possibility to adapt the tool to German language publications is crucial. Two tools that allow such retraining are CERMINE [10] and GROBID [7]. For ParsCit [8], an older version allows the adaption of the regular expressions that detect reference section headings and other relevant

---

[4] https://github.com/exciteproject/refext.

[5] https://github.com/exciteproject/amsd2017.

[6] http://www.ssoar.info/.

[7] https://github.com/exciteproject/ssoar-gold-standard.

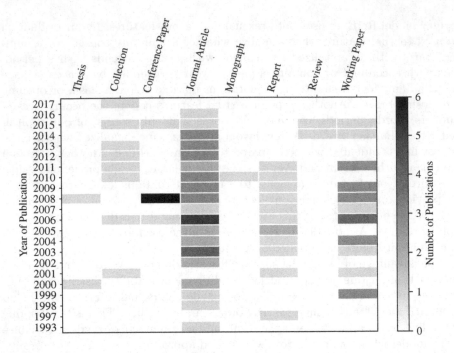

Fig. 2. Distribution of papers in the gold standard based on the publication type and year of publication.

headings such as appendices. Other tools that do not allow a retraining, such as PDFX [14] and pdfextract[8], were excluded due to their low performance on German language publications. The evaluation considers the performance on a line level based on the BIO notation in Table 1 and on a reference string level in Table 2 using the metrics macro precision, macro recall, and macro F1-score.

To compute macro metrics, a metric is first calculated on the individual publications and then averaged over all publications. Thereby, publications with a large amount of reference strings do not have a bigger impact on the final metric than publications with only a few reference strings. Further, in order to reduce the influence of the chosen split into training and testing data, the evaluation was performed using 10-fold cross-validation where each fold contains ten randomly chosen papers of the gold standard for testing and the remaining ninety papers for training. In the result tables, CERMINE, GROBID, and ParsCit are abbreviated with CER, GRO, and Pars, respectively. The suffixes D and T signalize whether the tool was using its default model or a model that was trained on the ninety publications from the gold standard. Pars-M represents version 101101 of ParsCit with modified regular expressions that match against German language section headings such as "Literatur" and "Anhang". Pars-D uses the latest ParsCit version as of May 31, 2017 which instead uses a trained model

---

[8] https://www.crossref.org/labs/pdfextract.

Table 1. Macro-metrics of BIO-annotated reference lines using 10-fold cross-validation on 100 German social science publications.

Metric	CER-D	CER-T	Pars-D	Pars-M	GRO-D	GRO-T	RefExt-T
B-REF Precision	0.719	0.734	0.683	0.769	0.692	0.871	**0.916**
B-REF Recall	0.600	0.557	0.620	0.688	0.789	0.865	**0.952**
B-REF F1-Score	0.616	0.589	0.616	0.689	0.712	0.861	**0.922**
I-REF Precision	0.729	0.755	0.577	0.678	0.664	0.857	**0.882**
I-REF Recall	0.340	0.313	0.809	0.843	0.839	0.871	**0.944**
I-REF F1-Score	0.432	0.415	0.647	0.716	0.703	0.855	**0.902**

Table 2. Macro-metrics of reference string extraction using 10-fold cross-validation on 100 German social science publications.

Metric	CER-D	CER-T	Pars-D	Pars-M	GRO-D	GRO-T	RefExt-T
Precision	0.296	0.303	0.558	0.617	0.627	0.847	**0.879**
Recall	0.233	0.220	0.552	0.595	0.718	0.839	**0.906**
F1-Score	0.245	0.235	0.542	0.590	0.650	0.837	**0.885**

for detecting reference sections. We were not able to retrain this model with the given source code and documentation. Further, we used CERMINE version 1.13 and GROBID version 0.4.1. Our approach, RefExt version 0.1.0, is based on the CRF models of MALLET [15] as well as the PDF text extraction and reading order detection of CERMINE. For the finite state transducer of MALLET we applied states in the "three-quarter" order. Thereby, the network contains a weight between every pair of adjacent labels as well as between the label and its corresponding features. Further, we applied a set of conjunctions that allow the usage of features of the previous two and following two lines. We found this network structure to perform similar to more complex structures while providing a reduced training time and a lower risk of overfitting. The learning was performed using the label log-likelihood with L1-regularization with a weight of 20.

The results show that RefExt is able to outperform the other tools on our gold standard. Over the 100 documents divided into ten folds, there were two publications[9] for which RefExt had a recall of zero in terms of reference strings. In both cases, the year number appeared at the end of the reference strings which is uncommon for the training corpus. In addition, one of the reference styles includes a line break after the listed authors in a reference string[10] which is also unusual. GROBID had a recall of zero in seven publications in terms of reference strings. Interestingly, the two publications that were problematic

---

[9] http://www.ssoar.info/ssoar/handle/document/32521 and http://www.ssoar.info/ssoar/handle/document/43525.

[10] Shown as the first example in Fig. 1.

for RefExt had a recall of 1.0 and 0.662 in GROBID, respectively. Thereby, a combined approach might be worthwhile.

5   Summary and Future Work

We have presented a novel approach to reference string extraction using line-based CRFs. The evaluation demonstrated that this approach outperforms existing tools when trained on the same amount of annotated data it the area of German language social sciences. Yet, there are several aspects that require further efforts. Having a precision and recall of around 0.9 is not sufficient for the usage in a productive system and it remains to be evaluated how the performance improves when extending the training data. Improvements might also be possible by adding more domain-specific features. Examples for such features are last name dictionaries or words that commonly appear in German language reference strings such as "Hrsg." and "Zeitschrift". Further, a number of journals in the German social sciences such as "Totalitarismus und Demokratie"[11] and "Südosteuropäische Hefte"[12] use a citation style where references are not grouped in a separate reference section but instead appear in the footnotes. This could present an interesting use case of our approach.

Acknowledgements. This work has been funded by Deutsche Forschungsgemeinschaft (DFG) as part of the project "Extraction of Citations from PDF Documents (EXCITE)" under grant numbers MA 3964/8-1 and STA 572/14-1. We would like to thank Dominika Tkaczyk for her support regarding the CERMINE tool as well as Alexandra Bormann, Jan Hübner, and Daniel Kostić for contributing to the gold standard that was used in this research.

References

1. Hienert, D., Sawitzki, F., Mayr, P.: Digital library research in action-supporting information retrieval in Sowiport. D-Lib Mag. **21**(3/4) (2015)
2. Moed, H.F.: Citation Analysis in Research Evaluation, vol. 9. Springer, Dordrecht (2005)
3. Körner, M.: Reference String Extraction Using Line-Based Conditional Random Fields. ArXiv e-prints (2017)
4. Peng, F., McCallum, A.: Information extraction from research papers using conditional random fields. Inf. Process. Manage. **42**(4), 963–979 (2006)
5. Cortez, E., da Silva, A.S., Gonçalves, M.A., Mesquita, F., de Moura, E.S.: FLUX-CiM: flexible unsupervised extraction of citation metadata. In: Proceedings of the 7th ACM/IEEE-CS Joint Conference on Digital Libraries, pp. 215–224. ACM (2007)
6. Groza, T., Grimnes, G.A., Handschuh, S.: Reference information extraction and processing using conditional random fields. Inf. Technol. Libr. (Online) **31**(2), 6 (2012)

---

[11] http://www.hait.tu-dresden.de/td/home.asp.
[12] http://suedosteuropaeische-hefte.org/.

7. Lopez, P.: GROBID: combining automatic bibliographic data recognition and term extraction for scholarship publications. In: Agosti, M., Borbinha, J., Kapidakis, S., Papatheodorou, C., Tsakonas, G. (eds.) ECDL 2009. LNCS, vol. 5714, pp. 473–474. Springer, Heidelberg (2009). doi:10.1007/978-3-642-04346-8_62

8. Councill, I.G., Giles, C.L., Kan, M.Y.: ParsCit: an open-source CRF reference string parsing package. In: Proceedings of LREC, vol. 2008, pp. 661–667 (2008)

9. Wu, J., Williams, K., Chen, H.H., Khabsa, M., Caragea, C., Ororbia, A., Jordan, D., Giles, C.L.: CiteSeerX: AI in a digital library search engine. In: AAAI, pp. 2930–2937 (2014)

10. Tkaczyk, D., Szostek, P., Fedoryszak, M., Dendek, P.J., Bolikowski, Ł.: CERMINE: automatic extraction of structured metadata from scientific literature. Int. J. Doc. Anal. Recogn. (IJDAR) **18**(4), 317–335 (2015)

11. Lafferty, J., McCallum, A., Pereira, F., et al.: Conditional random fields: probabilistic models for segmenting and labeling sequence data. In: Proceedings of the Eighteenth International Conference on Machine Learning, ICML, vol. 1, pp. 282–289 (2001)

12. Houngbo, H., Mercer, R.E.: Method mention extraction from scientific research papers. In: COLING 2012, 24th International Conference on Computational Linguistics, Proceedings of the Conference: Technical Papers, 8–15 December 2012, Mumbai, India, pp. 1211–1222 (2012)

13. Koller, D., Friedman, N.: Probabilistic Graphical Models: Principles and Techniques. MIT Press (2009)

14. Constantin, A., Pettifer, S., Voronkov, A.: PDFX: fully-automated PDF-to-XML conversion of scientific literature. In: Proceedings of the 2013 ACM symposium on Document engineering, pp. 177–180. ACM (2013)

15. McCallum, A.K.: MALLET: a machine learning for language toolkit (2002)

CEUR Make GUI - A Usable Web Frontend Supporting the Workflow of Publishing Proceedings of Scientific Workshops

Muhammad Rohan Ali Asmat[1,3(⊠)] and Christoph Lange[2,3]

[1] RWTH Aachen, Aachen, Germany
m.rohan.a.asmat@gmail.com
[2] University of Bonn, Bonn, Germany
math.semantic.web@gmail.com
[3] Fraunhofer IAIS, Sankt Augustin, Germany

Abstract. CEUR-WS.org is a widely used open access repository for computer science workshop proceedings. To publish a proceedings volume there, workshop organisers have to follow a complex, error-prone workflow, which mainly involves the creation and submission of an HTML table of contents. With ceur-make we had previously provided a command-line tool for partially automating this workflow. However, in a recent usability evaluation we confirmed that the tool is difficult to learn, highly dependent on other software, not portable and hard to use. We sought to solve these issues with a web-based user interface, which we present here. A usability evaluation of the latter proves significant improvements.

Keywords: Scholarly publishing · Open access · User experience · User interfaces

1 Introduction

Scientific social networks such as ResearchGate and free-of-charge open access repositories such as the Computing Research Repository (CoRR[1]) have significantly lowered the barrier for sharing research results in the form of individual papers. Open access repositories for complete proceedings of scientific events include the Proceedings of Machine Learning Research (PMLR) and the Electronic Proceedings in Theoretical Computer Science (EPTCS), addressing specific fields of computer science, and the CEUR Workshop Proceedings (CEUR-WS.org), addressing workshops from all over computer science.[2] Each of these employ an individual workflow for publishing, which proceedings editors and/or authors need to follow strictly to keep the effort low for those who run the service, usually volunteers. For example, PMLR requires editors to provide

---

[1] http://arxiv.org.

[2] http://proceedings.mlr.press/, http://www.eptcs.org/, http://ceur-ws.org.

© Springer International Publishing AG 2017
M. Kirikova et al. (Eds.): ADBIS 2017, CCIS 767, pp. 146–157, 2017.
DOI: 10.1007/978-3-319-67162-8_16

a BibTeX metadata database following specific rules[3], EPTCS acts as an overlay to CoRR, i.e. requires papers to be pre-published there, and CEUR-WS.org requires editors to provide an HTML table of contents following a certain structure[4]. Here, we focus on facilitating the latter by adding a web-based graphical user interface to a tool that auto-generates such tables of content, improving over the usability issues of the previous standalone command-line version of that tool.

Section 2 provides a more precise problem statement. Section 3 discusses related work. Section 4 presents the design and implementation of our web-based user interface. Section 5 evaluates the usability of the frontend in comparison to its command-line backend. Section 6 concludes with an outlook to future work.

2   Problem Statement

2.1   The Publishing Workflow

The HTML table of contents of a CEUR-WS.org workshop proceedings volume includes metadata about the workshop (title, date, venue, proceedings editors, etc.) and each of its papers (title, authors). This structure is prescribed[5]; around once a year, it has so far evolved a bit, e.g., in the form of more explicit semantic annotations to facilitate reuse of the metadata. Besides following the latest template and producing syntactically valid HTML, requirements for proceedings editors include following a consistent capitalisation scheme for paper titles, providing full names of authors, and using relative links to the full texts of the individual papers (typically PDF files). The HTML table of contents together with the full texts has to be submitted to CEUR-WS.org as a ZIP archive.

2.2   Automation of the Workflow with Ceur-Make

Traditionally, proceedings editors had to prepare the submission ZIP file manually. With ceur-make[6], the second author, technical editor of CEUR-WS.org, has provided a tool to automate part of this job – aiming at three objectives:

- Helping proceedings editors to learn more quickly how to create a table of contents, reducing their effort, and helping recurrent editors to cope with structural changes.
- Reducing the workload of the volunteers who carry out the subsequent publishing steps at CEUR-WS.org; so far, around one in ten submissions requires further communication with its editors to resolve problems, mainly rooted in the table of contents.
- Reducing the implications that subsequent improvements to the structure of the table of contents have on both proceedings editors and the CEUR-WS.org team by reducing their exposure to manual editing.

---

[3] http://proceedings.mlr.press/spec.html.
[4] http://ceur-ws.org/HOWTOSUBMIT.html#PREPARE.
[5] http://ceur-ws.org/Vol-XXX/.
[6] https://github.com/ceurws/ceur-make.

For ceur-make, the metadata about the workshop and its papers have to be provided in two XML files. ceur-make can auto-generate the latter XML file from the metadata that the widely used EasyChair submission and review management system exports in LNCS mode (cf. Sect. 3.1). From these two XML files, ceur-make auto-generates an HTML table of contents and finally a ZIP archive conforming with the CEUR-WS.org requirements. In addition, ceur-make can generate a BibTeX database to facilitate the citation of the papers in a proceedings volume, as well as a copyright form by which the authors agree to the publication of their papers with CEUR-WS.org.

2.3   Shortcomings of Ceur-Make

Shortcomings of ceur-make include that it depends on a Unix-style command line environment and a number of software packages that typically only developers have installed: the Make build automation tool[7], the Saxon XSLT processor and the Perl scripting language. Furthermore, it requires proceedings editors to edit one or two XML files manually, without validating their content with regard to all rules that editors should follow. It also requires them to follow certain conventions for naming and arranging files and directories; most importantly, the sources of ceur-make have to be downloaded to the same directory in which the proceedings volume is being prepared. These reasons may explain why ceur-make has so far only been used for less than one in ten proceedings volumes.

2.4   Research Objectives

The objectives of our research were 1. to assess the shortcomings of ceur-make in a more systematic way, and 2. to overcome them by providing a user-friendly web frontend to ceur-make.

3   Related Work

3.1   Conference Management Systems

The complex process of managing scientific events (conferences, workshops, etc.) is facilitated by a broad range of systems, of which we briefly review three representatives and their proceedings generation capabilities. Parra et al. have reviewed further systems without providing details on proceedings generation [6]. In computer science, **EasyChair**[8] enjoys particularly wide usage.[9] EasyChair features a special "proceedings manager" interface, which is initialised by adding all accepted papers and then supports the collection of the final ("camera ready") versions, including a printable form (usually PDF), editable sources (LaTeX,

---

[7] https://www.gnu.org/software/make/.

[8] http://www.easychair.org.

[9] EasyChair has so far been used to manage 53,739 events and has had 1,954,080 users (http://www.easychair.org/users.cgi, accessed 2017-04-18).

Word, or anything else, e.g., HTML, in a ZIP archive), and a copyright transfer form. Proceedings chairs can define an order of papers and add or edit additional documents such as a preface. Specific support for exporting all these files and their metadata is provided for events that publish in Springer's Lecture Notes in Computer Science (LNCS) series. Microsoft's **Conference Management Toolkit** (CMT[10]) assists with publishing accepted papers to CoRR. With a professional license, **ConfTool**[11] supports the export of metadata in multiple formats (Excel, XML and CSV) to facilitate proceedings generation.

3.2   Usability Evaluation of Command Line Vs. GUI

Comparing the usability of command-line (CLI) vs. graphical user interfaces (GUI) has been a long-standing research topic. Hazari and Reaves have evaluated the performance of students in technical writings tasks using a graphical word processor vs. a command-line tool [3]. Starting from the same level of background knowledge and given the same time for training, a significantly larger share of users felt comfortable using the GUI rather than the command line; also, their task-based performance was slightly higher with the GUI. Gracoli is an operating system shell with a hybrid user interface that combines GUI and CLI [12]. Its design is motivated by common drawbacks of CLIs, which are stated as follows: – the user can interact with the application in a limited way; – the output is hard to understand for the user;– the user does not easily get a clue of how to perform a task.

4   Design and Implementation of CEUR Make GUI

4.1   Architecture

The CEUR Make GUI is a graphical layer built on top of ceur-make. Figure 1 shows its three-layer architecture (Interface, Middleware, and Storage).

The **Interface Layer** consists of all the presentation elements. It displays visual elements, handles dependencies on external libraries for user interface elements, styles the web pages, validates forms and manages user interaction the web pages. It also initiates the communication with the Middleware Layer on user's request and displays the results from the Middleware. Technologies used on Interface Layer include standard web technologies used for front end clients (HTML 5, CSS, JavaScript), and the following libraries: Materialize CSS[12] is a JavaScript and CSS library based on Google's Material Design principles[13], used here to incorporate standard design patterns into the GUI. jQuery Steps[14] is used to create wizards for taking inputs.

---

[10] https://cmt.research.microsoft.com/cmt/.
[11] http://www.conftool.net/.
[12] http://materializecss.com.
[13] https://material.io/guidelines/.
[14] http://www.jquery-steps.com.

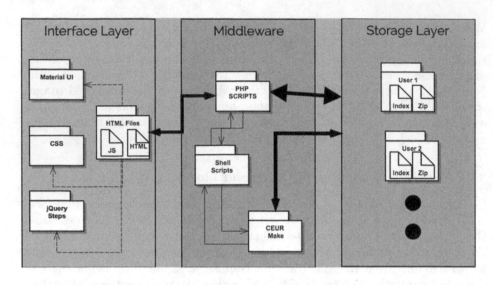

Fig. 1. System architecture of CEUR make graphical user interface

The **Middleware Layer** generates artifacts required for publishing at CEUR-WS.org. The Middleware Layer creates the files, as requested through the Interface Layer, by running ceur-make. It returns links to the artifacts stored at the Storage Layer to the Interface Layer, thus acting as a service provider.

The **Storage Layer** stores the files that are created temporarily on the server. It separates the files based on the user's identity and then also based on the workflow that the user chooses to create the artifacts for publishing (manual metadata input vs. EasyChair import).

4.2   Interface

We aim at providing an easy to use, task oriented interface. On the main screen, we give users the option of switching between four tasks: *viewing announcements, viewing published proceedings, publishing a proceeding* and *reporting an issue* through a navigational menu (cf. Fig. 2). Further, we separate the site navigation of the two proceedings publishing workflows using a Card design pattern [9], representing each workflow as an independent card (cf. Fig. 3a). We follow the Wizard design pattern [11] to collect workshop metadata input from users (cf. Fig. 3b). For the list of all proceedings volumes[15], we follow the style of the

Fig. 2. Navigational menu of CEUR make graphical user interface

---

[15] For now, we only implemented the list of proceedings volumes as a hard-coded mockup for the purpose of evaluating the usability of our user interface design.

(a) Workflows for Publishing Proceedings

(b) Workshop metadata input wizard

Fig. 3. Workflow screens

CEUR-WS.org user interface but make it more accessible by following standard design patterns. We applied the Pagination design pattern [10] to address the problem of the current CEUR-WS.org site that one has to scroll down a lot because all content is displayed at once; secondly, we applied the Autocomplete design pattern [8] to facilitate the task of finding proceedings volumes already published at CEUR-WS.org easier.

Source code, documentation and a working installation of the CEUR Make GUI are available at https://github.com/ceurws/ceur-make-ui.

5 Evaluation

5.1 Methodology

Participants. Twelve persons participated in the evaluation of the usability of the ceur-make CLI vs. the GUI. We chose nine participants with previous CEUR-WS.org publishing experience[16] and three participants without. The latter were trained to publish at CEUR-WS.org to avoid learning biases in our evaluation results.

Procedure. The participants were divided into two groups based on their availability. Those who were physically available participated in a Thinking Aloud test [7], and the other ones participated in a Question Asking test:

Thinking Aloud: Participants were provided with task definitions as explained below. They were asked to think aloud about their plans and their interaction with the system, particularly including problems they faced or unusual mental models, while the evaluator took notes. The task completion time was recorded for the purpose of comparison.

---

[16] Among them we would have preferred to have some with ceur-make experience, but this proved infeasible given the small number of people who had ever used it.

Question Asking: In a video conferencing setting with screen sharing (using Skype), the evaluator performed each task according to its definition. The participants were allowed to ask questions during the usability test, where the evaluator also asked questions to test the user's understanding. From an audio recording, the evaluator compiled a transcript of pain points afterwards.

Following a within-subject design setup[17], each participant first had to test the CLI and then the GUI. The participants were given four tasks to be performed in each system, designed to cover all major use cases of the system in a comparable way: 1. Initiate generation of a proceedings volume, 2. Generating workshop metadata, 3. Generating table of contents metadata, and 4. Search a proceedings volume. Each tasks were subdivided into smaller steps, e.g., as follows for Task 4:

Task 4 – Search a Proceedings Volume

1. Go to the proceedings page at http://ceur-ws.org (or in the GUI, respectively).
2. Search the proceedings volume that has the name "Cultures of Participation in the Digital Age 2015".

For the full list of task definitions, please see Appendix A and B of [1].

Usability tests were followed by a post study questionnaire for each user, which was created and filled using Google Forms[18]. The questionnaire was divided into following sections:

System Usability Scale (SUS [4]**),** a ten point heuristic questionnaire to evaluate general usability of the system on a Likert scale from 1 (strongly agree) to 5 (strongly disagree).

Question for User Interaction Satisfaction (QUIS [5]**),** a 27 point questionnaire to evaluate specific usability aspects of the system, covering overall reaction to the software, screen layout, terminology and system information, as well as learning and system capabilities, from a scale from 0 (lowest) to 9 (highest). The mean score was calculated for every user.

Dataset. All users used the same input data for both systems to ensure unbiased comparability of the content created and of completion times across users and systems. A full record of the data is provided in Appendix A and B of [1].

5.2   Results

This section summarizes the evaluation results; for full details see Appendix C and D of [1].

---

[17] https://web.mst.edu/~psyworld/within_subjects.htm.
[18] https://www.google.com/forms/about/.

Table 1. Quantitative usability evaluation results using thinking aloud

Tasks	CEUR Make (Min)	CEUR Make GUI (Min)
Task 1	0.13	0.10
Task 2	4.77	2.88
Task 3	2.40	1.46
Task 4	0.76	0.10

Quantitative Results (Completion Times). Table 1 shows the completion times per system and task - in detail:

1. **Task 1 (Initiate generation of a proceedings volume):** This required entering a terminal command for the CLI and pressing a button in the GUI. On average, this took less time in the CLI, but the difference is too marginal to be significant.
2. **Task 2 (Generating Workshop Metadata):** This required entering workshop metadata into the GUI input wizard, and using a text editor and the command line in the CLI. The difference in completion time is significant: completing the task using the GUI took only 60% of the time taken using the CLI, which emphasizes the user-friendliness of the GUI.
3. **Task 3 (Generating Table of Contents Metadata):** This required entering metadata of two papers similarly as for Task 2, with similar results.
4. **Task 4 (Search a proceedings volume):** This task took 7.6 times as long on the CEUR-WS.org homepage compared to the GUI. This result highlights the importance of using the autocomplete design pattern for searching in the graphical user interface, compared to just the "find in page" search built into browsers.

Overall, users took significantly less time to complete tasks with the GUI, which proves the usability improvement it provides over the CLI.

Qualitative Results. Notes recorded while performing the usability test were categorized in ten heuristics, i.e.: *Speed in performing a task*, *Documentation of the software*, *Ease in performing a* **Task**, **Learnability** *of the software*, *clear* **Navigation** *structure of the system*, **Portability** *of the system*, **Error** *correction chances, easy to use* **Interface**, **Dependency** *on other systems* and **Features** *to be added*. Table 2 shows the number of responses of the twelve participants for each qualitative heuristic, where "bad" means they were not comfortable using it, "good" means they liked the software, and "excited" means that the user is interested but would like to see more features to be implemented.

The total number of qualitative responses was 36 for the CLI and 34 for the GUI. 15 good responses were recorded for the CLI, regarding the heuristics *Speed*, *Documentation* and *Task*, whereas 21 bad responses were recorded

Table 2. Qualitative results for ceur-make and the CEUR make GUI

Heuristics	ceur-make	# of responses from users	CEUR make GUI	# of responses from users
Speed	Good	3	Neutral	–
Documentation	Good	4	Neutral	–
Task	Good	8	Neutral	–
Learnability	Bad	5	Good	5
Navigation	Bad	4	Good/Bad	5/2
Portability	Bad	2	Good	2
Error	Bad	2	Good	3
Interface	Bad	4	Good	8
Dependency	Bad	4	Good	3
Feature	Neutral	–	Excited	6
Total responses		36		34

regarding the heuristics *Learnability, Portability, Navigation, Error, Interface* and *Dependency*. For the GUI no response was recorded against the heuristics *Speed, Documentation* and *Task*, which were reported as good for the CLI. This was the case because users did not require documentation to operate the GUI as they never requested for it from the evaluator, speed was not an issue while using it as they never complained about it, and it enabled them to perform their respective tasks. 26 good responses were recorded for the GUI against the heuristics *Learnability, Portability, Navigation, Error, Interface* and *Dependency*, which were all recorded as bad in case of the CLI. This highlights the usability improvement provided by the GUI over the CLI. Only two bad responses were recorded for the GUI, against the heuristic *Navigation*, which means a slight improvement in navigation is required – as quoted by a user: *"ceur-make make things easier but has a complex setup, whereas the GUI is straightforward and requires no prior learning. With little improvement in the flow of screens it could be even better."*

Moreover, for the GUI, 6 responses were recorded as excited, against the heuristic *Feature*, which means users would like to use the software and would like additional features to be integrated.

Post Evaluation Questionnaire. The overall usability of the two systems was evaluated using **System Usability Scale**[19]. The SUS score for ceur-make was 41.25, which is below grade F (x-axis) as shown in Fig. 4. This rating demands immediate usability improvements. On the other hand, the SUS score of the GUI was 87.08, which is above grade A (x-axis). This means that the GUI has a good usability and its users would recommend it to others.

---

[19] http://www.userfocus.co.uk/articles/measuring-usability-with-the-SUS.html.

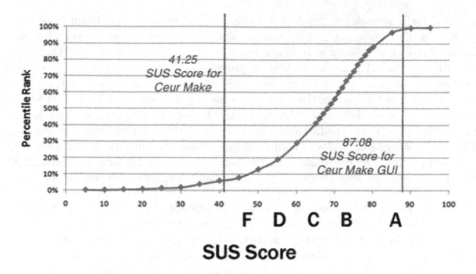

Fig. 4. SUS score: ceur-make vs CEUR make graphical user interface

Results of the **Questionnaire for User Interaction Satisfaction** reflect a high usability improvement of the GUI over the CLI. For the questions related to the *learnability of the system*, a visible difference in mean scores was recorded for *easy to remember the commands, learning to operate the system* and *trying new features by trial and error*. For these three questions, mean scores of the CLI were 3.75, 3.25, and 4.25 (all below average) and for the GUI they were 8.5, 8.25, and 8.0 (all above average). Likewise, mean scores for the GUI for the questions related to *information representation*, including *information organization, positioning of messages, highlighting of information to simplify task, prompts and progress* were 7.75, 8.0, 6, 6, and 6.25 (above average), whereas for the CLI they were 4, 4, 2.25, 4, and 3.25 − i.e.a notable difference. Another highlight was that users appreciated that the GUI was considered to be *designed for all levels of users* as backed by a mean score of 7.75, whereas the CLI was considered not to be designed for all levels of users as its mean score was just 3.

6   Conclusion

We aimed at automating a workflow for publishing scientific results with open access, focused on the CEUR-WS.org workshop proceedings repository. We developed a graphical user interface on top of the ceur-make command line tool and systematically evaluated the usability of both. Quantitative results on task completion time prove that the GUI is more efficient in performing common tasks. Qualitative evaluation suggests that on all heuristics where ceur-make performed badly, i.e., *learnability, navigation, portability, error, interface* and *dependency*, the GUI yielded good responses. In the post-evaluation questionnaires, a notable difference was recorded in the SUS scores of the two systems:

grade F for ceur-make vs. grade A for the GUI. 11 out of 27 QUIS questions of ceur-make had responses below average, and others were satisfactory, whereas for the GUI all responses were above average. Overall, results indicate that the usability of the GUI has noticeably improved over the command line. As our evaluation setup covered most typical tasks of proceedings editors, the results suggest that the GUI makes the overall process of publishing with CEUR-WS.org more effective and efficient and thus will attract a broad range of users. Thanks to the input validation of the metadata wizard and to the detailed explicit semantic RDFa annotations of tables of contents that ceur-make outputs, broad usage of the GUI will improve the quality of CEUR-WS.org metadata, largely eliminating the need for reverse-engineering data quality by information extraction (cf. [2]).

Future Work The next immediate step is to officially invite all CEUR-WS.org users to use the GUI for preparing their proceedings volumes. Partly inspired by feedback from the evaluation participants, we are planning to enhance the GUI with functionality addressing the following use cases (all filed as issues at https://github.com/ceurws/ceur-make-ui/issues/): **User Profiles** would help to automatically suggest information related to the editors while working on that section of the workshop metadata (Issue #1). Even without user profiles, building and accessing a **database** of previously published workshops' metadata would facilitate input, e.g., by auto-completing author names, and by reusing metadata of a workshop's previous edition. The RDF linked open data embedded into ceur-make generated tables of contents or extracted from old tables of contents (cf. [2]) can serve as such a database once collected in a central place (#5). A **Collaborative Space for Editors** would support multiple editors to work in parallel on a proceedings volume (#2). **Saving System State** would improve user experience and give users more control (#4). Currently, there is no way to restore the state of the interface, where one left in case the browser is accidentally closed or users want to complete the task later. **Extraction of Author Names, Titles and Page Numbers** from the full texts of the papers would further lower task completion time, as the system would automatically suggest most metadata (#3).

References

1. Asmat, M.R.A.: A Usable Web Frontend for Supporting the Workflow of Publishing Proceedings of Scientific Workshops. MA thesis. RWTH Aachen (2016). http://eis-bonn.github.io/Theses/2016/Rohan_Asmat/thesis.pdf
2. Dimou, A., et al.: Challenges as enablers for high quality linked data: insights from the semantic publishing challenge. In: PeerJ Computer Science: Semantics, Analytics, Visualisation: Enhancing Scholarly Data (2017)
3. Hazari, S.I., Reaves, R.R.: Student preferences toward microcomputer user interfaces. Comput. Educ. **22**, 225–229 (1994)
4. Measuring Usability with the System Usability Scale (SUS). MeasuringU (2017). https://measuringu.com/sus/. Accessed 17 Apr 2017
5. Norman, K.L., Shneiderman, B.: Questionnaire for User Interaction Satisfaction QUIS. http://www.lap.umd.edu/quis/. Accessed 25 Sep 2016

6. Parra, L., et al.: Comparison of Online Platforms for the Review Process of Conference Papers. In: CONTENT (2013)
7. Thinking Aloud: The Number 1 Usability Tool. Nielsen Norman Group (2017). https://www.nngroup.com/articles/thinking-aloud-the-1-usability-tool/. Accessed 10 Apr 2017
8. User Interaction Design Pattern Library: Autocomplete. UIPatterns (2017). http://ui-patterns.com/patterns/Autocomplete. Accessed 17 Apr 2017
9. User Interaction Design Pattern Library: Card. UIPatterns (2017). http://ui-patterns.com/patterns/cards. Accessed 17 Apr 2017
10. User Interaction Design Pattern Library: Pagination. UIPatterns (2017). http://ui-patterns.com/patterns/Pagination. Accessed 17 Apr 2017
11. User Interaction Design Pattern Library: Wizard. UIPatterns (2017). http://ui-patterns.com/patterns/Wizard. Accessed 17 Apr 2017
12. Verma, P.: Gracoli : a graphical command line user interface. In: CSCW (2013)

The 1st Workshop on Novel Techniques for Integrating Big Data (BigNovelTI 2017)

A Framework for Temporal Ontology-Based Data Access: A Proposal

Sebastian Brandt[1], Elem Güzel Kalaycı[2], Vladislav Ryzhikov[2],
Guohui Xiao[2(✉)], and Michael Zakharyaschev[3]

[1] Siemens CT, Munich, Germany
sebastian-philipp.brandt@siemens.com
[2] KRDB Research Centre for Knowledge and Data,
Free University of Bozen-Bolzano, Bolzano, Italy
{kalayci,ryzhikov,xiao}@inf.unibz.it
[3] Department of Computer Science and Information Systems,
Birkbeck, University of London, London, UK
michael@dcs.bbk.ac.uk

Abstract. Predictive analysis gradually gains importance in industry. For instance, service engineers at Siemens diagnostic centres unveil hidden knowledge in huge amounts of historical sensor data and use this knowledge to improve the predictive systems analysing live data. Currently, the analysis is usually done using data-dependent rules that are specific to individual sensors and equipment. This dependence poses significant challenges in rule authoring, reuse, and maintenance by engineers. One solution to this problem is to employ ontology-based data access (OBDA) that provides a conceptual view of data via an ontology. However, classical OBDA systems do not support access to temporal data and reasoning over it. To address this issue, we propose a framework of temporal OBDA. In this framework, we use extended mapping languages to extract information about temporal events in RDF format, classical ontology and rule languages to reflect static information, as well as a temporal rule language to describe events. We also propose a SPARQL-based query language for retrieving temporal information and, finally, an architecture of system implementation extending the state-of-the-art OBDA platform Ontop.

Keywords: Temporal logic · Ontology-based data access · Ontop

1 Introduction

Analysis of the logs sensor data is an important problem in industry, as it reveals crucial insights into the performance and conditions of devices. The outcomes of this analysis, known also as retrospective diagnostics, enables IT experts to improve the capabilities of real-time systems monitoring abnormal or potentially dangerous events developing in devices, i.e., the systems that perform predictive diagnostics. For complex devices (such as those we consider in the use-case below)

M. Kirikova et al. (Eds.): ADBIS 2017, CCIS 767, pp. 161–173, 2017.
DOI: 10.1007/978-3-319-67162-8_17

such events do not amount to simply measurable instances (say, the temperature above 100°C), but involve a number of measurements from sensors attached to a device, with each of them having a certain temporal duration and occurring in a certain temporal sequence.

Our central use-case is monitoring turbine conditions at Siemens, which maintains thousands of devices related to power generation, including gas and steam turbines. It provides operational support for these devices through a global network of more than 50 remote diagnostic centres. These centres are linked to a common database centre. Siemens wants to employ retrospective and predictive diagnostics in order to anticipate problems with turbines and take relevant countermeasures. A major challenge in this task is the access to *heterogeneous* data sources, since various turbine models have different schemas of underlying databases storing sensor measurements.

We propose a solution based on detaching the conceptual view of an event, such as 'HighRotorSpeed of the turbine tb01 in the period from 2017-06-06 12:20:00 to 2017-06-06 12:20:03' from a concrete database(s) that store the log of the sensors of that turbine. Thus, we work in the spirit of *ontology-based data access* (OBDA) [11,23], where only the high level conceptual layer of classes and properties is exposed to the end-users, while the complex structure of the underlying data sources is hidden from them. (In fact, those classes and properties are mapped to the data source schemas through a declarative specification.) In addition, an ontology is used to model the domain of interest by asserting conceptual relations (e.g., isA) between the classes and properties. There are several systems implementing the OBDA approach, some of which (Ontop[1] and Morph[2]) are freely available, whereas others (Stardog[3], Mastro[4] and Ultrawrap[5]) are distributed under commercial licences.

Unfortunately, none of the available OBDA systems supports access to temporal data well enough to work with the events relevant to our analysis. On the one hand, the common mapping languages are not tailored towards extracting validity time intervals of conceptual assertions. On the other hand—and this is a bigger problem—the supported ontology languages do not allow one to construct classes and properties whose temporal validity depends on the validity of other classes and properties, which is essential for defining complex temporal events. (In fact, the OBDA systems used industrially are based on the lightweight ontology languages, such as *OWL 2 QL* profile of the Web Ontology Language (OWL) and *DL-Lite* [11], a lightweight Description Logic, in order to guarantee maximally efficient query answering.) When limited to a classical ontology language, one approach to enable the extraction of temporal events is by extending the end-user query language with various temporal operators [5,7,15,17,18,20,22]. However, this leaves the burden of encoding the complex events in temporal

---

[1] http://ontop.inf.unibz.it/.
[2] https://github.com/oeg-upm/morph-rdb.
[3] http://stardog.com/.
[4] http://www.obdasystems.com/mastro.
[5] https://capsenta.com/.

queries to the end-user. In the Siemens scenario, this is a prohibitive limitation, since the end-users are the Siemens service engineers who are not trained in temporal logic.

Therefore, we pursue a more expressive setting, where the ontology language is extended by temporal operators that are capable of defining complex temporal events. In several works [3,4,6], non-temporal ontology languages were extended with *LTL*-based operators such as 'next time' or 'eventually'. Unfortunately, sensor data tend to come at irregular time intervals, which makes it impossible to adequately represent '10 s' or '1 min' in *LTL*. The same problem occurs if metric temporal logic is used with assumption of discrete time [14]. To overcome this limitation, a language *datalogMTL* based on metric temporal logic *MTL* over dense time was proposed [8]. This language also appears to be capable of capturing the events of interest for the diagnostic tasks at Siemens.

In this paper, using a running example from the Siemens use-case, we sketch our proposal of a framework for temporal OBDA with $datalog_{nr}MTL$, a non-recursive version of *datalogMTL*, as a rule language to describe events. At the same time, this framework supports the standard *OWL 2 QL* language to model static knowledge (such as a configuration of modules of a turbine), and extends it with non-recursive datalog rules to describe static knowledge of more complex structure. We outline the extension of the standard mapping language R2RML and the query language SPARQL to extract information on the validity intervals of conceptual assertions. Finally, we discuss an implementation of the proposed framework in the OBDA system Ontop.

2   A Framework of Temporal OBDA

In this section, we present our framework for temporal OBDA. We first consider the notion of OBDA specification, which is comprised of a set of mappings, static ontology, and static and temporal rules. Then we define a temporal OBDA instance and its specification through RDF triples. Finally, we present a query language for those instances based on τ-SPARQL [24].

2.1   Temporal OBDA Specification and Instance

A traditional *OBDA specification* is a triple $\langle \mathcal{O}, \mathcal{M}, \mathcal{S} \rangle$, where $\mathcal{O}$ is an ontology, $\mathcal{M}$ a set of mapping assertions between ontological entities and data sources, and $\mathcal{S}$ a database schema. As we want to build temporal OBDA on top of the standard atemporal one, we allow the use of existing specifications, calling $\mathcal{O}$ a *static ontology*, $\mathcal{M}$ a set of *static mapping assertions*, and $\mathcal{S}$ a *database schema*. In what follows we will extend the *static* vocabulary Σ_s of classes and properties occurring in $\mathcal{O}$ and $\mathcal{M}$ by a disjoint *temporal* vocabulary Σ_t. We now describe static ontologies in more detail using an example from the Siemens use case.

Static Ontology. At Siemens, the devices used for power generation are monitored by many different kinds of sensors reporting the temperature, pressure, vibration and other relevant measurements. Siemens uses an ontology in order to model the static knowledge regarding the machines and their deployment profiles, sensor configurations, component hierarchies and functional profiles. Example 1 below gives a snippet of the static conceptual model formalized in the syntax of the description logic *DL-Lite$_R$* [2].

Example 1. The signature Σ_s of the Siemens static ontology $\mathcal{O}$ contains the following ontological entities (concepts and roles):

Train, Turbine, GasTurbine, SteamTurbine, TurbinePart,

PowerTurbine, Burner, Sensor, RotationSpeedSensor,

TemperatureSensor, isMonitoredBy, isPartOf, isDeployedIn.

Some of the axioms (concept inclusions) of $\mathcal{O}$ are shown below:

$$GasTurbine \sqsubseteq Turbine, \qquad \exists isMonitoredBy \sqsubseteq TurbinePart,$$
$$SteamTurbine \sqsubseteq Turbine, \qquad \exists isMonitoredBy^- \sqsubseteq Sensor,$$
$$RotationSpeedSensor \sqsubseteq Sensor, \qquad \exists isPartOf \equiv TurbinePart,$$
$$TemperatureSensor \sqsubseteq Sensor, \qquad \exists isPartOf^- \sqsubseteq Turbine,$$
$$PowerTurbine \sqsubseteq TurbinePart, \qquad \exists isDeployedIn \sqsubseteq Turbine,$$
$$Burner \sqsubseteq TurbinePart, \qquad \exists isDeployedIn^- \sqsubseteq Train.$$

It turns out, however, that the language of *DL-Lite$_R$* is not able to capture all the static knowledge required in the Siemens use case. We complement this ontology with nonrecursive datalog *static rules*.

Static Rules. In the Siemens use case, we have turbine parts that are monitored by a number of different co-located sensors within the same turbine, for example, a temperature sensor and a rotation speed sensor. This situation can be readily described by a datalog rule with a ternary predicate and a complex body such as the one in the example below, but not by *DL-Lite$_R$* axioms.

Example 2. Static rules $\mathcal{R}$ can be of the form

$$ColocTempRotSensors(tb, ts, rs) \leftarrow isMonitoredBy(pt, ts), TemperatureSensor(ts),$$
$$isMonitoredBy(pt, rs), RotationSpeedSensor(rs),$$
$$isPartOf(pt, tb), Turbine(tb).$$

Note that we extend Σ_s to contain also the (static) symbols occurring in $\mathcal{R}$. We also require those rules to be *non-recursive* (that is, contain no cycles in the definitions of their predicates). Finally, observe that the static rules in $\mathcal{R}$ may contain concepts and roles from $\mathcal{O}$.

Temporal Rules. Siemens is interested in detecting abnormal situations in the working equipment as well as in monitoring running tasks in order to see whether they proceed ordinarily. A typical event that is crucial to monitor would be a *normal start of a turbine*. However, this complex event is composed of many subevents distributed in time, each of which is characterized by a temporal duration. One of these subevents is described in the following example.

Example 3. Purging is Over is a complex event of a certain *turbine* such that the following hold:

(*i*) there is a pair of sensors co-located in some part of a turbine, one of which is a rotor speed sensor and the other one a temperature sensor;
(*ii*) the temperature sensor detects that the main flame was burning for at least 10 s;
(*iii*) at the same time, the following occurred within preceeding 10 min:

 - the rotor speed sensor measures the speed of at most 1000 rpm for at least 30 s,
 - within preceding 2 min the rotor speed measured at least 1260 rpm for at least 30 s.

The described event is illustrated below:

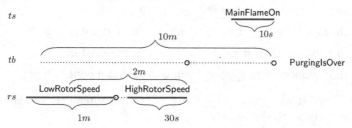

Temporal diagnostic patterns of this sort can be described by means of a $datalog_{nr}MTL$ program [8], which is a set of nonrecursive datalog rules with temporal operators of the metric temporal logic MTL [1]. The event 'purging is over' can be formalized by the following $datalog_{nr}MTL$ program $\mathcal{T}$.

Example 4. The program $\mathcal{T}$ includes the rules:

$$\mathsf{PurgingIsOver}(tb) \leftarrow \boxminus_{[0s,10s]} \mathsf{MainFlameOn}(ts),$$
$$\diamondsuit_{(0,10m]} \big[\boxminus_{(0,30s]} \mathsf{HighRotorSpeed}(rs),$$
$$\diamondsuit_{(0,2m]} \boxminus_{(0,1m]} \mathsf{LowRotorSpeed}(rs) \big],$$
$$\mathsf{ColocTempRotSensors}(tb, ts, rs).$$
$$\mathsf{HighRotorSpeed}(tb) \leftarrow \mathsf{rotorSpeed}(tb, v), v > 1260.$$
$$\mathsf{LowRotorSpeed}(tb) \leftarrow \mathsf{rotorSpeed}(tb, v), v < 1000.$$

In such programs, we require a (countably infinite) *temporal alphabet* Σ_t that is disjoint from the static Σ_s. Intuitively, the predicates in Σ_t hold on domain objects within some *validity time* intervals. This is in contrast to the predicates in Σ_s that are assumed to hold independently of the time (or, equivalently, at all time instants). In $datalog_{nr}MTL$ programs like $\mathcal{T}$, only predicates from Σ_t can occur in the head of the rules, whereas their bodies can contain predicates from both Σ_t and Σ_s. Moreover, as mentioned above, we require $\mathcal{T}$ to be non-recursive. Thus, intuitively, the temporal rules in $\mathcal{T}$ define temporal predicates in terms of both temporal and static ones. In the example above, Σ_t contains the predicates

PurgingIsOver, MainFlameOn, rotorSpeed, HighRotorSpeed, LowRotorSpeed.

Databases and Mappings. In our approach, we assume that databases have generic schemas. However, in temporal OBDA, we have to deal with temporal data. Therefore, we are particularly interested in databases with tables containing timestamp columns.

Example 5. An example data schema $\mathcal{S}$ for the Siemens data, including sensor measurements and deployment details, can look as follows (the primary keys of the tables are underlined):

$$tb\_measurement(\underline{timestamp}, \underline{sensor\_id}, value),$$

$$tb\_sensors(\underline{sensor\_id}, sensor\_type, mnted\_part, mnted\_tb),$$

$$tb\_deployment(\underline{turbine\_id}, turbine\_type, deployed\_in),$$

$$tb\_components(\underline{turbine\_id}, \underline{component\_id}, component\_type).$$

Snippets of data from the tables tb_measurement, tb_sensor and tb_components, respectively, are given below:

tb_measurement		
timestamp	sensor_id	value
2017-06-06 12:20:00	rs01	570
2017-06-06 12:21:00	rs01	680
2017-06-06 12:21:30	rs01	920
2017-06-06 12:22:50	rs01	1278
2017-06-06 12:23:40	rs01	1310
...		
2017-06-06 12:32:30	mf01	2.3
2017-06-06 12:32:50	mf01	1.8
2017-06-06 12:33:40	mf01	0.9
...		

tb_sensors			
sensor_id	sensor_type	mnted_part	mnted_tb
rs01	0	pt01	tb01
mf01	1	b01	tb01
...	...		

tb_components		
turbine_id	component_id	component_type
tb01	pt01	0
tb01	b01	1
...		

In classical OBDA, mapping assertions take the form $\phi(\boldsymbol{x}) \rightsquigarrow \psi(\boldsymbol{x})$, where $\phi(\boldsymbol{x})$ is a query over the schema $\mathcal{S}$ and $\psi(\boldsymbol{x})$ is an atom mentioning a symbol in Σ_s and variables $\boldsymbol{x}$.

Example 6. Given the static ontology $\mathcal{O}$, the alphabet Σ_s from Example 1, and the schema $\mathcal{S}$ from Example 5, the following are examples of mapping assertions:

> SELECT sensor_id AS X FROM tb_sensors
> > WHERE sensor_type = 1 $\rightsquigarrow$ TemperatureSensor(X)
>
> SELECT component_id AS X FROM tb_components
> > WHERE component_type = 1 $\rightsquigarrow$ Burner(X)
>
> SELECT mnted_part AS X, sensor_id AS Y
> > FROM tb_sensors $\rightsquigarrow$ isMonitoredBy(X, Y)

By applying these mapping assertions to the databases from Example 5, we retrieve the following facts:

> Burner(b01), TemperatureSensor(mf01),
>
> isMonitoredBy(pt01, rs01), isMonitoredBy(b01, mf01).

We call the mapping assertions that define concepts of Σ_s *static* mapping assertions and denote sets of them by $\mathcal{M}_s$.

On the other hand, we also require *temporal* mapping assertions of the form

$$\phi(\boldsymbol{x}, \text{begin}, \text{end}) \rightsquigarrow \psi(\boldsymbol{x})@\langle t_{\text{begin}}, t_{\text{end}}\rangle$$

where $\phi(\boldsymbol{x}, \text{begin}, \text{end})$ is a query over $\mathcal{S}$ such that the variables begin and end are mapped to values of the date/time format, t_{begin} is either a variable begin or a constant temporal value including ∞, $-\infty$ (and similarly for t_{end}), '$\langle$' is either '(', indicating that an interval is left-open, or '[', indicating that an interval is left-closed (and similarly for '$\rangle$'). Temporal mapping assertions are required to define predicates from Σ_t only. We denote sets of such mapping assertions by $\mathcal{M}_t$.

Example 7. Given Σ_t and $\mathcal{T}$ from Example 4 and the schema $\mathcal{S}$ from Example 5, the following is an example of a temporal mapping assertion:

> SELECT sensor_id, value,
> > timestamp AS begin, LEAD(timestamp, 1) OVER W AS end
> > FROM tb_measurement
> > WINDOW W AS (PARTITION BY sensor_id ORDER BY timestamp)
> > $\rightsquigarrow$ rotorSpeed(sensor_id, value)@[begin, end)

The temporal facts we retrieve from the database in Example 5 are as follows:

> rotorSpeed(rs01, 570)@[2017-06-06 12:20:00, 2017-06-06 12:21:00),
>
> rotorSpeed(rs01, 680)@[2017-06-06 12:21:00, 2017-06-06 12:21:30),
>
> rotorSpeed(rs01, 920)@[2017-06-06 12:21:30, 2017-06-06 12:22:50),
>
> rotorSpeed(rs01, 1278)@[2017-06-06 12:22:50, 2017-06-06 12:23:40).

For instance, the first fact above states that the rotor speed measured by the sensor rs01 was 570 in the specified interval. Note that the interval is left-closed and right-open, which reflects the logic of how turbine sensor outputs are produced: a sensor outputs a value when a measured value differs sufficiently from the previously returned value.

Temporal OBDA Specification. Thus, in the temporal OBDA framework, an OBDA *specification* is the octuple

$$\mathfrak{S} = \langle \Sigma_s, \Sigma_t, \mathcal{M}_s, \mathcal{M}_t, \mathcal{O}, \mathcal{R}, \mathcal{T}, \mathcal{S} \rangle,$$

where Σ_s (Σ_t) is a static (respectively, temporal) vocabulary, $\mathcal{M}_s$ ($\mathcal{M}_t$) a set of static (respectively, temporal) mapping assertions, $\mathcal{O}$ an ontology, $\mathcal{R}$ ($\mathcal{T}$) a set of static (respectively, temporal) rules, and $\mathcal{S}$ is a database schema. In the table below we clarify the intuition behind the components of $\mathfrak{S}$ and the specification languages.

Component	Defines predicates in	In terms of predicates in	Language
$\mathcal{M}_s$	Σ_s	$\mathcal{S}$	R2RML
$\mathcal{M}_t$	Σ_t	$\mathcal{S}$	R2RML
$\mathcal{O}$	Σ_s	Σ_s	$DL\text{-}Lite_{\mathcal{R}}\,/\,OWL\,2\,QL$
$\mathcal{R}$	Σ_s	Σ_s	non-recursive datalog
$\mathcal{T}$	Σ_t	$\Sigma_s \cup \Sigma_t$	$datalog_{nr}\,MTL$

Temporal OBDA Instance. A temporal OBDA *instance* $\mathfrak{J}$ is a pair $\langle \mathfrak{S}, D \rangle$ with a temporal OBDA specification $\mathfrak{S}$ and a database instance D compliant with the database schema $\mathcal{S}$ in $\mathfrak{S}$.

Recall that the semantics of a classical OBDA instance is given by the exposed RDF graph consisting of a set of triples of the form (subject, property, object); see, e.g., [19]. In temporal OBDA, we advocate the use RDF Datasets, that is, a set of named RDF graphs and a default graph, following the model of RDF stream proposed by the W3C RDF Stream Processing Community Group [13]. To model temporal facts, for each validity interval, we introduce a graph identifier and collect the triples within the interval into this graph. The details of intervals (namely, the beginning and the end) of these graph identifiers are described in the default graph using the vocabulary from the W3C TIME ontology [26]. The static triples also reside in the default graph. We leave a formal definition of our semantics to future work and only explain here the underlying intuition using the following example.

Example 8. The RDF dataset exposed by our running example includes the static facts in Example 6 and the temporal facts in Example 7. For instance, the static facts Burner(b01), isMonitoredBy(b01, mf01) are represented by two triples

$$(b01, a, \text{Burner}), (b01, \text{isMonitoredBy}, \text{mf01}).$$

The temporal fact

$$\text{rotorSpeed}(\text{rs01}, 570)@[2017\text{-}06\text{-}06\ 12\text{:}20\text{:}00,\ 2017\text{-}06\text{-}06\ 12\text{:}21\text{:}00)$$

is modeled as the named graph

$$\text{GRAPH } g_0 \ \{(\text{rs01}, \text{rotorSpeed}, 570)\}$$

and the following set of triples in the default graph:

$(g_0, a, \text{time:Interval})$,
$(g_0, \text{time:isBeginningInclusive}, \text{true})$, $(g_0, \text{time:isEndInclusive}, \text{false})$,
$(g_0, \text{time:hasBeginning}, b_0)$, $(b_0, \text{time:inXSDDateTimeStamp}, \text{'2017-06-06 12:20:00'})$,
$(g_0, \text{time:hasEnd}, e_0)$, $(e_0, \text{time:inXSDDateTimeStamp}, \text{'2017-06-06 12:21:00'})$.

2.2 Query Answering

The RDF datasets exposed by temporal OBDA instances are ready to be queried with a proper query language. We begin by an example of a query that is of interest in the context of the Siemens use-case.

Example 9. Suppose there is a train with the ID *T001* in which a gas turbine is deployed. An interesting query would be 'find all gas turbines deployed in the train with the ID *T001* and time periods of their accomplished purgings'.

Our query language is based on τ-SPARQL proposed in [24]. In τ-SPARQL, the atomic triple patterns, such as

$$\text{?person a : Person} \quad \text{or} \quad \text{?child : childOf ?parent}$$

can be prefixed by an expression $[?v1, ?v2]$ that represents a validity time interval of the pattern. In our case, the intervals have open and closed ends, therefore we extend the format of the timestamp to $\langle ?e1, ?v1, ?v2, ?e2 \rangle$, where $?e1$ is either 'true' or 'false' depending on whether the validity time interval of the statement is left-closed or left-open, and similarly for $?e2$. Moreover, we alter the notation of τ-SPARQL and write intervals in a prefix, as it is more consistent with our notation for OBDA instances. Thus, we formulate the query from Example 9 as:

```
PREFIX ss: <http://siemens.com/ns#>
PREFIX st: <http://siemens.com/temporal/ns#>
PREFIX obda: <https://w3id.org/obda/vocabulary#>

SELECT ?tb ?left_edge ?begin ?end ?right_edge WHERE {
?tb a ss:GasTurbine ; ss:isDeployedIn ss:train_T001 .

{?tb a st:PurgingIsOver}@<?left_edge,?begin,?end,?right_edge>
}
```

Note that we allow the use of @⟨?e1, ?v1, ?v2, ?e2⟩ only as a suffix of atomic patterns whose predicates are taken from Σ_t. In fact, the use of a SPARQL-based query language does not allow us to construct queries with predicates of arity more than 2 like ColocTempRotSensors from Example 2. However, this does not seems to be a major limitation, since such predicates as ColocTempRotSensors are auxiliary (the latter was needed for a convenient definition of the unary temporal predicate PurgingIsOver in Example 4).

Similarly to τ-SPARQL, our query language is a shorthand format for temporal queries and each query is translated into standard SPARQL using the GRAPH constructor. For example, the query above is translated as follows:

```
PREFIX xsd: <http://www.w3.org/2001/XMLSchema#>
PREFIX time: <http://www.w3.org/2006/time#>
PREFIX ss: <http://siemens.com/ns#>
PREFIX st: <http://siemens.com/temporal/ns#>
PREFIX obda: <https://w3id.org/obda/vocabulary#>

SELECT ?tb ?left_edge ?begin ?end ?right_edge WHERE {
  ?tb a ss:GasTurbine ; ss:isDeployedIn ss:train_T001 .

  GRAPH ?g { ?tb a st:PurgingIsOver.}

  ?g a time:Interval ;
     time:isBeginningInclusive ?left_edge ;
     time:hasBeginning [ time:inXSDDateTimeStamp ?begin ] ;
     time:hasEnd [ time:inXSDDateTimeStamp ?end ] ;
     time:isEndInclusive ?right_edge .
}
```

3   System Architecture and Implementation in Ontop

In this section, we describe a system architecture of temporal OBDA, including concrete languages for the inputs and a query answering algorithm, and also how to implement this architecture by extending the OBDA system Ontop.

Concrete Languages. Our principle when choosing concrete languages for the inputs is to be compliant with relevant existing standards whenever possible; we only extend and create new syntax/languages when it is absolutely necessary. Recall that, in classical OBDA systems (e.g., Ontop [9] and Mastro [10]), the adopted concrete ontology language is usually *OWL 2 QL* [21], the language for mappings is R2RML, and the query language is SPARQL 1.1 [16] under OWL 2 QL entailment regime [19,25]. Ontop also provides a more compact mapping language that is equivalent to R2RML.

In temporal OBDA, we continue to use *OWL 2 QL* for static ontologies. For the temporal rule language, we propose a new language since there is no existing language available. The proposed concrete syntax for $datalog_{nr}MTL$ is inspired by Datalog, SWRL, and SPARQL. We continue to use R2RML as a mapping language. Meanwhile, noticing that it is rather verbose to map all the temporal information in R2RML, we also extend the Ontop mapping language [9] to provide an alternative compact mapping language close to the one used in Example 7. As for the query language, we support both τ-SPARQL based language and plain SPARQL as discussed in Sect. 2.2.

Query Answering Algorithm. The core of a temporal OBDA system is the query answering algorithm. We propose a rewriting-based algorithm extending the structure proposed in [9]. The algorithm takes as inputs a temporal OBDA instance $(\mathfrak{S}, D)$ and a SPARQL query q, and returns the answers of q over $(\mathfrak{S}, D)$. The algorithm consists of two main stages: (1) an offline stage compiling $\mathfrak{S}$ into a set of mapping assertions $\mathcal{M}_\mathfrak{S}$ using the mapping saturation technique, (2) an online stage translating q into an SQL query Q with respect to $\mathcal{M}_\mathfrak{S}$, evaluating Q over D, and converting the SQL answers to SPARQL answers. We give some details of the components in this algorithm:

The offline stage has two sub-stages:

(i) The first sub-stage can be thought of as consisting of two phases: In phase (a) the system classifies the static ontology $\mathcal{O}$ as in [9] and in phase (b) it saturates the input mapping $\mathcal{M}_s$ with the classified ontology and obtains the saturated mapping $\mathcal{M}_s^{\mathcal{O}}$;

(ii) the second sub-stage also has two phases: In phase (a) the system saturates $\mathcal{M}_s^{\mathcal{O}}$ with static rules $\mathcal{R}$ as in [27] and obtains the mapping $\mathcal{M}_s^{\mathcal{O},\mathcal{R}}$; and in phase (b) it saturates $\mathcal{M}_t$ with $\mathcal{M}_s^{\mathcal{O},\mathcal{R}}$ and $\mathcal{T}$ w.r.t. [8], and obtains the final saturated mapping $\mathcal{M}_\mathfrak{S}$.

The online stage is for answering temporal queries. In step *(1)*, the ontology-mediated query is rewritten according to the classified ontology and normalized $datalog_{nr}MTL$ program. Then in step *(2)* the rewritten query is unfolded by substituting each triple occurring in the rewritten query by its saturated mapping specification $\mathcal{M}_\mathfrak{S}$. Obviously, it is not expected that the resulting unfolded query is efficient. So, in step *(3)*, the system applies some optimization techniques in order to eliminate redundant self-joins and replace joins of unions with unions of joins. Then, in step *(4)*, the system translates the optimized query into the target query languages, and in step *(5)* it delegates the evaluation of the translated query to DB engines. Finally in step *(6)*, the results retrieved from DB engines are transformed into SPARQL answer form.

Implementation in Ontop. Ontop is a state-of-the-art OBDA system developed at the Free University of Bozen-Bolzano. It is the core OBDA engine of the EU FP7 Optique project focused on end-user access to big data [12]. More recently, Ontop has also been integrated in Stardog to provide support for SPARQL end-user queries through a virtual RDF graph. Ontop supports the standard W3C recommendations related to OBDA (such as OWL 2 QL, R2RML, SPARQL, and the OWL 2 QL entailment regime in SPARQL). The system is available as a Protege plugin, and an extensible open-source Java library supporting OWL API and RDF4J.

We aim to support temporal OBDA in Ontop by taking advantage of the components already available in Ontop (e.g., *(i)*, *(5)*, *(6)*) for the query answering algorithm, implementing the new components (i.e., *(ii)*), and extending further some of the existing components (i.e., *(1)*, *(2)*, *(3)* and *(4)*).

Acknowledgements. This research has been partially supported by the project "Ontology-based analysis of temporal and streaming data" (OBATS), funded through the 2017 call issued by the Research Committee of the Free University of Bozen-Bolzano.

References

1. Alur, R., Henzinger, T.A.: Real-time logics: complexity and expressiveness. Inf. Comput. **104**(1), 35–77 (1993)
2. Artale, A., Calvanese, D., Kontchakov, R., Zakharyaschev, M.: The DL-Lite family and relations. J. Artif. Intell. Res. **36**(1), 1–69 (2009)
3. Artale, A., Kontchakov, R., Kovtunova, A., Ryzhikov, V., Wolter, F., Zakharyaschev, M.: First-order rewritability of temporal ontology-mediated queries. In: Proceedings of IJCAI 2015, pp. 2706–2712. AAAI Press (2015)
4. Artale, A., Kontchakov, R., Wolter, F., Zakharyaschev, M.: Temporal description logic for ontology-based data access. In: Proceedings of the 23rd International Joint Conference on Artificial Intelligence, IJCAI 2013. IJCAI/AAAI (2013)
5. Baader, F., Borgwardt, S., Lippmann, M.: Temporalizing ontology-based data access. In: Bonacina, M.P. (ed.) CADE 2013. LNCS, vol. 7898, pp. 330–344. Springer, Heidelberg (2013). doi:10.1007/978-3-642-38574-2_23
6. Basulto, V.G., Jung, J., Kontchakov, R.: Temporalized EL ontologies for accessing temporal data: complexity of atomic queries. In: Proceedings of the 25th International Joint Conference on Artificial Intelligence (IJCAI 2016). AAAI Press (2016)
7. Borgwardt, S., Lippmann, M., Thost, V.: Temporal query answering in the description logic DL-Lite. In: Proceedings of FroCoS 2013, pp. 165–180 (2013)
8. Brandt, S., Kalaycı, E.G., Kontchakov, R., Ryzhikov, V., Xiao, G., Zakharyaschev, M.: Ontology-based data access with a horn fragment of metric temporal logic. In: AAAI (2017)
9. Calvanese, D., Cogrel, B., Komla-Ebri, S., Kontchakov, R., Lanti, D., Rezk, M., Rodriguez-Muro, M., Xiao, G.: Ontop: answering SPARQL queries over relational databases. Semant. Web **8**(3), 471–487 (2017)
10. Calvanese, D., De Giacomo, G., Lembo, D., Lenzerini, M., Poggi, A., Rodriguez-Muro, M., Rosati, R., Ruzzi, M., Savo, D.F.: The mastro system for ontology-based data access. Semant. Web J. **2**(1), 43–53 (2011). Listed among the 5 most cited papers in the first five years of the Semantic Web Journal
11. Calvanese, D., Giacomo, G., Lembo, D., Lenzerini, M., Rosati, R.: Tractable reasoning and efficient query answering in description logics: the DL-Lite family. J. Autom. Reasoning **39**(3), 385–429 (2007)
12. Giese, M., Soylu, A., Vega-Gorgojo, G., Waaler, A., Haase, P., Jiménez-Ruiz, E., Lanti, D., Rezk, M., Xiao, G., Özçep, Ö.L., Rosati, R.: Optique - zooming in on big data access. IEEE Comput. **48**(3), 60–67 (2015)

13. R. S. P. C. Group. RDF stream processing: Requirements and design principles. W3C draft community group report, W3C (2016)
14. Gutiérrez-Basulto, V., Jung, J.C., Ozaki, A.: On metric temporal description logics. In: ECAI 2016, pp. 837–845 (2016)
15. Gutiérrez-Basulto, V., Klarman, S.: Towards a unifying approach to representing and querying temporal data in description logics. In: Krötzsch, M., Straccia, U. (eds.) RR 2012. LNCS, vol. 7497, pp. 90–105. Springer, Heidelberg (2012). doi:10.1007/978-3-642-33203-6_8
16. Harris, S., Seaborne, A., Prud'hommeaux, E.: SPARQL 1.1 query language. W3C recommendation, W3C (2013)
17. Kharlamov, E., Brandt, S., Jimenez-Ruiz, E., Kotidis, Y., Lamparter, S., Mailis, T., Neuenstadt, C., Özçep, O., Pinkel, C., Svingos, C., Zheleznyakov, D., Horrocks, I., Ioannidis, Y., Moeller, R.: Ontology-based integration of streaming and static relational data with optique. In: Proceedings of the 2016 International Conference on Management of Data, SIGMOD 2016, pp. 2109–2112. ACM, New York (2016)
18. Klarman, S., Meyer, T.: Querying temporal databases via OWL 2 QL. In: Proceedings of RR 2014, pp. 92–107 (2014)
19. Kontchakov, R., Rezk, M., Rodríguez-Muro, M., Xiao, G., Zakharyaschev, M.: Answering SPARQL queries over databases under OWL 2 QL entailment regime. In: Mika, P., et al. (eds.) ISWC 2014. LNCS, vol. 8796, pp. 552–567. Springer, Cham (2014). doi:10.1007/978-3-319-11964-9_35
20. Möller, R., Özçep, Ö., Neuenstadt, C., Zheleznyakov, C., Kharlamov, E.: D5.1: a semantics for temporal and stream-based query answering in an obda context. Optique project deliverable, FP7-318338, EU (2013)
21. Motik, B., Fokoue, A., Horrocks, I., Wu, Z., Lutz, C., Cuenca Grau, B.: OWL Web Ontology Language profiles. W3C Recommendation, World Wide Web Consortium (2009)
22. Özçep, Ö.L., Möller, R., Neuenstadt, C.: A stream-temporal query language for ontology based data access. In: Lutz, C., Thielscher, M. (eds.) KI 2014. LNCS, vol. 8736, pp. 183–194. Springer, Cham (2014). doi:10.1007/978-3-319-11206-0_18
23. Poggi, A., Lembo, D., Calvanese, D., Giacomo, G., Lenzerini, M., Rosati, R.: Linking data to ontologies. In: Spaccapietra, S. (ed.) Journal on Data Semantics X. LNCS, vol. 4900, pp. 133–173. Springer, Heidelberg (2008). doi:10.1007/978-3-540-77688-8_5
24. Tappolet, J., Bernstein, A.: Applied temporal RDF: efficient temporal querying of RDF data with SPARQL. In: Aroyo, L., Traverso, P., Ciravegna, F., Cimiano, P., Heath, T., Hyvönen, E., Mizoguchi, R., Oren, E., Sabou, M., Simperl, E. (eds.) ESWC 2009. LNCS, vol. 5554, pp. 308–322. Springer, Heidelberg (2009). doi:10.1007/978-3-642-02121-3_25
25. W3C. SPARQL 1.1 entailment regimes. Technical report, W3C, March 2013
26. W3C. Time ontology in OWL. W3C working draft, OGC & W3C (2017)
27. Xiao, G., Rezk, M., Rodríguez-Muro, M., Calvanese, D.: Rules and ontology based data access. In: Kontchakov, R., Mugnier, M.-L. (eds.) RR 2014. LNCS, vol. 8741, pp. 157–172. Springer, Cham (2014). doi:10.1007/978-3-319-11113-1_11

Towards Semantic Assessment of Summarizability in Self-service Business Intelligence

Luis-Daniel Ibáñez[1], Jose-Norberto Mazón[2(✉)], and Elena Simperl[1]

[1] University of Southampton, Southampton, UK
{l.d.ibanez,e.simperl}@soton.ac.uk
[2] DLSI, Universidad de Alicante, Alicante, Spain
jnmazon@dlsi.ua.es

Abstract. Traditional Business Intelligence solutions allow decision makers to query multidimensional data cubes by using OLAP tools, thus ensuring summarizability, which refers to the possibility of accurately computing aggregation of measures along dimensions. With the advent of the Web of Open Data, new external sources have been used in Self-service Business Intelligence for acquiring more insights through ad-hoc multidimensional open data cubes. However, as these data cubes rely upon unknown external data, decision makers are likely to make meaningless queries that lead to summarizability problems. To overcome this problem, in this paper, we propose a framework that automatically extracts multidimensional elements from SPARQL query logs and creates a knowledge base to detect semantic correctness of summarizability.

Keywords: OLAP · Data cube · Summarizability · Open data

1 Introduction

Traditional Business Intelligence (BI) rely on a well-controlled, consistent and certified *Data Warehouse* (DW) from which multidimensional cubes are designed in order to analyze data fast and accurately. Data cubes consists of measures or facts (representing events of interests for analysts) and dimensions (as different ways that data can be viewed, aggregated, and sorted). These Data Cubes (see Fig. 1 for a simplified example) are the basis of OLAP (OnLine Analytical Processing) tools that traditionally support the decision making process [5].

The advent of the Web of Data and the Open Data initiatives have made a tremendous amount of data to become available that can be combined with locally warehoused data, potentially improving the decision making process. The capability of (automatically) incorporating external data for supporting decision making processes gives rise to a novel branch of BI called *Self-Service BI* [1]. The goal of Self-Service BI is to accomplish the search, extraction, integration and querying of external data, while minimising the intervention of DW designers or programmers. Consequently, unlike the well-controlled and closed scenario

© Springer International Publishing AG 2017
M. Kirikova et al. (Eds.): ADBIS 2017, CCIS 767, pp. 174–185, 2017.
DOI: 10.1007/978-3-319-67162-8_18

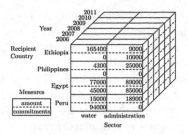

Fig. 1. Sample data cube for financial aid given by Finland, measured in *amount* and *commitments*, covering the *Recipient Country*, *Year* and *Sector* dimensions. Source: [7]

Table 1. Temperatures in cities versus temperatures in reactors

City	Temperature		Reactor	Temperature
London	14		R1	1400
Southampton	19		R2	1900
Manchester	11		R3	1100
Is it meaningful to sum temperatures across cities?			Is it meaningful to sum temperatures across reactors?	

of traditional BI, users of self-service BI do not necessarily know the nature of new external data to be analyzed [2], raising the issue of how to know if the multidimensional queries they ask on them are *meaningful* (i.e., semantically correct). This is especially significant when computing aggregations of measures with respect to dimensions, since *summarizability* problems may arise [11].

The simplified example on Table 1 illustrates the issue of semantics in summarizability. On the left side, a local dataset including cities in the UK is enriched with their temperatures, from external sources; on the right hand, a dataset containing chemical reactors in an industrial complex is enriched with temperature measures for each reactor. A syntactically-valid query for both datasets is to sum the temperature across cities/reactors. However, depending on the semantics, the query could be meaningless: while a meteorologist would argue that computing the sum of the temperatures across cities is not meaningful (instead, average could be more sensible), for the chief of the industrial complex, the sum of the temperatures across reactors being below or above a certain threshold might indicate a malfunction. Despite the syntactic correctness of both queries, a meaningful summarization also depends on the particular *semantic* of the involved multidimensional elements.

Previous works [6,9,11,15] propose categories of measures and dimensions and develop algorithms to check if a given query ensures summarizability. However, in all of them, the task of categorising measures and dimensions is left to the designer of the cube which disagrees with the self-service essence. Furthermore, when integrating Web Data on the fly (even curated sources of statistical open data like those coming from The World Bank[1] or Eurostat[2]), there is no

---

[1] http://data.worldbank.org/.

[2] http://ec.europa.eu/eurostat/data/database.

metadata on measures or dimensions available. At best, they provide a verbose explanation as part of their metadata[3]. Therefore, introducing manual annotations is required for categorizing MD elements, which hinders the ability to provide fast responses of self-service BI.

These issues lead us to argue that at Web-of-Data-Scale, the large amount of possible combinations of measure, dimensions and contexts makes very hard the development of an universal category of measures, thus, instead of trying to automate the mapping of measure and dimension types to one of the current categorisations, we propose to leverage the knowledge of query logs, made in possibly different contexts, by extracting semantic consistency information from queries made by others. The idea is to provide a hint on the semantic consistency of an aggregation query by looking if others have made the same or a similar query (in a crowdsourcing spirit). To improve the confidence on the hints, we consider the addition of provenance information to let users accept them or not depending on their own parameters (from the simple existence of a similar query, to more complex trust-based algorithms). Specifically, our contributions are as follows: (i) a framework to enable hinting of the summarizability of multidimensional queries based knowledge from queries issued by others, (ii) an extension to an existing vocabulary for describing Data Cubes in RDF for expressing the semantics of summarizability; and (iii) an algorithm to extract the multidimensional elements from SPARQL queries conforming to a multidimensional pattern.

The rest of the paper is organised as follows: preliminary definitions are provided in Sect. 2; Sect. 3 reviews the related work; Sect. 4 describe our general framework; Sect. 5 describes our extension to the QB4OLAP vocabulary; Sect. 6 describes our algorithm to extract semantics of multidimensional elements from SPARQL queries and reports our preliminary results; and finally, Sect. 7 concludes the paper and provides future work directions.

2   Preliminaries

The W3C recommends the use of the RDF Data Cube vocabulary [3][4] (QB) to publish multidimensional data on the web. Multidimensional data in QB is comprised by a description of the *observations* it holds and a set of statements about them. Each observation is characterized by a set of dimensions that define what the observation applies to, along with metadata describing what has been measured, how it was measured and how the observations are expressed.

Example 1.1 shows the SPARQL version (assuming the cube is in RDF format following QB) of the multidimensional query *What is the sum of commitment aid that the Philippines received from 2005 to 2008? Commitment* is the measure, *country* and *year* are dimensions, while *sum* is the aggregation function.

Example 1.1. Example SPARQL multidimensional query: What is the sum of commitment aid that the Philippines received from 2006 to 2008?

PREFIX qb: http :// purl . org / linked −data / cube#

---

[3] E.g., http://ec.europa.eu/eurostat/cache/metadata/en/hbs_esms.htm.

[4] https://www.w3.org/TR/vocab-data-cube/.

```
PREFIX  ex:  http://example−namespace.org/#
SELECT  sum(?commitment)  {
?aid  a  qb:Observation  .
?aid  ex:commitment  ?commitment  .
?aid  ex:recipient−country  ex:Philippines  .
?aid  ex:refYear  ?y  .
filter  (year  (?y)=2006  OR  year  (?y)=2008)  }
```

However, as pointed out in [18], QB vocabulary has no support for representing: (i) dimension hierarchies, since QB only allows representing relationships between dimension instances, (ii) aggregate functions, (iii) descriptive attributes of dimension hierarchy levels. To overcome these drawbacks, QB was extended in QB4OLAP [4] to fully support the multidimensional model, also enabling the implementation of typical OLAP operations, such as rollup, slice, dice, and drill-across using standard SPARQL queries. Unfortunately, although QB4OLAP includes the concept of aggregation functions and how they are associated to measures, they do not provide the means to describe summarizability.

Executing a multidimensional querying on a data cube [17] refers to obtain a set of measures indexed by one or more dimensions and applying an aggregation function (e.g., sum, count or average) to it. Lenz and Shoshani [11] established the importance of summarizability for queries on multidimensional data, since violations of this property may lead to erroneous conclusions and decisions. They show that summarizability is dependent on (i) types of measures and (ii) the specific dimensions under consideration. Moreover, they state three necessary conditions for summarizability. The first of these conditions, called disjointness, states that consecutive levels of a dimension hierarchy must have a many-to-one relationship. The second condition, called completeness, requires that all the elements of each level of a dimension hierarchy exist as well as that each element in a level is assigned to one element of the upper level. These two conditions are at the syntactic level and they should be solved by manipulating schema to create hierarchies that ensure these summarizability constraints [16]. The third condition, called *type compatibility*, ensures that the aggregate function applied to a measure is summarizable according to the type of the measure (e.g., stock, flow and value per-unit as defined by Lenz and Shoshani) and the type of the related dimensions (e.g., temporal, non-temporal). According to [11], this third condition ensures the semantic correctness of summarizability in a multidimensional query.

After elaborating a multidimensional query and before executing it, a user would like to know if the summarizability is *semantically ensured* according to the type compatibility condition. In other words, and recalling our Example 1.1: does it make sense to sum the commitments across recipient countries and years?. The problem of *semantic correctness of summarizability* can be formalised as follows:

Definition 1 (Semantic correctness of summarizability). *Given a multidimensional query Q with aggregation function F, measure M and set of dimensions $D = [D_1, ..., D_n]$, decide if the application of F to the set of observations* *characterized by M and D has types that are semantically compatible. If so, we say that M is* semantically summarizable *by F across D, and denote it as* $M \xrightarrow{F} D$ *and then* semantic correctness *of summarizability is ensured.*

Back to Example 1.1, if the *commitment* measure is categorised as summarizable across *country* and *year* following one of the approaches in [9,11,15], then one can use their proposed methods to decide if the query is meaningful or not. Importantly, this can be only done in traditional BI because multidimensional elements are known before analysis and they can be categorized in advance according to the expertise of DW designers, thus limiting users to meaningful queries. However, in self-service BI, multidimensional elements coming from external sources are unknown beforehand, and are uncategorised.

3   Related Work

An in-depth analysis of summarizability problems and a taxonomy of reasons why summarizability may not hold is presented by Horner and Song [6]. They distinguish schema problems (e.g., disjointness and completeness) from data problems (e.g., inconsistencies and imprecision) and computational problems (e.g., type compatibility in the sense of [11]), give typical examples for each problematic case, and suggest guidelines for their management. A survey on summarizability issues and solutions is in [13]. Categorisation of measures is the main approach for ensuring semantic correctness of summarizability. The most recent categorization is proposed by Niemi et al. [15]. They consider measures and categorize them, depending on their nature, into tally, semi-tally, reckoning, snapshot and conversion factor. Also, they propose an algorithm to decide summarizability according to this novel categorization. Main pitfall of these approaches is that they define a fixed categorization for measures in order to ensure semantic correctness of summarization queries. This is useful in traditional BI scenarios based only on querying internal and well-known data, but is has limitations in self-service BI scenarios that query unforeseen data from heterogeneous domains (see simplified example on Table 1).

Several efforts have advanced in the direction of realising self-service BI on the Web of Data based on Linked Data Cubes and the QB vocabulary. Kämpgen et al. [8] propose a mapping from OLAP operations to SPARQL based on RDF Data Cube. They define every multidimensional element but they do not automatically deal with summarizability problems and they manually create a specific measure for each possible aggregation function. Höffner [7] defines Question-Answering (in the sense of natural language questions) on top of Linked Data Cubes.

Varga et al. [?] propose a series of steps to automate the enrichment of QB data sets with specific QB4OLAP semantics; being the most important, the definition of aggregate functions and the detection of new concepts in the dimension hierarchy construction. They state that not every aggregate function can be applied to a measure and give a valid result. Therefore, techniques to automate the association between measures and aggregate functions should be provided. To do so, they create a mapping of measures to aggregate functions called *MAggMap*. However, they assume that the user must explicitly provide the *MAggMap* mapping since they state that the large variety of measure and aggregate function types makes the compatibility check a tedious task that can hardly be fully automated.

Nebot and Berlanga [14] consider semantic correctness of summarization in multidimensional queries on LOD in the context of exploring the potential analysis of LOD datasets. They propose a statistical framework to automatically discover candidate multidimensional patterns hidden in LOD. They consider different kind of aggregation semantics than the previous literature: (i) applicable to data that can be added together; (ii) applicable to data that can be used for average calculations; and (iii) applicable to data that is constant, i.e., it can only be counted. However, they assume the existence of a function that gives for each dimension type returns the compatible aggregation functions which should be manually established by a designer or an expert of the domain.

4   Detecting Semantic Correctness of Summarizability

Our proposal focuses on leveraging the information on query logs about aggregations made by decision makers, to provide them with hints on the summarizability of their multidimensional queries based on what others have done on their data. Figure 2 gives an overview of the framework. The flow of the framework is as follows:

1. A decision maker wants to execute a multidimensional query on a data cube without measure categorisation available (e.g. on a self-service BI scenario). Before issuing the query to the cube, the decision maker asks a *Summarizability Hint Module* (SHM) to get a hint on summarizability of the query.
2. A *Summarizability Knowledge Base* (SUM-KB) is previously constructed by extracting from query logs how others have used multidimensional queries that ensure summarizability on different cubes and knowledge bases.
3. The SHM takes the query Q and the knowledge from SUM-KB and produces a hint about the summarizability of Q (i.e., summarizability awareness).
4. If the decision maker accepts the hint, the query is executed on his/her data cube and this decision of considering that Q ensures summarizability is added to SUM-KB.

To realize this scenario, we need to solve three problems, namely (i) how to describe semantic correctness of summarizability, (ii) how to extract semantic correctness knowledge from query logs, and (iii) how to compute a hint from said information. The first problem is solved by extending the QB4OLAP vocabulary to consider summarizability (we refer reader to Sect. 5 for further details), while the two latter problems are formalized as follows.

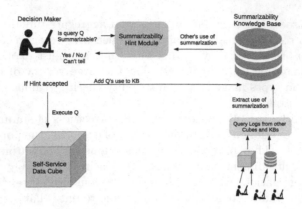

Fig. 2. Overview of framework for detecting semantic correctness of summarizability

Definition 2 (Extraction of multidimensional elements (EME)). *Given a multidimensional query string Q, extract the aggregation function F, the measure M and the dimensions D it uses. We say that Q summarizes M by F across D, and denote it as $M \overset{F}{\Rightarrow} D$.*

Section 6 proposes an algorithm to solve EME for SPARQL queries. Besides the multidimensional elements of queries, the hint process can benefit of having available metadata about their context or provenance, *e.g.*, who executed it, on which dataset, etc. We formalise this as a *Summarizability statement*.

Definition 3 (Summarizability statement). *Let $M \overset{F}{\Rightarrow} D$ be the EME of a query Q and C a set containing the context or provenance of Q. We call the tuple $(M \overset{F}{\Rightarrow} D, C)$ a summarizability statement.*

In layman terms, a summarizability statement encodes the fact that someone or something issued a multidimensional query with M, F and D and therefore considered it summarizable. In Sect. 5 we provide an extension of the QB4OLAP vocabulary that allows us to encode summarizability statements in RDF. Finally, we formalize the problem of providing a summarizability hint:

Definition 4 (Summarizability Hint). *Given a set S of summarizability statements and an instance of the semantic correctness of summarizability denoted as $M \overset{F}{\rightarrow} D$ (cf. Sect. 2), return one of the following three outputs:*

1. *$M \overset{F}{\rightarrow} D$ is hinted by the statements in S*
2. *$M \overset{F}{\rightarrow} D$ is not hinted by S*
3. *The knowledge in S is inconclusive to hint summarizability.*

5   QB4OLAP Extension for Constructing SUM-KB

We chose to extend QB4OLAP [4] to encode summarizability statements because it already considers aggregation functions, measures and dimensions. We want

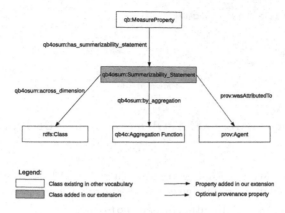

Fig. 3. Extension to QB4OLAP to encode summarizability statements

to express the following knowledge (recall 3): *The Measure M has been summarized by function F across dimensions $D_1, ..., D_n$ in context C*. Figure 3 shows a diagram describing our extension (we only show the classes from QB4OLAP linked to the classes and properties defined by us). We encode the n-ary relationship between a measure and the aggregation function, dimensions and context on which it was summarized by means of the *Summarizability Statement* (SS) class and three properties: *has_summarizability_statement* that relates a *qb:MeasureProperty* with a SS; *across_dimension* that relates a SS with a *rdfs:Class* and *by_aggregation* that relates a SS with a *qb4o:AggregationFunction*. We chose to relate SSs to *rdfs:Class* instead of *qb:DimensionProperty* because knowledge bases that are not data cubes or data cubes not in QB format usually regard dimensions as concepts or classes instead of properties. QB provides the mean to link a *qb:DimensionProperty* to the concept and/or class it represents through the *qb:concept* and *rdfs:Range* properties, therefore, a hint to a query on RDF Data Cube could be provided by examining the summarizability statements in SUM-KB refer to the same or similar classes or concepts. If available, context information can be linked to summarizability statements and be used as input for the hinting process. Figure 3 shows the sample statement linked to a *prov:agent* through *prov:wasAttributedTo* property of PROV ontology [10].

6   EME for SPARQL Queries

An algorithm to solve the EME (Extraction of Multidimensional Elements) problem (cf. Definition 2) for SPARQL queries has been developed. For queries executed on data cubes modelled with the QB vocabulary, e.g., Example 1.1, properties used are typed as *qb:DimensionProperty* or *qb:MeasureProperty*, making EME straightforward. Algorithm 1 shows EME for the class of queries on QB. This class is characterized as *multidimensional* queries centered around a variable of type *qb:Observation*.

```
Input: SPARQL Query Q on RDF Data Cube
                     Fj
Output: Set of {Mᵢ ⟹ Dₖ}
S ← ∅;
Aggs ← Q.getAggregateFunctions();
foreach F ∈ Aggs do
    aggVariable ← fun.getVariable();
    D ← ∅;
    M = Property of type qb:MeasureProperty of the BGP in Q having aggVariable as object ;
    foreach Property P in Q of type qb:DimensionProperty do
              F
        S ∪ {M ⟹ {P.getRDFSRange(), P.getQBConcept()}};
    end
end
```

Algorithm 1. EME for SPARQL queries using RDF Data Cube

Example 1.2. Example SPARQL aggregation query from USEWOD: Compute maximum, minimum and average runtime of movies starring Clint Eastwood

PREFIX: dbo:<http://dbpedia.org/ontology/>
PREFIX: dbp:<http://dbpedia.org/resource/>
SELECT ?movie max(?runtime) min(?runtime) avg(?runtime)
WHERE
{?movie dbo:runtime ?runtime .
?movie dbo:starring dbp:Clint_Eastwood . }
GROUP BY ?movie

However, for queries made on Knowledge Bases that are not QB data cubes, there is no explicit link to dimensions and measures. Therefore, we need to define an algorithm to perform EME. Our Algorithm 2 is divided in three main steps. We will illustrate it with the query depicted in Example 1.2.

1. Extract the aggregation functions and their aggregation variables. In the example, there are three aggregation functions: *max*, *min* and *avg*, all of them with *?runtime* as aggregation variable.
2. For each aggregation function, locate the Basic Graph Pattern (BGP) that contains the aggregation variable as object. We call this the *measure BGP* of the query. In our example, the measure BGP is *?movies dbo:runtime ?runtime*. The predicate of this pattern is the *Measure type* of the query. In our example *dbo:runtime*.
3. To extract the dimensions there are two cases:
   (a) If the query has a Group By clause, the GroupBy variables indicate the dimensions. To get their type, we check for each one if the query has a BGP explicitly stating their rdf:type. This is not the case in Example 1.2. Note that if the query had included the BGP *?movie rdf:type schema:Movie*, the algorithm would add *schema:Movie* as a dimension to the summarizability statement. If there is not such a BGP, we add as dimensions the rdfs:domain of the predicates of the BGPs having the group variable as subject, and the rdfs:range of the predicates of the BGPs having the group variable as subject (as in our running example). The algorithm identifies both BGPs in the query as containing the *?movie* variable and dereferences the predicates *dbo:runtime* and *dbo:starring* to get their domains. Both properties have as domain *dbo:Work*.

Input: SPARQL Query Q using aggregate functions
Output: Set of $\{M_i \xoverset{F_j}{\Rightarrow} D_k\}$
$S \leftarrow \emptyset$;
$Aggs \leftarrow Q.getAggregateFunctions()$;
foreach $F \in Aggs$ **do**
    $aggVariable \leftarrow fun.getVariable()$;
    $measureBGP \leftarrow Q.getBGPcontaining(aggVariable)$;
    $M \leftarrow measureBGP.getPredicate()$;
    if $Q.hasGroupBy()$ **then**
        foreach $GroupBy$ variable v **do**
            if $Q.hasBGP(v,rdf{:}type,a\_type)$ **then**
                $S \cup \{M \xRightarrow{F} \{a\_type\}\}$
            else
                $Dims \leftarrow \emptyset$ **foreach** BGP in Q having v as subject **do**
                    $Dims \cup \{BGP.getRDFSDomain()\}$
                end
                foreach BGP in Q having v as object **do**
                    $Dims \cup \{BGP.getRDFSRange()\}$
                end
                $S \cup \{M \xRightarrow{F} Dims\}$
        end
    else
        if $Q.hasBGP(measureBGP.getSubject(),rdf{:}type,a\_type)$ **then**
            $S \cup \{(M \xRightarrow{F} \{a\_type\})\}$
        else
            $Dims \leftarrow \emptyset$ **foreach** BGP in Q having $measureBGP.getSubject()$ as subject **do**
                $Dims \cup \{BGP.getRDFSDomain()\}$
            end
            foreach BGP in Q having $measureBGP.getSubject()$ as object **do**
                $Dims \cup \{BGP.getRDFSRange()\}$
            end
            $S \cup \{M \xRightarrow{F} Dims\}$
    end
end

Algorithm 2. EME for SPARQL queries

(b) Else, if there is no GroupBy clause, we take the variable that appears as subject in the *measure BGP* and apply the same procedure that we do to GroupBy variables: check if the rdf:type is explicit in the query, and if not, get the domain and ranges of the predicates involving the variable.

Complexity of Algorithm 2 is $O(numF \times numBGP)$ where $numF$ is the number of aggregation functions and $numB$ the number of Basic Graph Patterns.

From the EME and the context/provenance metadata of the query, it is straightforward to produce RDF summarizability statements following the extension to QB4OLAP described in Sect. 5. The following excerpt shows it for Example 1.2 for the *Max* aggregation function:

Example 1.3. Example of Summarizability Statement extracted from Example 1.2

```
PREFIX dbo:<http://dbpedia.org/ontology/>
PREFIX qb4o:<http://purl.org/qb4olap/cubes#>
PREFIX ex:<http://example.org/#>
PREFIX prov:<http://www.w3.org/ns/prov#>
dbo:runtime qb4o:Summarizability_Statement :_SS_instance
:_SS_instance qb4o:by_aggregation qb4o:Max
:_SS_instance qb4o:across_dimension dbo:Work
\----- Example Context/Provenance information -----/
:_SS_instance prov:wasAttributedBy ex:Jane_Doe
```

A simple way to provide a solution to the summarizability hinting problem (cf. Definition 4) is to ask the SUM-KB if a query using the same multidimensional elements exists. For example, given a query with $F = qb4o{:}Min$,

$M = dbo{:}runtime$ and $D = dbo{:}Work$, the following ASK query on SUM-KB returns TRUE if there is a summarizability statement using the same combination, and FALSE otherwise:

Example 1.4. SPARQL ASK query to solve Summarizability Hinting problem
```
PREFIX dbo:<http://dbpedia.org/ontology/>
PREFIX qb4osum:<http://purl.org/qb4sum/cubes#>
ASK {
dbo:runtime qb4osum:Summarizability_Statement ?ssinstance .
?ssinstance qb4osum:by_aggregation qb4o:Min .
?ssinstance qb4osum:across_dimension dbo:Work . }
```

7   Conclusion and Future Work

The capability of (automatically) incorporating external data for supporting decision making processes gives rise to a novel BI scenario labelled as *Self-Service BI* [1]. Within this scenario decision makers are likely to make meaningless queries that lead to summarizability problems. To overcome these problems, we propose a framework that contain (i) a *Summarizability Hint Module* (SHM) that allows decision makers to be aware on the potential summarizability problems of multidimensional queries; and (ii) a *Summarizability Knowledge Base* (SUM-KB) constructed by extracting multidimensional elements (EME), from multidimensional query logs, that contain information about semantic correctness of summarization queries. To proof-concept our framework, we generated a first version of SUM-KB from the DBpedia logs of the USEWOD 2016 research dataset [12] by filtering the queries that use aggregation functions (8946 queries out of 35124962). We applied Algorithm 2[5] to the remaining queries. Only in 0.5% of the queries the extraction was successful. Results suggest that the log of a public endpoint where many people come to learn SPARQL and the percentage of queries using aggregation is not significant. Therefore, although using the USEWOD 2016 dataset shows the feasibility of our approach works, as future work, we plan to extend our experimentation to be conducted on query logs from other endpoints (where more multidimensional queries are executed). Also, the algorithm that extracts multidimensional elements (EME) will be extended for considering a larger class of SPARQL queries.

Acknowledgments. Jose-Norberto Mazón is funded by grant number JC2015-00284 under "José Castillejo" research program from Spanish Government (Ministerio de Educación, Cultura y Deporte en el marco del Programa Estatal de Promoción del Talento y su Empleabilidad en I+D+i, Subprograma Estatal de Movilidad, del Plan Estatal de Investigación Científica y Técnica y de Innovación 2013–2016). This work is also funded by research project TIN2016-78103-C2-2-R from Spanish Government (Ministerio de Economía, Industria y Competitividad).

---

[5] Available at https://github.com/ldibanyez/eme-extractor.

References

1. Abelló, A., Darmont, J., Etcheverry, L., Golfarelli, M., Mazón, J.N., Naumann, F., Pedersen, T., Rizzi, S.B., Trujillo, J., Vassiliadis, P., Vossen, G.: Fusion cubes: towards self-service Business Intelligence. Int. J. Data Warehous. Min. **9**(2), 66–88 (2013)
2. Abello, A., Romero, O., Pedersen, T.B., Berlanga, R., Nebot, V., Aramburu, M.J., Simitsis, A.: Using semantic web technologies for exploratory OLAP: a survey. IEEE Trans. Knowl. Data Eng. **27**(2), 571–588 (2015)
3. Cyganiak, R., Reynolds, D., Tennison, J.: The RDF data cube vocabulary. Technical report, W3C (2014). https://www.w3.org/TR/vocab-data-cube/
4. Etcheverry, L., Vaisman, A., Zimányi, E.: Modeling and Querying Data Warehouses on the Semantic Web Using QB4OLAP. In: Bellatreche, L., Mohania, M.K. (eds.) DaWaK 2014. LNCS, vol. 8646, pp. 45–56. Springer, Cham (2014). doi:10.1007/978-3-319-10160-6_5
5. Golfarelli, M., Rizzi, S.: Data Warehouse Design: Modern Principles and Methodologies. McGraw-Hill Inc., New York (2009)
6. Höffner, K., Lehmann, J., Usbeck, R.: CubeQA - question answering on RDF data cubes. In: International Semantic Web Conference (ISWC) (2016)
7. Horner, J., Song, I.-Y.: A Taxonomy of Inaccurate Summaries and Their Management in OLAP Systems. In: Delcambre, L., Kop, C., Mayr, H.C., Mylopoulos, J., Pastor, O. (eds.) ER 2005. LNCS, vol. 3716, pp. 433–448. Springer, Heidelberg (2005). doi:10.1007/11568322_28
8. Kämpgen, B., O'Riain, S., Harth, A.: Interacting with statistical linked data via OLAP operations. In: Simperl, E., Norton, B., Mladenic, D., Della Valle, E., Fundulaki, I., Passant, A., Troncy, R. (eds.) ESWC 2012. LNCS, vol. 7540, pp. 87–101. Springer, Heidelberg (2015). doi:10.1007/978-3-662-46641-4_7
9. Kimball, R., Ross, M.: The Kimball Group Reader: Relentlessly Practical Tools for Data Warehousing and Business Intelligence. Wiley, Indianapolis (2010)
10. Lebo, T., Sahoo, S., McGuinness, D.: PROV-O: the PROV ontology. Technical report, W3C (2013). https://www.w3.org/TR/prov-o
11. Lenz, H.J., Shoshani, A.: Summarizability in OLAP and statistical data bases. In: Proceedings of the Ninth International Conference on Scientific and Statistical Database Management, pp. 132–143. IEEE Computer Society,., January 1997
12. Luczak-Roesch, M., Aljaloud, S., Berendt, B., Hollink, L.: Usewod 2016 research dataset. doi:10.5258/SOTON/385344
13. Mazón, J.N., Lechtenbörger, J., Trujillo, J.: A survey on summarizability issues in multidimensional modeling. Data Knowl. Eng. **68**(12), 1452–1469 (2009)
14. Nebot, V., Berlanga, R.: Statistically-driven generation of multidimensional analytical schemas from linked data. Knowl. Based Syst. **110**, 15–29 (2016)
15. Niemi, T., Niinimäki, M., Thanisch, P., Nummenmaa, J.: Detecting summarizability in OLAP. Data Knowl. Eng. **89**, 1–20 (2014)
16. Pedersen, T.B.: Managing Complex Multidimensional Data. In: Aufaure, M.-A., Zimányi, E. (eds.) eBISS 2012. LNBIP, vol. 138, pp. 1–28. Springer, Heidelberg (2013). doi:10.1007/978-3-642-36318-4_1
17. Rafanelli, M., Bezenchek, A., Tininini, L.: The aggregate data problem: a system for their definition and management. ACM Sigmod Rec. **25**(4), 8–13 (1996)
18. Varga, J., Etcheverry, L., Vaisman, A.A., Romero, O., Pedersen, T.B., Thomsen, C.: Qb2olap: Enabling olap on statistical linked open data. In: 32nd International Conference on Data Engineering (ICDE), pp. 1346–1349. IEEE (2016)

Theta Architecture: Preserving the Quality of Analytics in Data-Driven Systems

Vasileios Theodorou[1(✉)], Ilias Gerostathopoulos[2], Sasan Amini[2],
Riccardo Scandariato[3], Christian Prehofer[4], and Miroslaw Staron[3]

[1] Intracom SA Telecom Solutions, Athens, Greece
theovas@intracom-telecom.com
[2] Technische Universität München, Munich, Germany
Gerostat@in.tum.de, sasan.amini@tum.de
[3] University of Gothenburg, Gothenburg, Sweden
{riccardo.scandariato,Miroslaw.Staron}@cse.gu.se
[4] Fortiss GmbH, Munich, Germany
prehofer@fortiss.org

Abstract. With the recent advances in Big Data storage and processing, there is a real potential of data-driven software systems, i.e., systems
that employ analysis of large amounts of data to inform their runtime
decisions. However, for these decisions to be trustworthy and dependable,
one needs to deal with the well-known challenges on the data analysis domain: data scarcity, low-quality of data available for analysis, low
veracity of data and subsequent analysis results, data privacy constraints
that hinder the analysis. A promising solution is to introduce flexibility
in the data analytics part of the system enabling optimization at runtime
of the algorithms and data streams based on the combination of veracity,
privacy and scarcity in order to preserve the target level of quality of the
data-driven decisions. In this paper, we investigate this solution by providing an adaptive reference architecture and illustrate its applicability
with an example from the traffic management domain.

Keywords: Big Data · Reference architecture · Data veracity

1 Introduction

Modern systems collect raw data from the users and their personal devices (like
health trackers), raw data from the environment and its smart objects (smart
thermostats, home automation devices), as well as higher-level data coming
from information providers like social platforms, open-data sites (e.g., Open-
StreetMap), and other silos of information. Beyond functionality, the success of
such systems is tied to the availability of the information that is processed as
well as its quality. Functionality is often centered around the analysis of data to
extract useful information (e.g. make user-specific recommendations, adapt to
user habits to make an application more ergonomic, etc.).

© Springer International Publishing AG 2017
M. Kirikova et al. (Eds.): ADBIS 2017, CCIS 767, pp. 186–198, 2017.
DOI: 10.1007/978-3-319-67162-8_19

We believe that we are moving from traditional software systems, which are functionality- and software-centric, towards systems that are data-centric and where the functionality is driven by the availability of data and the decisions drawn from them. This is true not only for systems that perform offline analysis (e.g. business intelligence systems), but also, and even more so, for systems that employ real-time analysis of data to inform their runtime decisions (personalized advertising, traffic control).

In this paper, we focus on systems that make decisions at runtime based on Big Data analytics. For these decisions to be trustworthy and dependable, one needs to deal though with the well-known challenges on the data analysis domain: data scarcity, low-quality of data available for analysis, low veracity of data and subsequent analysis results, data privacy constraints that hinder the analysis. Indeed, we argue that unless a data-centric system deals with the above issues effectively at runtime, it will not matter whether it can process terabytes or petabytes at very high rates (which, in itself, is a noteworthy achievement of current Big Data systems)—as the resulting decision recommendation cannot be trusted.

Promising solutions already exist in the data analysis domain, where several data cleaning and data preparation methods have been proposed [7]. While these are certainly relevant and important, this paper takes a different (yet complementary) approach and proposes to introduce flexibility to change data streams in runtime. This will allow the system architects to depart from the current, more rigid model where the data analytics schema is planned ahead of deployment (which data sources to use, which information to mine, which algorithms to use) and kept fixed afterwards, at run time. The main drawbacks of such systems are that (i) the quality of the decisions will decay over time in case the quality of the analyzed data drops, and (ii) the system cannot opportunistically adapt or even improve in light of new operating conditions (e.g., when a new, richer, better data source becomes available).

In this work we introduce a novel reference architecture for adaptive data-driven systems (called Theta architecture) that can change used data sources and data analysis algorithms at runtime to preserve the target quality of its outcome. To achieve our ambitious goal we have set up an agenda consisting of three items: (i) identify the need for adaptive Big Data analytics in the context of data-driven systems via concrete examples, (ii) propose a high level architecture for adaptive Big Data analytics in data-driven systems, (iii) evaluate our architecture via applying it in concrete systems identified in (i). In this paper, we present our first steps towards fulfilling (i) and (ii) from above. In particular, we identify and describe an example of a data-driven system where adaptive Big Data analytics are of high value. The example comes from the vehicular traffic management domain and is presented in Sect. 2. We present the Theta architecture in Sect. 3 and illustrate its use on the running example. Finally, after presenting an overview of the related work in Sect. 4, we provide the concluding remarks in Sect. 5.

2   Real-Time Freeway Traffic Control System

In this section, we present an example of a data-driven system to showcase that adapting the data analytics processes at runtime is beneficial. This is used for both motivating and exemplifying our approach.

The main objective of traffic management on freeways is to increase the freeway capacity, i.e. to maximize the vehicular flow per time unit:

$$flow\,(veh/h) = speed\,(km/h) * density\,(veh/km)$$

In order to increase the capacity of a freeway, existing control measures involve a combination of dynamically changing the speed limit, opening or closing freeway lanes and recommending alternative routes through in-car navigation devices and Variable Message Signs (VMS).

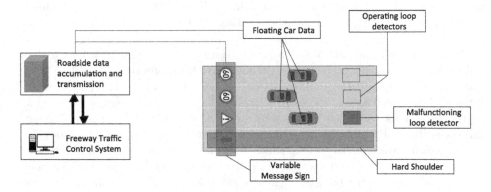

Fig. 1. Data sources in real-time freeway control system.

To know when to make the decision of applying the control measures, a freeway traffic control system (FTCS) can rely on data coming from different sources (Fig. 1). Data sources include inductive loop detectors installed on the pavement, surveillance cameras, cars transmitting their position, speed, etc. (referred to as *floating car data* in the Intelligent Transport System domain), Doppler radars, and ultrasonic and passive infrared sensing devices. In addition, environmental detectors are commonly used for freeway traffic control such as scatter measurements—measuring visibility range—and precipitation detectors—measuring thickness of water film on the road. Data from these different sources are transmitted to a central traffic control center where they are analyzed (this typically involves also visualizing the data).

We envision a fully automated real-time FTCS. In this case, a human operator simply configures the FTCS upon startup, e.g. by setting the thresholds on calculated flow for opening and closing the hard shoulder. The FTCS performs the calculation of traffic flow in predefined time windows (e.g. every 30 s) and, based on these calculations, makes autonomous decisions to apply (a combination of) the available control measures. Applying the opening of a hard shoulder, for instance, results in changing the signs on the VMS gantries and disseminating the information about the opening/closing of a lane or a different speed limit to the in-vehicle navigation devices.

A challenge in order to achieve a fully automated FTCS is to be able to deal with non-veracious (untrue) data that may creep in the analysis and ultimately influence the control logic, which resides in the control center. Data can have low veracity (i.e., little correspondence to reality) due to several reasons, e.g., sensor inaccuracy, sensor faults, sensor tampering, communication errors, to name a few. For example, dirt could block the view of a camera or water spray from the passing vehicles may influence the visibility range at the location of the scatter measurements. A loop detector may start malfunctioning due to pavement cracking or moving, inadequate sealant application, electronics unit failure, breakdown of wire insulation, electrical surges, etc.

Ideally, FTCS should be able to detect that it is using non-veracious data and adapt its analytics logic by choosing different data sources or a different analytics algorithm (e.g., one that makes weaker assumptions on the veracity of incoming data).

3 Theta Architecture

We have already pointed out the need for introducing flexibility in the data analytics as part of our running example in Sect. 2. We believe this need extends to other systems that need to make correct and reliable real-time decisions based on large numbers of collected data. In this section, we propose a high level architecture to support this.

Our proposed architecture, termed *Theta* architecture, is depicted in Fig. 2. It consists of three main sub-systems: Business subsystem, Analytics subsystem and Adaptation subsystem. We describe each one in turn and exemplify them by applying them to our running example (Fig. 3).

Business Subsystem. This part represents any software-intensive cyber-physical system that needs to be controlled at runtime based on analysis of data collected from its execution, optionally enriched with external data.

Interfaces. *Business* communicates with the other two subsystems through two interfaces, the *Data & metadata* and the *Adaptation* interface. The first is used by the *Analytics* to collect data in a pull (e.g., polling a RESTfull API [26]) or a push (e.g., publishing Kafka [15] messages) fashion. Together with the actual data values (e.g., speed, precipitation, etc.), data points may also contain metadata such as reported accuracy of sensors, ownership, etc.

The second interface is used by the *Adaptation* for requesting the *Business* to adapt. It is the task of the system designer to define how adaptation requests from *Adaptation* are translated into actual changes in *Business*. We note that a change can affect the cyber part of the system (e.g., setting a parameter of a route planner software to a new value) or the physical one (e.g., switching a traffic light).

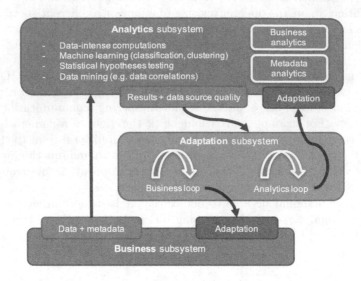

Fig. 2. Theta architecture.

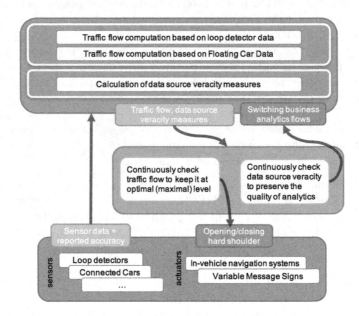

Fig. 3. Application of Theta architecture to the running example.

Example. In our running example (Fig. 3), *Business* is the traffic control system deployed in the freeway, consisting of a number of different types of sensors, e.g., inductive loop detectors, cars, cameras, radars, etc. Each sensor provides different data according to its type. For instance, a loop detector provides data on the number of vehicles that passed in a time interval, together with its average occupancy in the interval. A car provides its position and speed, while a camera provides either live feed or how many cars have been detected at a time interval. *Business* includes also two actuators, in-vehicle navigation systems and variable message signs, which are used to implement the "open/close hard shoulder" directive from *Adaptation*.

Analytics Subsystem. It is responsible for generating actionable information by large-scale data analysis. It should be able to provide different types of data analysis, from simple but data-intense computations to statistical hypothesis testing and machine learning and data mining.

In terms of the analytics processes that are run in *Analytics*, we distinguish between processes that support the data-driven functionality of the business system (business analytics) and processes that support the reasoning of the quality of input sources (metadata analytics). Metadata analytics not only use and summarize the metadata on provided accuracies of sensed data (e.g., from confidence interval values accompanying GPS measurements or weather data-based quality estimations on camera feeds), but also generate inferred metadata about the data sources based on the actual data values. Flexible formalisms can be employed for maintaining metadata about quality measurements, such as the Data Quality Vocabulary (DQV[1]).

Technically, *Analytics* can be implemented as a Big Data processing system following one of the existing lambda or kappa architectures or forks of them and the corresponding Big Data stacks (e.g., HDFS and Hive/Pig, Kafka and Spark/Flink).

Interfaces. *Analytics* exposes two interfaces for its interaction with the other two subsystems, the *Results & data source quality* and the *Adaptation* interface. The first interface is used by *Adaptation* in order to access the results of the business analytics processes, as well as to retrieve quality information about the available and utilized data sources. An interesting approach would be for the *Adaptation* to receive provided and inferred metadata about the data sources in an event-driven fashion, e.g., with notifications being triggered when veracity of some source is estimated to fall behind a predefined threshold.

The *Adaptation* interface is used by *Adaptation* for imposing adaptations on *Analytics*. The adaptations we currently consider take the form of switching the running business analytics processes and/or their input data sources. We also envision more fine-grained adaptations such as parameter configuration of the analytics algorithms embodied in the running processes.

Example. In our running example, *Analytics* comprises (a minimum of) three analytics processes. Two of them are business analytics; they compute the traffic

---

[1] https://www.w3.org/TR/vocab-dqv/.

flow based on different data sources and computations algorithms. The computation may involve also near-future prediction of the traffic flow (e.g., based on regression using historical data). A third one is metadata analytics; it provides an estimation of the veracity of data sources such as loop detectors or cars. In case of loop detectors, the veracity calculation process can involve performing some well-known plausibility checks about loop detector data from the traffic domain [13].

Adaptation Subsystem. It is the subsystem responsible for decision making. This is performed in two independent adaptation loops, the business and the analytics loop. The first is responsible for planning and executing adaptations on *Business* based on the business analytics results of *Analytics*. The second is responsible for planning and executing adaptations on *Analytics* based on the metadata analytics results of *Analytics*. To this end, data profiling [2] and metadata extraction [27] play a crucial role. A notable adaptation in the latter loop is the substitution of running business analytics processes with others that have either the same input data or different ones (a sensible option when the veracity of previously used input data gets high).

By making these two loops explicit in Theta architecture, we achieve separation between the concern of performing reliable and robust business analytics—by adapting *Analytics*—and using them to inform business decisions—by adapting *Business*).

Interfaces. *Adaptation* communicates with the other two subsystems with three interfaces (not depicted in Figs. 2 and 3). Via the first one, it receives the results of both the business and metadata analytics. The second and third ones are used to propagate the adaptation directives to both *Analytics* and *Business*.

Example. In our running example, the two loops in Adaptation take the following form. The business loop continuously checks the computed values of traffic flow (current values or near-future predictions) and issues the adaptation requests of opening or closing the hard shoulder in order to maximize the flow. In this case, decision making is based on well-known, empirically-validated rules on when to open/close a hard shoulder in a freeway based on the flow [11].

The analytics loop continuously checks the data source veracity measures calculated in *Analytics*. If the measures for the active business analytics processes indicate low veracity of input data, the analytics loop looks for alternative processes that can calculate the same result with different inputs to substitute the running processes. For example, the analytics loop switches the process that calculates the traffic flow based on loop detector data to an equivalent one that uses Floating Car Data, when a number of loop detectors start malfunctioning.

4   Overview of Related Work

Our approach builds upon existing works in Big Data analytics (relevant for the Analytics subsystem), self-adaptive systems (relevant for the Adaptation subsystem), and data veracity definition and management (relevant for the Meta-data Analytics inside the Analytics subsystem and the Analytics loop inside the Adaptation subsystem), which we overview below.

4.1 Big Data Analytics

Analytics on Big Data refer to data analysis approaches able to scale to large amounts of data (e.g. petabytes or exabytes) which are typically unstructured or semi-structured [30], i.e. they do not follow a particular schema. Big Data are considered to encompass the "5-V" properties: volume, velocity, variety, veracity, and value [21]. Here, the value is of particular importance, since Big Data analytics is ideally concerned with deriving insightful information, or even actionable results, from raw data by applying techniques from statistics, machine learning, and data mining. Big Data analytics has been used in a diverse set of applications, from analyzing user clicks on websites to analyzing the results of high-energy physics experiments. It has also been used in the analysis of software artifacts such as source code, execution traces, and runtime logs (software analytics [23]).

On the tooling side, Hadoop [1] is an open-source ecosystem that has become the de facto standard. A distinction is typically made between analyzing historical data in batch mode (full-data analytics) and analyzing data as they come in stream mode ("tuple-at-at-time" analytics) [22]. Important tools in the first category are Hadoops MapReduce, Hive, and Pig; Flink and Kafka Streams are representative of the second category. Spark is a hybrid system that combines batch and stream processing. Stream processing in Spark relies on analysis of micro-batches, rather than on analyzing one tuple at a time. Spark comes also with extensive support in machine learning algorithms (its MLlib library).

When architecting a Big Data analytics system, two main approaches exist, namely lambda and kappa architecture. Lambda architecture [22] combines both batch and stream analytics in two layers called batch and speed layer, respectively. The results of the two layers are combined at a third layer, the serving layer (a NoSQL database). Essentially, stream processing complements batch processing by analyzing the data that comes in the system since the start of batch processing (which typically has high latency). An alternative to lambda is kappa architecture, where stream processing is used for both the batch and stream processing. This builds on the idea that stream processing, when applied for large windows that can fit in all the historical data, essentially corresponds to batch processing. We note here that both lambda and kappa architectures focus on solving performance issues (e.g. by balancing throughput and latency), and not on quality issues of data and data analysis results—which is our focus.

4.2 Self-adaptive Systems

Self-adaptation refers to the ability of a system to change its structure and/or behavior at runtime in response to external stimuli and changes in internal state. Self-adaptation is typically achieved in three fundamental ways: (i) by relying on a detailed application model, e.g., Markov Decision Processes [12], and employing simulations or other means of state-space traversal to infer the best response of the system, (ii) by identifying control parameters and employing feedback-based control techniques from control theory [8], and (iii) by reconfiguring architecture models, typically with the help of Event-Condition-Action rules [10].

Self-adaptation techniques typically follow the MAPE-K loop [14], which divides self-adaptation into four phases: *Monitoring* activities, *Analyzing* run-time metrics, *Planning* strategies, and *Execute* planall based on a shared *Knowledge* base. Self-adaptation strategies are expressed as actions involving particular architecture reconfigurations; they are applicable under certain conditions in the presence of certain events or situations [4]. Actions can be associated with the satisfaction of one or more system goals, typically quantified via fitness or utility functions [28].

Although adapting a system based on analysis of data collected from its execution is a well-researched idea, there is a vacuum of approaches that use Big Data analytics in self-adaptation. In our recent work, we have proposed an approach to do so [29], since we believe that as the amount of data collected from a running system and its environment increases, Big Data analytics become relevant for self-adaptation.

In our approach, we employ two self-adaptation loops: one that controls the data-driven system itself (by employing Big Data techniques in its Analysis phase), and one that controls the analytics subsystem (by switching data sources and analysis algorithms).

4.3   Data Veracity

Historically, the notion of veracity is derived from the area of sociology and its major popularity lies in the area of criminology—the ability to detect whether a witness is veracious or not [17,20]. In that particular context, the term veracity is used both in relation to actors (e.g. witnesses) and their statements [3]. The latter refers to judging the truthfulness of a statement and is in scope for our purposes as well.

In our context, we consider the definition of veracity as quoted by Krotofil [16] who defines the veracity as *the property that an assertion truthfully reects the aspect it makes a statement about.* We can see a direct relation to the field of criminology and also see the challenges related to automated assessment of the veracity in the context of software systems.

For instance, the veracity of the data can be violated by: (i) non-adequate measurement of a physical property by a sensor because of the inappropriate design of the sensor (ii) non-adequate measurement caused by a faulty sensor during the operation (iii) non-adequate measurement caused by an obstructed sensor (iv) faulty data caused by a malicious agents tempering with the sensor data.

Lukainova and Rubin recognize data veracity as the sum of a number of quality attributes such as correctness, accuracy, free from biases and free from errors [19]. Based on this work and our previous work [31], we can see that veracity can be modelled as a composition of elements. To be able to perform such modelling, we need to first decide upon which elements are relevant (e.g. correctness, free or errors) and how they relate to each other.

In our approach, we first need to model and assess the veracity of the data that is being used in Big Data analytics In our running example, a rudimentary way to do is to apply existing plausibility checks on the sensor readings (such checks are e.g. well-known for inductive loop detector readings). We then propose to switch between different data sources and corresponding data analysis algorithms (i.e. Big Data jobs) at runtime when the veracity of the used data drops below a threshold. In other words, we intend to use simple threshold-based adaptation rules to adapt the analytics subsystem in order to increase the quality of its results.

4.4 Data Source Evaluation

Data source selection has received interest since the advent of the Internet. Florescu et al. [9] use probabilistic knowledge on a mediation schema that quantitatively determines the probability of information being covered by specific data source and provide corresponding data source ranking in cases of overlapping data sources (i.e., sources containing same documents). Naumann et al. [25] additionally put into play data source quality and propose a methodology that weights different data sources with regards to their information quality, considering quantified quality criteria (e.g., relevance to specific query) and cost and thus formulating linear programs. Mihaila et al. [24] introduce the use of metadata to maintain in XML format content and data quality information about data sources and by relaxing accuracy requirements, they propose a methodology for efficient source selection and ranking.

More recently, Dong et al. [6] have dealt with the problem of selecting the subset of data sources that maximize quality gain and minimize cost. In their work, they showcase the peculiar behavior of information gain as a result of utilizing multiple data sources of varying information coverage and accuracy. This work poses a pragmatic view on data source selection and can provide a stepping stone for conducting analysis on integrating data sources in presence of errors and false values.

Examining the quality of web sources, Dong et al. [5] use an information extraction system to employ aside from exogenous signals, inference about probability of correctness of facts which they define as accuracy of a web source. They introduce a novel methodology for assessing source and extracted data correctness, which can pave the way for veracity inference in case of multiple available data sources. Finally, data source quality assessment approaches have been proposed [18, 32] that can deal with high variety and variability of available sources.

In our approach, we combine explicit metadata derived from the Business subsystem with statistical methods, to determine the veracity of data sources and to detect anomalies. This analysis provides feedback for decisions on data source selection and switching, which aim at maximizing veracity while minimizing cost.

5   Discussion

In this section, we conclude by providing a discussion that reflects on our proposed architecture:

1. Essentially, if we disregard *Analytics* from Theta architecture, the resulting architecture degenerates into the classic MAPE-K loop, comprised of a controlled subsystem (*Business*) and a controller (*Adaptation*).
2. Both the business and the analytics loops in *Adaptation* can be arbitrarily complex. In our first attempts based on Apache Kafka and Python for traffic management of a simulated freeway, we have successfully considered only simple adaptation logics (e.g. based on a short number of rules); however, Theta architecture imposes no restrictions to the complexity of the adaptation logics.
3. Cluster-based approaches at the Analytics subsystem are only necessary if data size is large enough to render single-node approaches impractical.
4. We note here that although data veracity is the primary concern in our running example, our architecture can be used for adapting between data sources based on other concerns such as data privacy and confidentiality.

References

1. Apache Hadoop (2017). http://hadoop.apache.org/
2. Abedjan, Z., Golab, L., Naumann, F.: Data profiling: a tutorial. In: Proceedings of the 2017 ACM International Conference on Management of Data, SIGMOD 2017, pp. 1747–1751 (2017)
3. Carey, P.W., Mehler, J., Bever, T.G.: Judging the veracity of ambiguous sentences. J. Verbal Learn. Verb. Behav. **9**(2), 243–254 (1970)
4. Cheng, S.W., Garlan, D., Schmerl, B.: Stitch: a language for architecture-based self-adaptation. J. Syst. Softw. **85**(12), 1–38 (2012)
5. Dong, X.L., Gabrilovich, E., Murphy, K., Dang, V., Horn, W., Lugaresi, C., Sun, S., Zhang, W.: Knowledge-based trust: estimating the trustworthiness of web sources. Proc. VLDB Endow. **8**(9), 938–949 (2015)
6. Dong, X.L., Saha, B., Srivastava, D.: Less is more: selecting sources wisely for integration. In: Proceedings of the 39th International Conference on Very Large Data Bases, PVLDB 2013, pp. 37–48. VLDB Endowment (2013)
7. Dustdar, S., Pichler, R., Savenkov, V., Truong, H.L.: Quality-aware service-oriented data integration: requirements, state of the art and open challenges. SIGMOD Rec. **41**(1), 11–19 (2012)
8. Filieri, A., et al.: Software engineering meets control theory. In: Proceedings of SEAMS 2015, pp. 71–82. IEEE, May 2015
9. Florescu, D., Koller, D., Levy, A.Y.: Using probabilistic information in data integration. In: Proceedings of the 23rd International Conference on Very Large Data Bases, VLDB 1997, Athens, Greece, pp. 216–225, 25–29 August 1997
10. Garlan, D., Cheng, S.W., Huang, A.C., Schmerl, B., Steenkiste, P.: Rainbow: architecture-based self-adaptation with reusable infrastructure. Computer **37**(10), 46–54 (2004)

11. Geistefeldt, J.: Operational experience with temporary hard shoulder running in Germany. Transp. Res. Rec. J. Transp. Res. Board **2278**(6), 67–73 (2012)
12. Ghezzi, C., Pinto, L.S., Spoletini, P., Tamburrelli, G.: Managing non-functional uncertainty via model-driven adaptivity. In: Proceedings of ICSE 2013, pp. 33–42. IEEE (2013)
13. Gladbach, B.: Bundesanstalt fr Straenwesen: Merkblatt fr die Ausstattung von Verkehrsrechnerzentralen und Unterzentralen (MARZ). Technical report, Ausgabe 1999 (1999)
14. Kephart, J., Chess, D.: The vision of autonomic computing. Computer **36**(1), 41–50 (2003)
15. Kreps, J., Narkhede, N., Rao, J., et al: Kafka: a distributed messaging system for log processing. In: Proceedings of the 6th International Workshop on Networking Meets Databases (NetDB 2011), pp. 1–7 (2011)
16. Krotofil, M., Larsen, J., Gollmann, D.: The process matters. In: Proceedings of the 10th ACM Symposium on Information Computer and Communications Security. Association for Computing Machinery (ACM) (2015)
17. Levine, T.R., Park, H.S., McCornack, S.A.: Accuracy in detecting truths and lies: documenting the "veracity effect". Commun. Monogr. **66**(2), 125–144 (1999)
18. Li, Q., Li, Y., Gao, J., Zhao, B., Fan, W., Han, J.: Resolving conflicts in heterogeneous data by truth discovery and source reliability estimation. In: Proceedings of the 2014 ACM SIGMOD International Conference on Management of Data, pp. 1187–1198. ACM (2014)
19. Lukoianova, T., Rubin, V.L.: Veracity roadmap: is Big Data objective, truthful and credible? (2014)
20. Mann, S., Vrij, A.: Police officers' judgements of veracity tenseness, cognitive load and attempted behavioural control in real-life police interviews. Psychol. Crime Law **12**(3), 307–319 (2006)
21. Marr, B.: Big Data: the 5 vs. everyone must know. https://www.linkedin.com/pulse/20140306073407-64875646-big-data-the-5-vs-everyone-must-know
22. Marz, N., Warren, J.: Big Data: Principles and Best Practices of Scalable Realtime Data Systems, 1st edn. Manning Publications Co., Greenwich (2015)
23. Menzies, T., Zimmermann, T.: Software analytics: so what? IEEE Softw. **30**(4), 31–37 (2013)
24. Mihaila, G.A., Raschid, L., Vidal, M.: Using quality of data metadata for source selection and ranking. In: Proceedings of the Third International Workshop on the Web and Databases, pp. 93–98 (2000)
25. Naumann, F., Freytag, J.C., Spiliopoulou, M.: Quality driven source selection using data envelope analysis. In: Third Conference on Information Quality (IQ 1998), pp. 137–152 (1998)
26. Pautasso, C., Zimmermann, O., Leymann, F.: Restful web services vs. "Big" web services: making the right architectural decision. In: Proceedings of the 17th International Conference on World Wide Web, WWW 2008, pp. 805–814. ACM, New York (2008)
27. Quix, C., Hai, R., Vatov, I.: Metadata extraction and management in data lakes with GEMMS. CSIMQ **9**, 67–83 (2016)
28. Salehie, M., Tahvildari, L.: Self-adaptive software: landscape and research challenges. ACM Trans. Auton. Adapti. Syst. **4**(2), 1–40 (2009)

29. Schmid, S., Gerostathopoulos, I., Prehofer, C., Bures, T.: Self-adaptation based on big data analytics: a model problem and tool. In: Proceedings of the 12th International Symposium on Software Engineering for Adaptive and Self-Managing Systems (SEAMS 2017), pp. 102–108. IEEE Press, Piscataway (2017). https://doi.org/10.1109/SEAMS.2017.20

30. Srinivasa, S., Bhatnagar, V. (eds.): BDA 2012. LNCS, vol. 7678. Springer, Heidelberg (2012)

31. Staron, M., Scandariato, R.: Data veracity in intelligent transportation systems: the slippery road warning scenario. In: 2016 IEEE Intelligent Vehicles Symposium (IV), pp. 821–826. IEEE (2016)

32. Zhang, Y., Wang, H., Gao, H., Li, J.: Efficient accuracy evaluation for multi-modal sensed data. J. Comb. Optim. **32**(4), 1068–1088 (2016)

Spatio-Temporal Evolution
of Scientific Knowledge

Goce Trajcevski$^{(\boxtimes)}$, Xu Teng, and Shailav Taneja

Department of Electrical Engineering and Computer Science,
Northwestern University, Evanston, USA
{goce,shailavt1,xuteng2}@eecs.northwestern.edu

Abstract. In this work we take a first step towards the problem of integrating the content and the spatio-temporal aspects of the evolution of the (published) scientific knowledge. A lot of research has been invested in developing tools and search engines that will enable more efficient querying of relevant medical (and broader scientific) data from various perspectives, spanning from retrieval of similar documents/images to HCI-based flexible query-answering systems. Variety of methodologies have been developed, founded on knowledge-bases, statistics, semantic similarity, etc. and quite a few systems are available (e.g., Medline). Parallel to this, another body of research works has emerged over the past couple of decades, targeting the efficient management of mobility and spatio-temporal data. What motivates this work is the observation that fusing the data (and corresponding techniques) developed in these two broad research fields could enable novel categories of queries that can be used to investigate various evolving spatio-temporal relationships between particular scientific topics.

We present a novel model and a formalization of this confluence, in what we call *Knowledge-Evolution Trajectories* (KET). We also provide a preliminary proof-of-concept implementation that enables answering novel categories of queries pertaining to KET data with a few initial observations regarding the impact of different data-representation approaches.

1 Introduction and Motivation

Shortly after the co-emerging of the fields of spatial [1,13,23] and temporal [8] data management, the miniaturization of computing devices and advances in Global Positioning Systems (GPS) technology have spurred a plethora of applications that demanded some type of a Location-Awareness (LBS) [22]. This, originated the fields of Spatio-Temporal Databases (STDb) [9] and Moving Objects Databases (MOD) [10], addressing various aspects of managing such data – from modelling, through indexing and query processing [7,9,18,20,28]. The peculiar feature of the popular query-categories – e.g., *range, (k) Nearest-Neighbor ((k-)NN)* [26,29,30]

Research supported by NSF – CNS 1646107 and III 1213038, and ONR – N00014-14-1-0215.

M. Kirikova et al. (Eds.): ADBIS 2017, CCIS 767, pp. 199–210, 2017.
DOI: 10.1007/978-3-319-67162-8_20

in MOD settings is that they are typically: (1) *continuous* (i.e., their answers may have to be re-evaluated based on the changes in the motion of the entities); and/or (2) *persistent* (i.e., their answers may need to be re-evaluated based both on the changes of the motion as well as the history of such changes) [7,18]. More recently, researchers have turned their attention to modelling, representing, and querying spatio-temporal trajectories which also have some kind of annotated information associated with the location and/or time – bringing about the concept of semantic (resp. symbolic, spatio-textual) trajectories [19].

Complementary to these developments, the need to reduce (or even eliminate) labor-intensive process associated with retrieval of textual documents matching particular criteria, along with the contemporary advances in information systems, have spurred the field of Information Retrieval (IR) [16,21,24] starting in the middle of the XX century. To enable effective retrieval relevant documents by various IR strategies, the documents are typically transformed into a suitable representation, often accompanied by preparation of suitable indexes [24]. In the subsequent decades, a plethora of research works[1] followed, enabling literature searches [6,17], detection of various semantic correlations among (topics) in existing publications [14], etc. In addition, several prototype systems and publicly available search engines have been generated over the years (e.g., MED-LARS [6], Medline (https://www.nlm.nih.gov/bsd/pmresources.html)).

At the heart of the motivation for this work is the observation that despite the rather long co-existence of the two fields (IR and STDb/MOD) and their respective rich histories, not much has been done in terms of exploiting the possible confluence of the two – which, as we will discuss shortly, could enable novel categories of queries of relevance for various entities, from researchers, to government agencies. For example, most of the works related to spatio-temporal aspects of medical phenomena pertain to: – modelling the temporal, spatial, and evolutionary nature of subject's conditions [4]; exploring the spreading of different chemicals, or respondses/reactions to particular stimuli (e.g., [2,12].

Our goal in this work is to provide a foundation for addressing the problem of modelling the spatio-temporal aspects of the evolution of scientific knowledge and take a step towards enabling novel queries. Specifically, we aim at answering queries such as:

Q1: *Retrieve the authors and institutions located in Eastern Europe, who have published results related to the topic of heterocyclic compounds between 2005 and 2010.*
Q2: *Retrieve all the topics that were published by an institution in Pennsylvania between 2008 and 2012.*

---

[1] We do not aim at presenting a comprehensive overview of the vast body of works from the well-established field of IR in this paper (cf. [6,21]. The purpose of this section is to provide a motivation for the research presented here.

Our main contributions are two-fold:

1. We introduce the concept of *Knowledge-Evolution Trajectory* (KET) as a model to formalize the fusion between the *spatial, temporal,* and *content-based* aspects of scientific publications.
2. We provide a preliminary proof-of-concept implementation that demonstrates the feasibility of the proposed model. Specifically, we created a SQL Server database that: (a) contains the data pertaining to medical publications, fetched from PubMed; (b) We generated the geospatial information about each publication by using Google Map API to convert the institution name into (Lat, Lon), and then used ArcGIS to generate the values to be used by the Geometry type of SQL Server.

We also conducted some preliminary experiments which, in addition to demonstrating the feasibility of our objectives, also point out some interesting research challenges.

In the rest of this paper, Sect. 2 introduces the KET model, followed by Sect. 3 in which we describe the current case-study implementation and the experimental observations. We summarize and outline directions for future work in Sect. 4.

2 Modelling Spatio-Temporal Evolution of Scientific Literature

We now present the main aspects of the KET model. We firstly introduce the concept of symbolic trajectories.

Semantic (synonymously, Symbolic or Enriched or Spatio-textual) *Trajectories* [3,5,11,15,19] embed contextual and/or situational knowledge into location-in-time data. In a MOD [10] setting, a trajectory is modelled as a structure of the form $Tr_i = [o_{ID}, (x_{i1}, y_{i1}, t_{i1}), \ldots, (x_{ik}, y_{ik}, t_{ik})]$, where x_{ij} and y_{ij} ($1 \leq j \leq k$) are the coordinates of the location ($l_{ij} = (x_{ij}, y_{ij})$) of the object with a unique oID, obtained at time instant t_{ij}. In-between two consecutive updates, the object's motion is approximated in accordance with some kind of an interpolation. STs attempt also to describe the kinds of activities associated with a particular location and time – e.g., "stop", "move", "walk", "eat", etc. Formally (cf. [5,19]), a semantic trajectory ST_i is a sequence of so-called, semantic episodes $se_{i,m}$ as follows:

$ST_i = [se_{i1}, se_{i2}, se_{i3}, \ldots se_{im}]$, and each semantic episode is a tuple of the form:

$$se_{ij} = (da_{ij}, sp_{ij}, x_{ij}^{in}, y_{ij}^{in}, t_{ij}^{in}, x_{ij}^{out}, y_{ij}^{out}, t_{ij}^{out}, tagList_{ij})$$

where:

- da_{ij} = defining annotation; typically expressing an activity (verb) such as stop or move.
- sp_{ij} = semantic location/position of the activity, like a POI (e.g., a museum, restaurant, zoo), home, work, etc.

- t_{ij}^{in} and t_{ij}^{out} = entry/exit times of a semantic position.
- $x_{ij}^{in}, y_{ij}^{in}, x_{ij}^{out}, y_{ij}^{out}$ = entry/exit coordinates of a semantic position.
- tagList$_{ij}$ = any additional semantic information, like transportation mode, additional activity description (e.g., eat), etc.

constitute the j-th semantic episode of the i-th semantic trajectory.

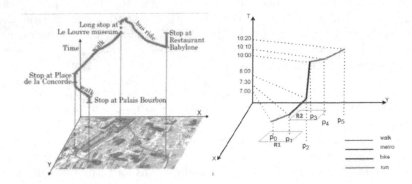

Fig. 1. Semantic trajectory and spatio-temporal range querying

Left portion of Fig. 1 illustrates the concept of a semantic trajectory (cf. [19]), and the right portion (cf. [15]) illustrates yet another way of visualizing a semantic trajectory (i.e., color-coded activity) along with the semantics of processing the query *Retrieve all the individuals who were running in the region R1, between 7:00 and 8:00 AM.*

There are two fundamental observations when it comes to the existing model of semantic trajectories, and the KET model that we are proposing:

1. There is no concept of a "motion", as commonly perceived in MOD trajectories (even when augmented with annotation). The scientific publications are associated with spatial data – e.g., the locations of the institution of the participating authors, and those are not mobile entities. However, there is an evolution over the temporal dimension that, in part, is associated with spatial values.
2. The scientific publications have a lot more contextual attributes, and those attributes are composite/richer. Namely, considering a typical meta-data[2], they contain:

- *Title*;
- *Category* (in accordance with an established nomenclature of the corresponding field; possibly a *set* of such categories, augmented with a set of *keywords*).
- A *set* of *authors names* and the corresponding *institution* of his affiliation.

---

[2] Aside from the main body of the text of the respective publications, or other attributes associated with, e.g., publishers, forum/venue, etc...

It is precisely the elements of the *institution* that contain an *implicit spatial value*, which we use for enabling the novel categories of queries.

Based on **(1)** and **(2)** above, we define the concept of *Scientific Publication Point* (SPP) as follows:

$SPP_i =$

$(P_{ID}, Title, category, [(author_{1i}, inst_{1i}, x_{1i}, y_{1i}), \ldots, (author_{ki}, inst_{ki}, x_{ki}, y_{ki})], T_{pub})$

We re-iterate that, in practice, most of the bibliographical data sources contain the name of the institution (along with its postal-address) for each author – however, not the actual coordinates for the corresponding address. This, in turn, eliminates the possibility of asking any queries containing predicates ranging over spatial domains. However, such queries may provide insight into data/trends that could influence both government funding as well as institutional/individual collaboration plans, as exemplified by the query: *Retrieve all the institutions within 100 Km from the coastal line in China, which have received more than 10M renminbi research funding in the last 4 years for medical research, but have published less than 20 articles on the topic of cytostatics.*

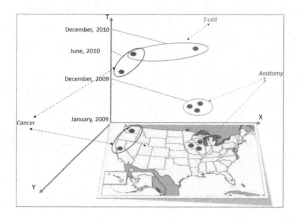

Fig. 2. KET for Geo-constrained Query

Assuming that the temporal value in each SPP_i (i.e., $SPP_i.T_{pub}$) is represented at a uniform particular level of granularity (e.g., *(month, year)*), we now present formally the model for a KET (*Knowledge-Evolution Trajectory*) defined as follows:

Definition 1. *A Knowledge-Evolution Trajectory is a sequence*
$[\alpha_1(SPP^{(1)}), \alpha_2(SPP^{(2)}), \ldots, \alpha_n(SPP^{(n)})]$ *where:*

- α_i *denotes an operator from relational algebra or a spatial predicate, applied to SPP_i*
- *For any pair $SPP^{(i)}$, $SPP^{(j)}$, $(i < j) \Rightarrow SPP^{(i)}.T_{pub} < SPP^{(j)}.T_{pub}$*

Given the definition of SPP_i, the role of α_i in Definition 1 is to extract the proper content of interest, the evolution of which needs to be queried. Thus, for example, we can focus on a particular author by applying $(SPP_i).author =$ $'Jones'$. However, the main benefit of the proposed model is that we can also apply spatial predicates such as: $(SPP_i).(x_{mi}, y_{mi})IN'Pennsylvania'$.

Clearly, the KET model generalizes the traditional model of a trajectory used in MOD literature and, for that matter, also generalizes the semantic trajectories.

Figure 2 shows an example of a KET corresponding to an answer of the query *Retrieve all the collaborative works between Midwest-based and California-based institutions, between January 2009 and December 2010.* As can be seen, instead of a traditional (x, y) point, we now have collection of Geo-points that constitute each one of the trajectories of the answer. Moreover, we can also see that a particular Geo-point can participate in > 1 KET – for as long as the constraints of the query are satisfied. Thus, we have a collaboration between an institution from California and Illinois on a publication related to T-cell studies, in December of 2010.

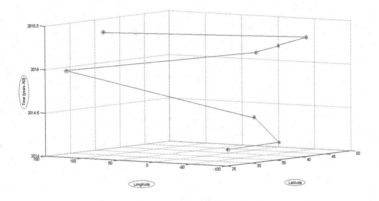

Fig. 3. KET for participation constraint query

Figure 3 illustrates another example of a KET – which visually (and type-wise) has a highest resemblance of a traditional moving point-object trajectory. However, it shows an example of an actual answer from our implementation, corresponding to the query *Retrieve all the institutions that have published an article on cytostatics in which all the authors were from a same institution, between January, 2014 and July 2015.*

We close this section with another example-query the answer of which, in some sense, does resemble spatio-temporal trajectories, but yet it has its own distinct semantics of the temporal evolution.

Figure 4 shows the answer to the query *Retrieve which topics/categories had most publications, for the researchers from the SouthWest University, as well as for the ones from Arizona State University, between 2008 and 2012.*

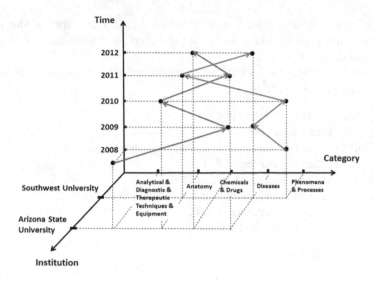

Fig. 4. Following most popular topics per institution

3   Case-Study: Spatio-Temporal Evolution in PubMed

We now describe in detail the current implementation of our proof-of-concept system for which the context is restricted to the PubMed data, pertaining to the various medical publications available on MedLine. We used SQL Server 2014 and we wrote python scripts to extract the data from PubMed and populate the tables[3]. As mentioned earlier, the PubMed data does not contain geo-spatial information[4] – whereas SQL Server provides two different geospatial data types: `Geometry` type and `Geography` type. Thus, in our prototype system, we selected [Publish ID], [Title], [Author] (including their names and institutions), [Publish Date] from the returned XML records from PubMed – however, in addition, we populated the entries for the corresponding `Geometry` type in which the coordinate system is World Geodetic System (WGS) 1984. To populate the corresponding values for the geospatial information, we used a two-step procedure:

(1) Google Map API was used to convert the institution name into (Lat, Lon) pair of values.
(2) Subsequently, ArcGIS was used to convert from (float Lat, float Lon) into the corresponding `Geometry` type.

---

[3] We note that all the data, code for the queries, as well as the scripts used to generate the values for the spatial attributes, is publicly available at https://github.com/ ShailavTaneja/PubMedDerivedDataAndImplementation.

[4] The description of the standard meta-data used in PubMed is available at https://www.ncbi.nlm.nih.gov/books/NBK3828/#publisherhelp.Example_of_a_ Standard_XML.

This is what enabled us to specify queries that can capture the spatio-temporal evolution of the knowledge represented in scientific – in this case, medical – publications.

In the first iteration, we had a "naïve" representation of the data residing in a single table, with attributes:

- `[Publish ID] varchar(50),`
- `[Title] varchar(500),`
- `[Publish Date] date,`
- `[Author Name] varchar(50),`
- `[Institution Name] varchar(150),`
- `[Geo Information] geometry,`
- `[Subcategory] varchar(150),`
- `[Category] varchar(150)`

The table had 409702 rows in total.

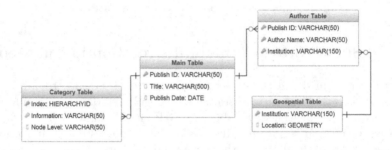

Fig. 5. Normalized database schema

Subsequently, to eliminate the redundancy, we normalized the naïve table, and the schemata that we used in the implementation is shown in Fig. 5, corresponding to the following tables:

- **Main** – with three attributes: [Publish ID] varchar(50), [Title] varchar(500), [Publish Date] date. 32768 row in total
- **Author** – with three attributes: [Publish ID] varchar(50), [Author Name] varchar(50), [Institution] varchar(150). 188786 rows in total.
- **Category** is a table capturing the hierarchy of the categories. It has three attributes: [Index] hierarchyid, [Information] varchar(150), [Node Level] varchar(50). 68652 rows in total. **Hierarchyid** is a special data type and works as the key of Category table. / represents root, /*/ represents the first level and so on
- **Geospatial** – with two attributes: [Institution] varchar(150), [Location] geometry. 1316 rows in total.

In the sequel, we describe an example of the difference between evaluating queries in the naïve representation and the normalized one. Consider the following:

Q_{topic}: *Retrieve the* KET *for publications addressing the topic of diagnosis, in the period of October 1, 2012 – September 1, 2016.*

The query returns a set of temporally annotated points (geo locations) for the topic, during that period.

The corresponding SQL statement for the naïve implementation is:

```
SELECT DISTINCT [Publish ID], [Publish Date], Location.STX, Location.STY
FROM NativeTable
WHERE ([Publish Date] between '2012-10-01' and '2016-09-01') and
      ([Subcategory] = 'diagnosis')
ORDER BY [Publish Date]
```

The SQL implementation for the same query in the normalized version is:

```
DECLARE @catLevel hierarchyid;
SET @catLevel = (SELECT Index
                         FROM Category
                         WHERE Information = 'diagnosis')

SELECT DISTINCT MainTable.[Publish ID], MainTable.[Publish Date],
                             Location.STX, Location.STY
FROM MainTable, Author, Geospatial
WHERE (MainTable.[Publish Date] between '2012-10-01' and '2016-09-01') and
(MainTable.[Publish ID] in
  (SELECT Information
     FROM Category
     WHERE Index.IsDescendantOf(@catLevel) = 1 and [Node Level] = 'Publish ID'))
            and
        (MainTable.[Publish ID] = Author.[Publish ID]) and
        (Author.Institution = Geospatial.Institution)
ORDER BY MainTable.[Publish Date]
```

To illustrate the impact of the different database representation, we first illustrate the benefits in terms of eliminating the redundancy via the normalized representation:

As shown in Fig. 6, the naïve implementation requires approximately five times more space than the normalized one.

When it comes to the efficiency of the execution with each implementation we report the corresponding measurements observed when executing Q_{topic} (labeled **Q1** in Fig. 7) and the query

Q2: *Retrieve the works jointly co-authored by Masaki Mori and Yuichiro Doki, between 2009–2010 and 2012–2013*

on a Windows 10 machine, with Intel(R) Core(TM) i5-5257U CPU (2.70 GHz, 2701 Mhz), with 2 Cores (4 Logical Processors) and 8 GB of RAM.

As shown in Fig. 7, the normalized implementation also yields a much more efficient execution than the naïve one.

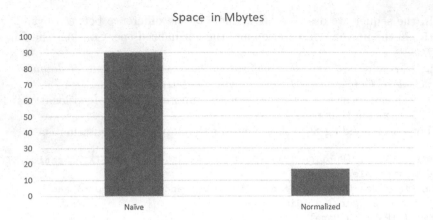

Fig. 6. Space requirements

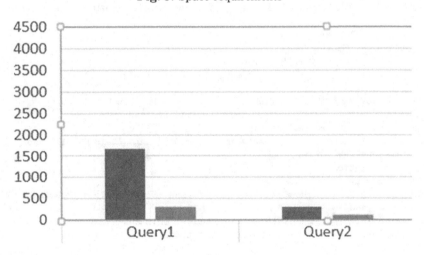

Fig. 7. Efficiency of execution (in milliseconds)

4   Conclusions and Future Work

We addressed the problem of modelling and querying the spatio-temporal evolution of the knowledge recorded in scientific publications, and took the first steps towards providing a formalism to capture such evolution across the (geo)spatial domain, as well as other contextual attributes. We proposed the concept of KET (Knowledge-Evolving Trajectories) as a possible unification between two fields (Information Retrieval and Spatio-Temporal/Moving Objects Databases). This type of unification provides opportunities for novel query categories which, to our knowledge, have not been formally treated in the literature. In addition, we provided a proof-of-concept implementation of our approach for the data available in PubMed, and compared the impacts of a naïve design of the database vs. a normalized one – albeit for limited set of queries. Our initial evaluations have demonstrated that in addition to reducing the space requirements, the normalized approach also enables faster execution of the KET-based queries.

Related Works: There are plethora of works in each of the fields of IR [6, 14, 17, 21] and STDb/MOD [7, 9, 10, 18, 20, 23, 28] – to mention but a few. The novelty of our proposed approach is to provide a formalism for fusing these works, along with the integration of the respective existing datasets, enabling novel query-categories. The closest formalism to our proposed approach is the one of semantic/symbolic trajectories [19]. However, as we argued, this approach: (1) has too simplistic model of the spatio-temporal evolution; and (2) is lacking the - dimensionality of the contexts typically associated with scientific publications.

We believe to have scratched the surface of a direction that may be of interest in many applications of societal relevance and, moreover, can pose interesting challenges. As part of our future work, we are planning to extend the KET model, and augment the current implementation so that it can incorporate different publications' data sources. In addition, although we have provided some preliminary discussions related to the efficiency, a challenging problem is to address the efficient processing different types of KET-based queries.

Last, but not the least, it seems rather intuitive that investigations along the direction of warehousing spatio-temporal evolution of scientific publications along with further semantic similarity searches [25, 27] data may yield novel categories of analytical queries.

References

1. Bedard, Y., Merrett, T., Han, J.: Fundamentals of spatial data warehousing for geographic knowledge discovery. Geogr. Data Min. Knowl. Discov. **2**, 53–73 (2001). Taylor and Francis
2. Bilgen, M., Abbe, R., Liu, S.J., Narayana, P.A.: Spatial and temporal evolution of hemorrhage in the hyperacute phase of experimental spinal cord injury: in vivo magnetic resonance imaging. Magn. Ressonance Med. **43**(4), 594–600 (2000)
3. Bogorny, V., Renso, C., de Aquino, A.R., de Lucca Siqueira, F.: Constant - A conceptual data model for semantic trajectories of moving objects. GIS **18**(1), 66–88 (2014)
4. Chu, W.W., Cardenas, A.F., Taira, R.T.: Kmed: A knowledge-based multimedia medical distributed database system. Inf. Sci. **20**(2), 75–96 (1995)
5. Damiani, M.L., Güting, R.H.: Semantic trajectories and beyond. In: Proceedings of IEEE - MDM, pp. 1–3. Brisbane, Australia (2014)
6. Dee, C.R.: The development of the medical literature analysis and retrieval system (medlars). J. Med. Library Assoc. **94**(5), 416–425 (2007)
7. Ding, H., Trajcevski, G., Scheuermann, P.: Towards efficient maintenance of continuous queries for trajcectories. GeoInformatica **12**(3), 255–288 (2008)
8. Etzion, O., Jajodia, S., Sripada, S. (eds.): Temporal Databases: Research and Practice. LNCS, vol. 1399. Springer, Heidelberg (1998)
9. Güting, R.H., Böhlen, M.H., Erwig, M., Jensen, C.S., Lorentzos, N., Schneider, M., Vazirgiannis, M.: A foundation for representing and queirying moving objects. ACM TODS **25**, 1–42 (2000)
10. Güting, R.H., Schneider, M.: Moving Objects Databases. Morgan Kaufmann, San Francisco (2005)

11. Güting, R.H., Valdés, F., Damiani, M.L.: Symbolic trajectories. ACM Trans. Spat. Algorithms Syst. **1**(2), 7:1–7:51 (2015)
12. Hirano, Y., Stefanovic, B., Silva, A.C.: Spatiotemporal evolution of the fmri response to ultrashort stimuli. J. Neurosci. **31**(4), 1440–1447 (2011)
13. Hjaltason, G.R., Samet, H.: Distance browsing in spatial databases. ACM Trans. Database Syst. **24**(2), 265–318 (1999)
14. Hristovski, D., Kastrin, A., Dinevski, D., Rindflesch, T.C.: Constructing a graph database for semantic literature-based discovery. In: MEDINFO 2015: eHealth-enabled Health - Proceedings of the 15th World Congress on Health and Biomedical Informatics, p. 1094 (2015)
15. Issa, H.: Spatio-textual trajectories: models and applications. PhD thesis, Universita degli studi di Milano (2017)
16. Korfhage, R.: The impact of personal computers on library-based information systems. SIGIR Forum **12**(4), 10–13 (1978)
17. Lowe, H.J., Barnett, G.O.: Understanding and using the medical subject heading (mesh) vocabulary to perform literature searches. J. Am. Med. Assoc. **271**(14), 1103–1108 (1994)
18. Mokbel, M.F., Aref, W.G.: SOLE: scalable on-line execution of continuous queries on spatio-temporal data streams. VLDB J. **17**(5), 971–995 (2008)
19. Parent, C., Spaccapietra, S., Renso, C., Andrienko, G.L., Andrienko, N.V., Bogorny, V., Damiani, M.L., Gkoulalas-Divanis, A., de Macêdo, J., Pelekis, N., Theodoridis, Y., Yan, Z.: Semantic trajectories modeling and analysis. ACM Comput. Surv. **45**(4), 42 (2013)
20. Pelanis, M., Saltenis, S., Jensen, C.S.: Indexing the past, present, and anticipated future positions of moving objects. ACM Trans. Database Syst. **31**(1), 255–298 (2006)
21. Salton, G.: Automatic Text Processing. Addison Wesley, Massachusetts (1989)
22. Schiller, J.H., Voisard, A. (eds.): Location-Based Services. Morgan Kaufmann, San Francisco (2004)
23. Shekhar, S., Chawla, S.: Spatial Databases: A Tour. Prentice Hall, New Jersy (2003)
24. Taine, S.I.: New program for indexing at the national library of medicine. Bull. Med. Libr. Assoc. **47**(2), 117 (1959)
25. Trajcevski, G., Donevska, I., Vaisman, A.A., Avci, B., Zhang, T., Tian, D.: Semantics-aware warehousing of symbolic trajectories. In: Proceedings of the 6th ACM SIGSPATIAL International Workshop on GeoStreaming, IWGS 2015, pp. 1–8, 3–6 November 2015, Bellevue, WA, USA (2015)
26. Trajcevski, G., Tamassia, R., Cruz, I., Scheuermann, P., Hartglass, D., Zamierowski, C.: Ranking continuous nearest neighbors for uncertain trajectories. VLDB J. **20**(5), 767–791 (2011)
27. Vaisman, A.A., Zimányi, E.: Data Warehouse Systems: Design and Implementation. Data-Centric Systems and Applications. Springer, Heidelberg (2014)
28. Xing, X., Mokbel, M.F., Aref, W.G., Hambrusch, S.E., Prabhakar, S.: Scalable spatio-temporal continuous query processing for location-aware services. In: International Conference on Scientific and Statistical Database Management (SSDBM) (2004)
29. Xiong, X., Mokbel, M.F., Aref, W.G.: Sea-cnn: Scalable processing of continuous k-nearest neighbor queries in spatio-temporal databases. In: ICDE, pp. 643–654 (2005)
30. Yu, X., Pu, K.Q., Koudas, N.: Monitoring k-nearest neighbor queries over moving objects. In: ICDE, pp. 631–642 (2005)

The 1st International Workshop on Data Science: Methodologies and Use-Cases (DaS 2017)

Parallel Subspace Clustering Using Multi-core and Many-core Architectures

Amitava Datta[1], Amardeep Kaur[1], Tobias Lauer[2($\boxtimes$)], and Sami Chabbouh[2]

[1] School of Computer Science and Software Engineering,
University of Western Australia, Perth, Australia
[2] Department of Electrical Engineering and Information Technology,
Offenburg University of Applied Sciences, Offenburg, Germany
`tobias.lauer@hs-offenburg.de`

Abstract. Finding clusters in high dimensional data is a challenging research problem. Subspace clustering algorithms aim to find clusters in all possible subspaces of the dataset where, a subspace is the subset of dimensions of the data. But exponential increase in the number of subspaces with the dimensionality of data renders most of the algorithms inefficient as well as ineffective. Moreover, these algorithms have ingrained data dependency in the clustering process, thus, parallelization becomes difficult and inefficient. SUBSCALE is a recent subspace clustering algorithm which is scalable with the dimensions and contains independent processing steps which can be exploited through parallelism. In this paper, we aim to leverage, firstly, the computational power of widely available multi-core processors to improve the runtime performance of the SUBSCALE algorithm. The experimental evaluation has shown linear speedup. Secondly, we are developing an approach using graphics processing units (GPUs) for fine-grained data parallelism to accelerate the computation further. First tests of the GPU implementation show very promising results.

Keywords: Data mining · Subspace clustering · Multi-core architectures · Many-core architectures · GPU computing

1 Introduction

The growing size and dimensions of data these days have set new challenges for data mining research. Clustering is a data mining process of grouping similar data points into clusters without any prior knowledge of the underlying data distribution. Due to the curse of dimensionality, data group together differently under different subsets of dimensions, called subspaces. Subspace clustering algorithms attempt to find clusters in all possible subsets of dimensions of a given data set [1,2].

The area of subspace clustering is of critical importance with diverse applications [1]. But due to the exponential search space with the increase in dimensions, subspace clustering becomes computationally very expensive. With the

© Springer International Publishing AG 2017
M. Kirikova et al. (Eds.): ADBIS 2017, CCIS 767, pp. 213–223, 2017.
DOI: 10.1007/978-3-319-67162-8_21

wide availability of multi-core processors and the spread of many-core coprocessors such as GPUs, parallelization seems to be an obvious choice to reduce this computational cost.

SUBSCALE is a recent subspace clustering algorithm to find the subspace clusters without enumerating the data points or computing any redundant clusters [3,4]. This algorithm combines the dense set of points across all single dimensions of the data to find non-trivial subspace clusters. Although SUBSCALE algorithm scales well with the dimensions and performs faster than other subspace clustering algorithms, it is still compute intensive due to the generation of combinatorial 1-dimensional dense set of points. However, the compute time can be reduced by parallelizing the computation of the dense units.

In this paper, we aim to parallelize the SUBSCALE algorithm in two ways and investigate its runtime performance. First, we exploit current multi-core architectures with up to 48 processing cores using OpenMP. The experimental evaluation demonstrates the speedup of up to a factor of 27. This modified algorithm is faster and scalable for high dimensional large data sets. Second, we use many-core graphics processing units to exploit data parallelism on a fine-granular level, with a significant speedup, especially for larger computations. The latter work is ongoing and results are expected to further improve with optimizations in the implementation.

In the next section we discuss the current related literature. Section 3 explains subspace clustering and the algorithm we parallelize. In Sect. 4, we describe our multi-core parallelization and analyse the performance of the parallel implementation. Section 5 describes our current work of massive parallelization using GPUs with preliminary results. The paper is concluded in Sect. 6.

2   Related Work

Over the past few years, there has been extensive research on clustering algorithms [2]. The underlying premise that data group together differently under different subsets of dimensions opened the challenging domain of subspace clustering algorithms [1,5].

However, the increase in the dimensions of data impedes the performance of clustering algorithms which are known to perform very well with low dimensions. Moreover, most of the subspace clustering algorithms have less obvious parallel structures [6,7]. This is partially due to the data dependency during the processing sequence [8].

SUBSCALE [3,4] is a recent subspace clustering algorithm and requires only k database scans to process a k-dimensional dataset. Also, this algorithm is scalable with dimensions and its structure contains the computation of independent tasks which can be parallelized. In the next section, we briefly discuss the SUBSCALE algorithm and our modifications for multi-core parallel implementation.

3   Subspace Clustering

This section provides the basics and definitions of subspace slustering and a brief description of SUBSCALE [4], the algorithm which we aim to make more efficient through parallelization.

Given an $n \times k$ set of data points, a point P_i is a k-dimensional vector $\{P_i^1, P_i^2, \ldots, P_i^k\}$ such that, P_i^d is the projection of a point P_i on the dth dimension. A *subspace* is a subset of k dimensions. A subspace cluster $C_i = (P, S)$ is a set of points P, such that the projections of these points in subspace S, are dense.

According to the Apriori principle [6], a dense set of points in a subspace S of dimensionality a, is dense in all of 2^a projections of S. Thus, it is sufficient to find a cluster in its maximal subspace. A cluster $C_i = (P, S)$ is called a **maximal subspace cluster**, if there is no other cluster $C_j = (P, S')$ such that $S' \supset S$. The SUBSCALE Algorithm finds such maximal subspace clusters by combining the dense points from single dimensions and without computing the redundant non-maximal clusters.

3.1   SUBSCALE Algorithm

The main idea behind the SUBSCALE algorithm is to find the dense sets of points (also called *density chunks*) in all of the k single dimensions, generate the relevant *signatures* from these density chunks, and *collide* them in a hash table ($hTable$) to directly compute the maximal subspace clusters as explained below.

Density Chunks. Based on two user defined parameters ϵ and τ, a data point is *dense* if it has atleast τ points within ϵ distance. The neighbourhood $N(P_i)$ of a point P_i in a particular dimension d is a set of all the points P_j such that $L_1(P_i^d, P_j^d) < \epsilon$, $P_i \neq P_j$. L_1 is the distance metric. Each dense point along with its neighbours, forms a *density chunk* such that each member of this chunk is within ϵ distance from each other.

The smallest possible dense set of points is of size $\tau + 1$, called a *dense unit*. In a particular dimension, a density chunk of size t can have $\binom{t}{\tau+1}$ possible combinations to form the dense units. Some of these dense units may or may not contain projections of the higher dimensional maximal subspace clusters. Without any prior information of the underlying data distribution, it is not possible to know the promising dense units in advance. Only viable solution is to check which of these dense units from different dimensions contain identical points.

Signatures. The SUBSCALE algorithm proposed a novel way to match the dense units by assigning signatures to them. To create signatures, each of the n data points is mapped to a random, unique and large integer key. The sum of the mapped keys of the data points in each dense unit is termed as its *signature*.

According to observations 2 and 3 in the SUBSCALE paper [3], *two dense units with equal signatures would have identical points in them with extremely high probability*. Thus, collisions of the signatures across dimensions $d_r, \ldots, d_s$ implies that, the corresponding dense unit exists in the maximal subspace, $S = \{d_r, \ldots, d_s\}$. We refer our readers to the extended version of the original paper [4] for the detailed explanation. Each single dimension may have zero or more dense chunks, which in turn generate different number of signatures in each dimension. Some of these signatures will collide with the signatures from the other dimensions to give a set of dense points in the maximal subspace.

Hash Table. The SUBSCALE algorithm uses a hash table data structure $hTable$ to store collision information about each signature. An $hTable$ has a fixed number of slots and each slot can store one or more signatures, depending upon the implementation (Fig. 1). When a signature Sig is generated in a dimension d, it is mapped to a slot in the $hTable$. In this paper, we used *modulo numSlots* for mapping of signatures to a slot. If a slot already contains a signature Sig' such that $Sig = Sig'$, then d is appended to the dimension-list attached to Sig.

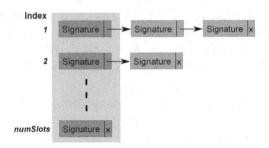

Fig. 1. $hTable$ data structure.

Since the size of each dense unit is $\tau + 1$, the value of a signature generated from a dense unit will lie in the range $R = [(\tau+1) \cdot min_K, (\tau+1) \cdot max_K]$, where min_K and max_K are the smallest and the largest keys respectively. Also, if $numSig_d$ is the number of total signatures in a dimension d, then the total number of signatures in a k-dimensional data set will be $total_{sig} = \sum_{d=1}^{k} numSig_d$. If memory is no onstraint a hash table with $|R|$ slots can easily accommodate $total_{sig}$, as typically, $total_{sig} \ll R$. Since memory is a constraint, the range R can be split into multiple slices and each slice can be processed independently using a separate and smaller hash table. The computations for each slice is not dependent of other slices. The split factor called sp determines the number of splits of R and its value can be set according to the available working memory. Also, the cluster quality is not affected by splitting of hash table computations. The clusters are formed by combining the dense units in each maximal subspace. The total number of dense units are decided by the density chunks created through epsilon neighbourhoods in single dimensions. As long as all dense units are processed, same clusters will be generated through sequential or parallel methods.

4 Multi-core Parallelization Using OpenMP

We used the OpenMP platform with C to parallelize the SUBSCALE algorithm. OpenMP is a set of complier directives and callable runtime library routines to facilitate shared-memory parallelism [9].

4.1 Processing Dimensions in Parallel

The generation of signatures from the density chunks in each single dimension is independent of other dimensions. Thus, the dimensions can be divided among the available processing cores to be run in parallel using threads. The hash table $hTable$ is shared among threads. However, the problem of thread contention arises when multiple threads try to get mutually exclusive access of the same slot of $hTable$ to update or store the signature information. Without mutually exclusive access, two threads with the same signatures generated from two different dimensions, would overwrite the same slot of $hTable$. The maximal subspace of a dense unit can only be found by having the information about which of the dimensions generated this dense unit. We discuss the results from this method in Sect. 4.3.

4.2 Processing Slices in Parallel

The other approach to avoid thread contention is to utilise the splitting of the range R of expected signature values as proposed by the SUBSCALE algorithm. The slices created through the splitting can be processed in parallel as each slice generates signatures from different range compared to other slices. Each slice requires a separate hash table. Though this approach helps to achieve faster clustering performance from the SUBSCALE algorithm, the memory required to store all of the hash tables can still be a constraint. Since R denotes the whole range of computation sums that are expected during the signature generation process, we can bring these slices into the main working memory one by one. Each slice is again split into sub-slices to be processed with multiple threads. The total number of signatures can be pre-calculated from the dense chunks in all dimensions. The results and their evaluation are discussed in the next section.

4.3 Results and Analysis

The experiments were carried out on the IBM Softlayer Server Quad Intel Xeon E7-4850, 2 GHz, with 48 cores, 128 GB RAM and Ubuntu 15.04 kernel. Hyperthreading was disabled on the server so that each thread could run on a separate physical core and parallel performance could be measured fairly. The parallel version of the SUBSCALE algorithm was implemented in C using OpenMP directives. Also, we used 14-digit non-negative integers for the key database.

The two main datasets for this experiment: 4400×500 *madelon* dataset [10] and 3661×6144 *pedestrian* dataset [11,12], are publicly available. These datasets were also used by authors of the SUBSCALE algorithm.

Multiple Cores for Dimensions. We used the *madelon* data set with $\epsilon = 0.000001$, $\tau = 3$ and experimented with three different number of slots in the shared $hTable$: 0.1 million, 0.5 million and 1 million. Figure 2a shows the runtime performance of the *madelon* data set by using multiple threads for dimensions. We observe that performance improves slightly by processing dimensions in parallel but as discussed before, thread contention due to mutually exclusive access to the same slot in the shared hash table results in performance degradation.

Multiple Cores for Slices. To avoid this memory contention due to shared $hTable$, we split the hash table computations into slices according to the SUBSCALE algorithm and distribute these slices among multiple cores. Figure 2b shows the results of the runtime versus the number of threads used for processing the slices of the *madelon* dataset. The hash computation was sliced with different values of split factor sp ranging between 200 and 2000. We can see the performance boost by using more threads. The speed up is shown in Fig. 2c, which becomes linear as the number of slices increases.

Scalability with Dimensions. The 6144 dimensional *pedestrian* dataset is used to study the speed up with the increase in dimensions. With $\epsilon = 0.000001$ and $\tau = 3$, 19860542724 total signatures are expected which would require $\sim 592\,GB$ of working memory to store the hash tables. To overcome this huge memory requirement, we can split these signature computations twice. We used a split factor of 60 to bring down the memory requirement for total hash tables. Each of these 60 slices were further split into 200 subslices to be run on 48 cores.

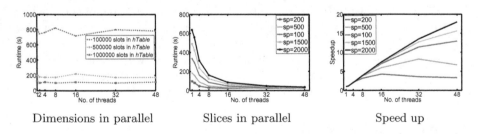

Fig. 2. *madelon* dataset: $\epsilon = 0.000001$ and $\tau = 3$. In (a), the total dimensions are divided among available cores using threads. Due to thread contention, the runtime fails to improve. In (b), the slices of hash computation (sp denotes the split factor) are distributed among multiple cores and runtime improves with number of threads.

The number of slots in each $hTable$ are fixed using $\frac{total\_signatures}{sp}$. The execution time decreases drastically with increase in the number of threads. It took around 26 h to finish processing all of the 60×200 slices. The sequential version of SUBSCALE clustering algorithm takes about ~ 720 h.

5 Fine-Grained Parallelization Using GPUs

In this section, we describe an alternative way of parallelization with a much finer-grained task structure, suitable for parallelization on graphics processing units (GPUs). This work is ongoing, the results are preliminary.

5.1 Levels of Granularity

Finer-grained parallelism than the one described so far can be achieved on two levels of granularity. First, we observe that within a dimension all density chunks can be processed independently, since they are necessarily disjoint. Hence, if a dimension contains k dense chunks, k independent tasks can be executed in parallel by k threads. The downside of this approach is that the number and sizes of density chunks in a dimension is not known a priori, thus resulting in a vastly varying number of active threads with varying workloads during the process, which is not efficient and not very suitable for GPU parallelization.

However, we note that the processing within a single density chunk can also be parallelized. Recall that this computation consists of computing the signatures of all dense units, i.e. all possible combinations of $\tau + 1$ elements, in that chunk. Signature computation only requires read access to the points in the respective dense unit, so even for non-disjoint units there are no thread conflicts. Furthermore, since all the dense units of the same dimension result in different signatures (with very high probability, cf. Sect. 3.1), hash collisions are very unlikely during the parallel computation within one dimension. Hence, one notable advantage of this task structure is that there will be no thread contention for accessing the hash table, if different dimensions are processed sequentially. (Note, however, that this does not preclude parallel computation of dimensions.)

5.2 Parallel Task Structure

In this parallel approach, a task consists of computing the signature value for one single dense unit and hashing it. All such tasks can be executed in parallel.

```
1  for each dense unit du[i] of length tau in parallel
2      signature := 0
3      for j = 0 to tau
4          signature := signature + key(du[i].point[j])
5      htable.insert(signature)
```

Algorithm 1. Code for parallel computation of signatures

Since a density chunk of t elements has $\binom{t}{\tau+1}$ dense units, this approach results in a large number of small and almost identical tasks. As τ is constant within one execution of the algorithm and since there are no branches, all tasks execute the same sequence of instructions, but on different data. This type of data parallelism is well-suited for implementation using GPUs. We have developed an implementation using Nvidia's CUDA architecture and programming model [13]. This model enables data parallelism by allowing scalable grids of threads, depending on the size of the data to be processed. Each thread is identified by its ID and can use this ID, e.g. to determine memory locations for reading input and writing output data. In our case, the ID is required to identify the dense unit to process.

5.3   Computing Subsets Efficiently

Algorithm 1 presupposes that the thread with ID i knows how to retrieve the i-th dense unit, i.e. the subset $\{P_{i_0}, , P_{i_\tau}\}$ of projected points whose signature it computes. In a sequential scenario, this is not a problem, as the subsets of size $\tau + 1$ can be enumerated one after another, using an ordering in which it is computationally cheap to calculate the lexicographically next subset from a given one [14]. In our parallel scenario, however, each thread needs to identify its relevant subset independently, without reference to other results, i.e., the i-th thread must find the i-th subset without access to the $(i-1)$-th subset. Calculating directly the i-th subset is significantly more complex than advancing to the next subset from a given one. Using the combinatorial representation (or, *combinadics*) of index i allows for a relatively efficient computation of the corresponding subset [15], but still involves $O(\tau \cdot \log t)$ calculations of binomial coefficients. Using a table of pre-calculated binomial coefficients can improve efficiency at the cost of extra memory usage.

An alternative solution – which is used in our current implementation – is precomputing an array containing the lexicographic enumeration of all $\binom{t}{\tau+1}$ dense units within a density chunk of size t, i.e. the i-th position of the array contains a representation of the i-th dense unit. A straightforward and space-efficient encoding of subsets of size $\tau + 1$ of a set with t elements is a bit string of length t with exactly $\tau + 1$ bits set to 1. The precomputation of the array is sequential but uses a very efficient implementation to compute the lexicographically next bit permutation [16]. Calculating the dense unit array of length ~ 500000 for a density chunk of size $t = 60$ and dense units of size $\tau + 1 = 4$ takes about $12\,\text{ms}$ on an Intel Core i7-4720HQ @2.6 GHz. Computing an array of ~ 50 million permutations ($t = 60, \tau + 1 = 6$) takes $1252\,\text{ms}$.

We are also working on a possible parallelization of this precomputation, similar to the idea of parallel prefix calculation [17].

5.4   GPU-Based Hashing

Hashing the calculated signatures into *htable* can also be carried out on the GPU. GPU-based hashing has been extensively studied by Alcantara, who proposed several efficient hashing schemes [18]. Our approach, based on the implementation used in [19], is currently being implemented and hence, not part of the evaluation presented here. Note that GPU memory is a limiting factor for the hash table size. State of the art GPUs come with up to 16 GB of RAM, which is sufficient to accommodate each partial table of the slicing approach described in Sect. 4.2.

5.5 Performance Evaluation

Our current implementation of the GPU approach is a first step. It has not been optimized regarding the GPU's memory hierarchy and hence does not benefit from caching effects. Also, it does currently not use more intricate CUDA functions such as, for instance, the SHFL (shuffle) command, which might be interesting for combinatoric tasks like subset enumeration.

The performance of the GPU algorithm was tested on an Intel Core i7-4720HQ @2.6 GHz machine equipped with an Nvidia GeForce GTX 950M GPU hosting 640 processing units (CUDA cores) and 4 GB of GPU RAM, against the sequential CPU algorithm for computing signatures, run on the same machine. The results are shown in Table 1. They do not include the time for precomputation of subsets and for hashing the signatures, but include all transfer times between GPU and host memory required for the GPU computations.

Table 1. Performance of CPU and GPU algorithms for computing signatures.

#Signatures computed	Time CPU (ms)	Time GPU (ms)	Speedup factor
1,770	0.4	1.0	0.4
34,220	11.5	1.9	6.1
487,635	148.8	13.8	10.8
5,461,512	1770.0	135.8	13.0
50,063,860	17692.5	1162.2	15.2

For smaller numbers of signatures, GPU is slower than CPU. This was to be expected as there is always a small but non-negligible ramp-up cost for GPU kernels Note that the speedup factor increases with the amount of calculations. Note that the GPU used for this preliminary evaluation is a relatively small model; high-performance Tesla GPUs contain thousands of CUDA cores and achieve significantly higher processing power.

6 Conclusion

In this paper, we have presented two independent ways of parallelizing the SUB-SCALE algorithm. First, we have described the use of a common shared memory multi-core architecture. Achieving the parallelization by assigning CPU cores to slices of the hash table for store signatures and finding larger-dimensional dense units, the results have shown linear speedup with the number of cores.

The second approach uses finer-granular data parallelism and can be implemented efficiently on graphics processing units (GPUs). First performance tests show very promising results, especially for larger data sets. This part of the work is ongoing; we are currently implementing the full functionality including GPU-based parallel hashing of the signatures.

The two approaches do not exclude each other. In fact, they can complement each other, using multi-core parallelism for coarse-grained tasks (processing of dimensions or has table slices) and many-core data parallelism for finer-grained subtasks (such as individual signature computation). Future work includes this combination of both approaches, making the best possible use of the different processing resources.

References

1. Parsons, L., Haque, E., Liu, H.: Subspace clustering for high dimensional data: a review. ACM SIGKDD Explor. Newsl. **6**(1), 90–105 (2004)
2. Aggarwal, C.C., Reddy, C.K.: Data Clustering: Algorithms and Applications, 1st edn. Chapman & Hall/CRC, Boca Raton (2013)
3. Kaur, A., Datta, A.: Subscale: fast and scalable subspace clustering for high dimensional data. In: 2014 IEEE International Conference on Data Mining Workshop (ICDMW), pp. 621–628 (2014)
4. Kaur, A., Datta, A.: A novel algorithm for fast and scalable subspace clustering of high-dimensional data. J. Big Data **2**(1), 17 (2015)
5. Sim, K., Gopalkrishnan, V., Zimek, A., Cong, G.: A survey on enhanced subspace clustering. Data Min. Knowl. Disc. **26**(2), 332–397 (2013)
6. Agrawal, R., Gehrke, J., Gunopulos, D.: Automatic subspace clustering of high dimensional data for data mining applications. In: Proceedings of the ACM SIGMOD International Conference on Management of Data, pp. 94–105 (1998)
7. Kailing, K., Kriegel, H.P., Kroger, P.: Density-connected subspace clustering for high-dimensional data. In: SIAM International Conference on Data Mining, pp. 246–256 (2004)
8. Zhu, B., Mara, A., Mozo, A.: CLUS: parallel subspace clustering algorithm on spark. In: Morzy, T., Valduriez, P., Bellatreche, L. (eds.) ADBIS 2015. CCIS, vol. 539, pp. 175–185. Springer, Cham (2015). doi:10.1007/978-3-319-23201-0_20
9. Dagum, L., Menon, R.: OpenMP: an industry standard API for shared-memory programming. IEEE Comput. Sci. Eng. **5**, 46–55 (1998)
10. Bache, K., Lichman, M.: UCI Machine Learning Repository (2013)
11. Geiger, A., Lenz, P., Stiller, C., Urtasun, R.: Vision meets robotics: the KITTI dataset. Int. J. Rob. Res. **32**(11), 1231–1237 (2013)
12. Zhu, J., Liao, S., Lei, Z., Yi, D., Li, S.Z.: Pedestrian attribute classification in surveillance: database and evaluation. In: ICCV Workshop on Large-Scale Video Search and Mining (LSVSM 2013), Sydney (2013)
13. Nvidia: CUDA home page. http://www.nvidia.com/object/cuda_home_new.html. Accessed 26 May 2017

14. Loughry, J., van Hemert, J., Schoofs, L.: Efficiently enumerating the subsets of a set (2000). applied-math.org/subset.pdf
15. McCaffrey, J.: Generating the mth lexicographical element of a mathematical combination. MSDN Library (2004)
16. Anderson, S.E.: Bit Twiddling Hacks compute the lexicographically next bit permutation. http://graphics.stanford.edu/~seander/bithacks.html#NextBitPermutation. Accessed 26 May 2017
17. Harris, M., Sengupta, S., Owens, J.D.: Parallel prefix sum (scan) with CUDA. GPU gems **3**(39), 851–876 (2007)
18. Alcantara, D.A.F.: Efficient hash tables on the GPU. Ph.D. thesis, University of California Davis (2011)
19. Strohm, P.T., Wittmer, S., Haberstroh, A., Lauer, T.: GPU-accelerated quantification filters for analytical queries in multidimensional databases. In: Bassiliades, N., Ivanovic, M., Kon-Popovska, M., Manolopoulos, Y., Palpanas, T., Trajcevski, G., Vakali, A. (eds.) New Trends in Database and Information Systems II. AISC, vol. 312, pp. 229–242. Springer, Cham (2015). doi:10.1007/978-3-319-10518-5_18

Discovering High-Utility Itemsets at Multiple Abstraction Levels

Luca Cagliero[(⊠)], Silvia Chiusano, Paolo Garza, and Giuseppe Ricupero

Dipartimento di Automatica e Informatica, Politecnico di Torino, Turin, Italy
{luca.cagliero,silvia.chiusano,paolo.garza,giuseppe.ricupero}@polito.it

Abstract. High-Utility Itemset Mining (HUIM) is a relevant data mining task. The goal is to discover recurrent combinations of items characterized by high profit from transactional datasets. HUIM has a wide range of applications among which market basket analysis and service profiling. Based on the observation that items can be clustered into domain-specific categories, a parallel research issue is generalized itemset mining. It entails generating correlations among data items at multiple abstraction levels. The extraction of multiple-level patterns affords new insights into the analyzed data from different viewpoints. This paper aims at discovering a novel pattern that combines the expressiveness of generalized and High-Utility itemsets. According to a user-defined taxonomy items are first aggregated into semantically related categories. Then, a new type of pattern, namely the Generalized High-utility Itemset (GHUI), is extracted. It represents a combinations of items at different granularity levels characterized by high profit (utility). While profitable combinations of item categories provide interesting high-level information, GHUIs at lower abstraction levels represent more specific correlations among profitable items. A single-phase algorithm is proposed to efficiently discover utility itemsets at multiple abstraction levels. The experiments, which were performed on both real and synthetic data, demonstrate the effectiveness and usefulness of the proposed approach.

Keywords: High-Utility Itemset Mining · Generalized itemset mining · Data mining · Knowledge discovery

1 Introduction

Frequent itemset mining is an exploratory data mining technique which focuses on discovering recurrent combinations of items (of arbitrary size) that occur in potentially large transactional data [1]. Frequent itemsets have been used in many research contexts, among which market basket analysis [1], service profiling [3], to discover correlations between multiple data items. For example, in the context of market basket analysis, each row of the dataset (transaction) represents a different market basket. Transactions contain the subsets of purchased items. Frequent itemsets represent sets of items that customers frequently purchased together. For instance, itemset {*Coke, bread*} indicates that customers

© Springer International Publishing AG 2017
M. Kirikova et al. (Eds.): ADBIS 2017, CCIS 767, pp. 224–234, 2017.
DOI: 10.1007/978-3-319-67162-8_22

who purchased coke frequently purchased bread as well. Since generating all the possible combinations of items in a transactional dataset is computationally intractable [1], frequent itemset mining entails discovering only the combinations of items whose frequency of occurrence (support) is above a given threshold. However, the traditional itemset mining problem relies on three (potentially unreliable) assumptions:

(A) Items appear at most once in each transaction (e.g., we disregard the amounts of purchased items within each basket).
(B) Items have all the same importance in the analyzed data (e.g., the unit profit is assumed to be the same for all the items in the market).
(C) The semantic relationships between items are ignored (e.g., the co-occurrences of items belonging to the same product group within the same basket are considered as uncorrelated with each other).

To overcome limitations (A) and (B), the concept of High-Utility Itemset (HUI) has been proposed [10]. HUIs represent sets of frequently co-occurring items that are characterized by averagely high utility within the analyzed data. To mine HUIs, items in the transactional dataset are enriched with both per-transaction weights (hereafter denoted as *internal utilities*) and global weights (denoted as *external utility*). For example, in the context of market basket analysis internal utilities represent per-item amounts (e.g. the customer purchased 3 bottles of coke), while external utilities indicate unit profits (bottles of coke cost 5 USD each). Utility itemsets represent sets of items whose total yield is above a given (user-specified) threshold. This knowledge may be exploited to perform cross-selling, to plan promotions, or to effectively arrange items on the shelves.

To overcome limitation (C), correlations between items at higher abstraction levels can be analyzed [11]. Based on a taxonomy built on top of the analyzed data, items are aggregated into semantically related groups (e.g. items *Coke* and *Water* into group *Beverage*). Then, generalized itemsets, which represent correlations among data items at different abstraction levels (e.g., not only {*Coke, bread*} but also {*Beverage, Food*}), can be extracted.

Many algorithms have been proposed to efficiently extract HUIs [7–9,12] and to select compact subsets of HUIs, e.g., the closed HUIs [5] and the top-k HUIs [13]. Existing solutions are efficient in term of temporal and spatial scalability, but, to the best of our knowledge, they are unable to cope with multiple-level data (limitation (C)). On the other hand, parallel works addressed the generalized itemset mining problem by performing bottom-up [2,3,11] or top-down [6] taxonomy visits during candidate itemset generation. However, since they do not consider item utilities, they still suffer from limitations (A) and (B).

This paper aims at bridging the gap between HUI mining and generalized itemset mining. To this purpose, it proposes a new type of pattern, namely the Generalized High-utility Itemsets (GHUIs). The proposed approach allows both multiple appearances of the same item within each transaction and different per-item profits. Unlike traditional HUIs, GHUIs are extracted from a transactional dataset enriched with a taxonomy, which describes the semantic is-a

Table 1. Transactional dataset

Transaction id	Items and internal utility
TId1	(Coke, 2), (Bread, 2), (Steak, 1)
TId2	(Water, 3), (Pasta, 2), (Steak, 1)
TId3	(Water, 2), (Bread, 2)
TId4	(Coke, 1), (Bread, 2)

Table 2. External utilities of items in dataset

Item	External utility
Water	1
Coke	5
Bread	1
Pasta	2
Steak	10

Fig. 1. Taxonomy on items in the dataset

relationships between data items. These relationships are exploited to drive the process of knowledge generalization thus generating profitable combinations of items at multiple abstraction levels. To extract GHUIs we proposed ML-HUI Miner, which extends a state-of-the-art HUI mining algorithm to cope with data enriched with taxonomies. The newly proposed algorithm integrates taxonomy information into the utility itemset mining process to mine GHUIs in a single-phase mining session (Table 4).

Preliminary experiments performed on both real retail data and benchmark datasets show the efficiency and effectiveness of the proposed approach.

The paper is organized as follows. Sections 2 and 3 introduce preliminary concepts and formalizes the newly proposed pattern, respectively. Section 4 presents the mining algorithm used to discover the newly proposed pattern. In Sect. 5 we summarized the experiments performed on real datasets, while Sect. 6 draws conclusions and discusses future works.

2   Preliminaries

A transactional dataset is a set of transactions [1], where each transactions is a set of data items (i.e., objects identified by literals). Herafter, let us denote as $\mathcal{I}$ the set of all possible items and as $t_j \in \mathcal{I}$ the j-th transaction of a transactional dataset $\mathcal{D}$. Items are characterized by (i) Internal Utility, denoted as $iu(i, t_j)$ which indicates the relative importance of item $i \in \mathcal{I}$ in transaction t_j, and (ii) External Utility, denoted as $eu(i)$, which indicates the relative importance of item i in $\mathcal{D}$ with respect to all the other items in the dataset.

Table 1 reports an example of market basket dataset consisting of 4 transactions (identified by Tids 1–4)).

Table 3. High-Utility Itemsets.
minutil (itemsets) = 17

Itemset	Utility
{Steak}	20
{Steak, Coke}	20
{Coke, Bread}	19
{Steak, Coke, Bread}	22

Table 4. Generalized High-Utility Itemsets.
minutil (generalized itemsets) = 30

Generalized Itemset	Utility
{Food}	30
{Food, Beverage}	50

The utility of item i in transaction t_j, hereafter denoted as $u(i, t_j)$, is computed as $eu(i) \cdot iu(i, t_j)$. In the running example, it indicates the total income related to an item appearing in the market basket (e.g., the price of all the bottles of coke in a given market basket).

Itemsets are sets of items of arbitrary size. We will denote as k-itemset a set of k items. The utility of itemset I in transaction t_j is the sum of the utilities of all the corresponding items, i.e., $u(I, t_j) = \sum_{i \in I} u(i, t_j)$. The utility of itemset I in the transactional dataset $\mathcal{D}$ is obtained by summing the utilities of the itemset in all the dataset transactions, i.e., $u(I, \mathcal{D}) = \sum_{t_j \in \mathcal{D}} u(I, t_j)$, where we assume that $u(I, t_j) = 0$ if $I \not\subseteq t_j$.

A notable type of itemset is the High-Utility Itemset. Given user-specified minimum utility threshold *minutil*, an itemset mined from dataset $\mathcal{D}$ is an High-Utility Itemset (HUI) if and only if $u(I, \mathcal{D}) > minutil$. Given a transactional dataset $\mathcal{D}$ and a minimum utility threshold *minutil*, the High-Utility Itemset Mining (HUIM) problem entails discovering all the HUIs in $\mathcal{D}$.

The HUIs extracted from the dataset in Table 1 by enforcing a minimum utility threshold $minutil = 20$ are enumerated in Table 3, where the corresponding utility value is given too.

Example. {Coke, Bread} is HUI because the utility values of the 2-itemset for each transaction are: 12 ($5 \times 2 + 1 \times 2$) in TId1, 0 in TId2 and TId3 (no matches), and 7 ($5 \times 1 + 1 \times 2$) in TId4.

To efficiently extract HUIs, the Transaction-Weighted Utilization (TWU) has been introduced [12]. It is an over-estimate of the utility of the itemset, which can be exploited to prune the search space because it satisfies the following downward closure property: given two itemsets I_1 and I_2 such that $I_1 \subset I_2$, if the TWU of I_1 is below the utility threshold *minutil* even the TWU of I_2 does. The TWU of an itemset I, denoted as $twu(I)$, is defined as the sum of the transaction utilities of all the transactions containing I, where the transaction utility of a transaction is the sum of the utility values of all its items. A formal definition of the TWU measure of itemset I follows: $twu(I) = \sum_{t_j \in D | I \subseteq t_j} \sum_{i \in t_j} u(i, t_j)$.

3   Generalized High-Utility Itemsets

Our goal is to discover HUIs that incorporate knowledge at multiple granularity levels. To this aim, we generalize items at different abstraction levels.

Let $\mathcal{T}$ be a taxonomy (i.e., a is-a hierarchy), which aggregates items in $\mathcal{I}$ into higher-level concepts, hereafter denoted as *generalized items*. Generalized items represent higher-level categories which group individual items based on their semantic meaning. Let $\mathcal{G}$ be the set of generalized items in $\mathcal{T}$. For the sake of simplicity, hereafter we will assume that in the given taxonomy $\mathcal{T}$ each item $i \in \mathcal{I}$ is aggregated into exactly one generalized item $g \in \mathcal{G}$ (i.e., each item belongs to a specific higher-level category). Generalized items can be further generalized as other generalized items at higher granularity levels.

For each generalized item $g \in \mathcal{G}$, $\mathrm{Desc}(g, \mathcal{T}) \in \mathcal{I}$ denote the subset of descendant items of g according to the given taxonomy. For our purposes, we formalize the concept of *level* of a generalized item $g \in \mathcal{G}$ in the taxonomy, hereafter denoted as $\mathrm{l}(g, \mathcal{T})$, as the length of the shortest path between g and any leaf node in the taxonomy. Note that, by construction, the level of non-generalized itemsets is zero, while the maximum level of an item corresponds to the taxonomy height (i.e., the length of the longest path from any node in $\mathcal{T}$ to a leaf node).

Example. Figure 1 depicts an example of taxonomy built on items occurring in the running example dataset (see Table 1). For instance, items *Coke* and *Water* are aggregated into the generalized item *Beverage*. The level of *Beverage* and is one, whereas the level of *Coke*, and *Water* is zero.

A generalized itemset is a set of generalized items in $\mathcal{G}$. Similar to [3,6,11] we focus our analyses on the combinations of generalized items having the same level, because they compactly represent information at a given abstraction level. Herafter we will denote as *level* of a generalized itemset the level of its items.

For our purposes, we extend the concept of utility to generalized items and itemsets. Specifically, the utility of a generalized item g in a transaction t_j, hereafter denoted as $\mathrm{u}(g, t_j)$, is the sum of the utility values of all the descendant items, while the utility of g in a transactional dataset is the sum of all the per-transaction utilities. More formal definitions follow.

Definition (Utility of a generalized item). Let g be a generalized item, $\mathcal{D}$ a transactional dataset, and $\mathcal{T}$ be a taxonomy. The utility of g in a transaction $t_j \in \mathcal{D}$ is defined as $\mathrm{u}(g, t_j) = \sum_{i \in Desc(g, \mathcal{T})} \mathrm{eu}(i) \cdot \mathrm{iu}(i, t_j)$. The utility of generalized item g in $\mathcal{D}$ is calculated as $\mathrm{u}(g, \mathcal{D}) = \sum_{i \in Desc(g, \mathcal{T})} u(i, \mathcal{D})$. $\square$

Similar definitions hold on itemsets (i.e., sets of items). The utility of a generalized itemset in a transactional dataset indicates the overall profit of a combination of item categories.

Definition (Utility of a generalized itemset). Let GI be a generalized itemset and let $\mathcal{D}$ a transactional dataset, and $\mathcal{T}$ be a taxonomy. The utility of a generalized itemset GI in transaction t_j is defined as $u(GI, t_j) = \sum_{g \in GI} u(g, t_j)$. The utility of a generalized itemset GI is in the transactional dataset $\mathcal{D}$ is defined as $u(GI, \mathcal{D}) = \sum_{t_j \in \mathcal{D}} u(GI, t_j)$, where we assume $u(GI, t_j) = 0$ if $GI \notin t_j$.   □

Example. Let us consider again the market basket dataset reported in Table 1 and the taxonomy in Fig. 1. The per-transaction utility of generalized item *Beverage* is 10 in transaction with TId1, 3 in transaction with TId2, 2 in transaction with TId3, and 5 in transaction with TId4. Hence, the utility of *Beverage* in the dataset is 20 (10 + 3 + 2 + 5).

In this work we address the extraction of a selection of generalized itemsets, called Generalized High-Utility Itemsets (GHUIs).

Definition (Generalized High-Utility Itemset). Let *minutil* be a (user-specified) minimum utility threshold, let $\mathcal{D}$ be a transactional dataset, let $\mathcal{T}$ be a taxonomy, and let GI be a generalized itemset. A generalized itemset GI is a Generalized High-Utility Itemset (GHUI) in $\mathcal{D}$ if and only if $u(GI, \mathcal{D}) > minutil$.   □

Example. The GHUIs mined from the dataset in Table 1 by enforcing a minimum utility threshold for generalized itemsets equal to 30 for GHUIs are enumerated in Table 4.

Given a transactional dataset $\mathcal{D}$, a taxonomy $\mathcal{T}$, and a minimum utility threshold *minutil*, the Generalized High-Utility Itemset Mining (GHUIM) problem addressed by this work entails discovering all the HUIs and GHUIs.

Per-level utility thresholds. The utility of a generalized itemset incorporates those of all of its descendant itemsets. Hence, itemsets at higher abstraction levels are more likely to satisfy a fixed minimum utility threshold than lower-level ones. To overcome this issue, we adapt the utility threshold to the level of generalization of the considered itemsets. The key idea is to set higher utility thresholds for itemsets including items at higher abstraction levels. We set the same utility threshold for all itemsets having the same level. Specifically, given a user-specified least minimum utility threshold *minutil* associated with non-generalized itemsets (level = 0), the minimum utility threshold thr(l) associated with level-l generalized itemsets is *minutil* if $l = 0$, thr(l) = $\alpha(l)$ · *minutil* otherwise, where $\alpha(l)$: $\mathbb{N} \rightarrow [1, +\infty)$ is a monotonically increasing (user-specified) function.

4   The ML-HUI Miner Algorithm

To extract Generalized High-Utility Itemsets we propose a new algorithm, namely Multiple-Level High-Utility Itemset Miner (ML-HUI Miner). The main features of GHUI-Miner are summarized below.

(a) *Taxonomy-driven HUI mining.* ML-HUI Miner supports transactional data enriched with taxonomy information. This allows us to extract patterns at different abstraction levels.

Algorithm 1. The ML-HUI Miner algorithm

Input: transactional dataset $\mathcal{D}$, taxonomy $\mathcal{T}$, minimum utility threshold $minutil$, function $\alpha(l)$
Output: $\mathcal{O}$, the set of High-Utility Itemsets and Generalized High-Utility Itemsets satisfying the per-level utility thresholds
{**Initializations**}
1: $\mathcal{I} \leftarrow$ set of items in $\mathcal{D}$
2: $\mathcal{GI} \leftarrow$ set of generalized items in $\mathcal{T}$
{**Preparation**}
3: scan $\mathcal{D}$ and $\mathcal{T}$ to compute Transaction-Weighted-Utility (TWU) of items in $\mathcal{I}$
4: Compute Transaction-Weighted-Utility of items in $\mathcal{GI}$
5: $\mathcal{I}^* \leftarrow$ set of items in $\mathcal{D}$ such that TWU is above $minutil$(level=0)
6: $\mathcal{GI}^* \leftarrow$ set of generalized items g in $\mathcal{T}$ such that TWU is above $\alpha \cdot minutil(l(g, \mathcal{T}))$
7: Build utility list and Estimated Utility Co-occurrence Structure of (generalized) items in $\mathcal{I}^* \cup \mathcal{GI}^*$
{**Recursive depth-first search**}
8: $\mathcal{O} \leftarrow$ Recursive generation of the combinations of (generalized) items in $\mathcal{I}^*$ and $\mathcal{GI}^*$ whose items share the same level and selection of all combinations satisfying the utility threshold $\alpha(l) \cdot minutil$

(b) *Single-step extraction of generalized and non-generalized HUIs.* The proposed algorithms explores the dataset and the taxonomy to generate multiple-level patterns in a single phase (i.e., without the need for multiple runs).

(c) *Prevent the generation of uninteresting combinations of items.* Similar to [6, 11], we focus on extracting itemsets including only items with the same level. Thus, GHUI-Miner prevents the generation of itemsets consisting of items with different levels in the taxonomy.

A high-level pseudo-code of the ML-HUI Miner algorithm is given in Algorithm 1. First, ML-HUI Miner scans the dataset and the taxonomy to identify the single non-generalized and generalized items whose Transaction-Weighted-Utility (TWU) is above the per-level utility threshold (Lines 3–6 in Algorithm 1). To this aim, the taxonomy is explored in a bottom-up fashion. The dataset and the accessory structures are properly adapted to prevent the generation of combinations of mixed-level items (according to Point (c). To compute the per-level utility thresholds, the user-specified threshold $minutil$ is adjusted using function α according to the level of each generalized item. Then, we compute the utility list associated with all non-generalized and generalized items whose TWU satisfies the per-level utility threshold (Line 7 in Algorithm 1). The utility list is a compact data structure that contains for each (generalized) item i (i) the list of transactions t_j such that $i \in t_j$, (ii) the utility values of the (generalized) item in each transaction $u(i, t_j)$, and (iii) the sum of the utilities of the remaining items with the same taxonomy level within each transaction, i.e., $\sum_{q \in t_j \wedge q \neq i \wedge l(q, \mathcal{T}) = l(i, \mathcal{T})} U(q, t_j)$. For our purposes, we extended the utility list proposed in [8] to integrate information about the generalized items at the same taxonomy level appearing in the taxonomy. The utility list is provided as input to the depth-first recursive procedure (Line 8 in Algorithm 1), which does not need to access neither the dataset nor the taxonomy. Specifically, starting from single items (generalized and not) the recursive procedure computes their exact utility value and then explores all their extensions using a depth-first strategy based on the utility list. To avoid generating itemsets consisting of items with different level, extensions are selectively generated. The recursive procedure is similar to the one adopted by FHM [8].

5   Experiments

To evaluate the performance of the ML-HUI Miner algorithm, we conducted experiments on four benchmark UCI datasets coming from different domains (*Connect, Mushroom, Chess, Retail* [4]), which have already been used to evaluate the performance of recently proposed High-Utility Itemset mining algorithms (e.g., FHM [8]). The number of transactions per dataset varies from 3,196 (*Chess*) to 67,557 (*Retail*). The analyzed datasets show different characteristics (e.g., number of distinct items *Connect* 129, *Mushroom* 119, *Chess* 75, *Retail* 2,741). The external utilities in the datasets range from 1 to 1,000. They were synthetically generated by using a log-normal distribution. Internal utilities are uniformly distributed between 1 and 5. To set the per-level utility threshold values ($thr(l)$), we defined $\alpha(l) = \gamma \cdot f(l)$, where $f(l)$ is the average number of level-0 descendants per level-l item. To generalize items at higher abstraction levels, we generated taxonomies on top of data items. Specifically, on the *Retail* dataset, which stores real sales of an online store, we generated a 2-level taxonomy by aggregating products into the corresponding group provided by the online store. The 2,741 available products are clustered into 38 produt groups (e.g., *Kitchen, Toy*). Products are evenly distributed across product groups. For instance, *Kitchen* clusters 1,513 products, while *Toy* 236. On the other datasets we generated a synthetic taxonomy where each generalized item aggregates, on average, 10 randomly selected products. The experiments were performed on a 2.67 GHz Intel Xeon workstation with 32 GB of RAM, running Ubuntu 12.04.

5.1   Performance Analysis

Figure 2 shows the number of mined (non-generalized) HUIs and GHUIs by varying the values of *minutil* and γ on a representative dataset (i.e., *Chess*). For both types of patterns, the number of mined itemsets is inversely proportional to the *minutil* value. By decreasing the values of *minutil* and γ the total number of mined patterns super-linearly increases. The number of GHUIs is two orders of magnitude lower than those of HUIs due to the significantly lower number of possible item combinations at higher taxonomy levels. The increase in the number of candidate itemsets due to taxonomy integration implies also a time complexity increase. Figure 3 shows the time spent by the ML-HUI Miner algorithm on the analyzed datasets with decreasing *minutil* values. Since the value of γ affects only the extraction of HGUIs, whose number is orders of magnitude lower than the number of HUIs, varying γ value slightly affects the execution time. For this reason, we report the execution time only for the representative value $\gamma = 0.2$ on all datasets. Despite a larger number of combinations were explored, the extraction times remain acceptable (few ms) on all the analyzed datasets. We compared also the execution times of the newly proposed ML-HUI

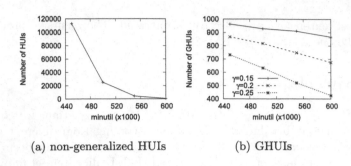

(a) non-generalized HUIs   (b) GHUIs

Fig. 2. Chess: number of mined patterns

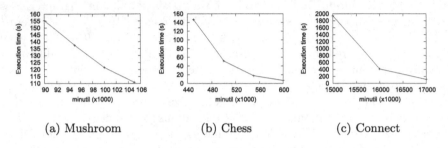

(a) Mushroom   (b) Chess   (c) Connect

Fig. 3. Execution time ($\gamma = 0.2$)

Miner with those of the FHM algorithm [8], which extracts only non-generalized HUIs. ML-HUI Miner execution time approximately doubles that of FHM, even when more than two aggregation levels are integrated in the taxonomy.

5.2 Knowledge Discovery

We present some example GHUIs mined from the *Retail* dataset (*minutil* = 10000, $\gamma = 0.2$). The GHUIs with length 1 reveal the most profitable product groups. For example, *Kitchen* is the category with maximal utility. *Toy* ranked second (utility gap from *Kitchen* to *Toy* 87%), *Home* is the third (92%), *Office product* ranked 4th (96%) and *Law & Patio* 5th (96%). For all the remaining categories the utility gap with respect to *Kitchen* was above 97%. The GHUIs with length greater than 1 point out combinations of groups that yielded high profits when the respective products were sold together. Let us consider, for instance, GHUI {Musical-Instruments, Home}. It indicates that the jointly sale of products in these categories provided a high income. Among GHUIs including group *Musical-Instruments*, $GHUI_i = \{$*Musical-Instruments, Kitchen*$\}$ is the one with highest utility. The other GHUIs in order of decreasing utility are: {Musical-Instruments, Toy} (utility −75% w.r.t. $GHUI_i$), {Musical Instruments, Home} (utility −91% w.r.t. $GHUI_i$), and {Musical-Instruments, Office-Product} (utility −94% w.r.t. $GHUI_i$). These patterns can can be exploited to figure out which categories of products should be promoted in the same advertising campaign, e.g., while planning a campaign on products belonging to *Musical Instrument*, products of *Kitchen*, *Toy*, *Home* or *Office-Product category* should be advertised as well. The utility value provided us a ranking of the most appealing groups of products to advertise together with products of *Musical Instrument*. GHUIs provide high-level knowledge related to single products that do not satisfy the utility threshold. Let us consider again group *Musical Instrument*. Only two non-generalized HUIs were extracted, i.e., {Red-Harmonica} (utility 26,205) and {Blue-Harmonica} (utility 10,271). However, the utility of the GHUI {Musical-Instruments} is 40,741. Hence, the utility value of {Musical-Instruments} is not only due to the {Red-Harmonica} and {Blue-Harmonica} products, but even to other products of the same group that do not satisfy the utility threshold. Considering GHUI {Musical-Instruments} allows us to consider also the contribution of the other products, even if their single profits are averagely low. Moreover, GHUIs represent correlations among products that can not be easily inferred while considering only HUIs. For instance, even if correlations between products of group {Musical-Instruments} and products of other groups have not been extracted, the high-lever correlations between *Musical Instrument* and other groups were extracted and can be analyzed.

6 Conclusions and Future Work

We proposed a new pattern, called GHUI, which represents sets of items groups, each one characterized by a high total profit. The significance of the proposed pattern and the performance of the proposed GHUI mining algorithm have been evaluated on retail data, with the goal of planning advertising campaigns of retail products. Future extensions of the work will address the efficient extraction of significant subsets of GHUIs (e.g., closed or minimal GHUIs).

References

1. Agrawal, R., Imielinski, T., Swami, A.: Mining association rules between sets of items in large databases. In: ACM SIGMOD 1993, pp. 207–216 (1993)
2. Baralis, E., Cagliero, L., Cerquitelli, T., D'Elia, V., Garza, P.: Expressive generalized itemsets. Inf. Sci. **278**, 327–343 (2014)
3. Cagliero, L.: Discovering temporal change patterns in the presence of taxonomies. IEEE Trans. Knowl. Data Eng. **25**(3), 541–555 (2013)
4. Fournier-Viger, P., Gomariz, A., Gueniche, T., Soltani, A., Wu, C.-W., Tseng, V.S.: SPMF: a Java open-source pattern mining library. J. Mach. Learn. Res. **15**(1), 3389–3393 (2014)
5. Fournier-Viger, P., Zida, S., Lin, J.C., Wu, C., Tseng, V.S.: Efficient closed high-utility itemset mining. In: Proceedings of the 31st Annual ACM Symposium on Applied Computing, Pisa, Italy, pp. 898–900, 4–8 April 2016
6. Han, J., Fu, Y.: Discovery of multiple-level association rules from large databases. In: VLDB Conference, pp. 420–431 (1995)
7. Krishnamoorthy, S.: Pruning strategies for mining high utility itemsets. Expert Syst. Appl. **42**(5), 2371–2381 (2015)

8. Lin, J.C., Fournier-Viger, P., Gan, W.: FHN: an efficient algorithm for mining high-utility itemsets with negative unit profits. Knowl. Based Syst. **111**, 283–298 (2016)
9. Liu, J., Wang, K., Fung, B.C.M.: Direct discovery of high utility itemsets without candidate generation. In: 12th IEEE ICDM Conference, pp. 984–989, December 2012
10. Liu, Y., Liao, W., Choudhary, A.: A two-phase algorithm for fast discovery of high utility itemsets. In: Ho, T.B., Cheung, D., Liu, H. (eds.) PAKDD 2005. LNCS (LNAI), vol. 3518, pp. 689–695. Springer, Heidelberg (2005). doi:10.1007/11430919_79
11. Srikant, R., Agrawal, R.: Mining generalized association rules. In: VLDB 1995, pp. 407–419 (1995)
12. Tseng, V.S., Shie, B.-E., Wu, C.-W., Yu, P.S.: Efficient algorithms for mining high utility itemsets from transactional databases. IEEE Trans. Knowl. Data Eng. **25**(8), 1772–1786 (2013)
13. Tseng, V.S., Wu, C.W., Fournier-Viger, P., Yu, P.S.: Efficient algorithms for mining top-k high utility itemsets. IEEE TKDE **28**(1), 54–67 (2016)

Automatic Organization of Semantically Related Tags Using Topic Modelling

Iman Saleh[(✉)] and Neamat El-Tazi

Faculty of Computers and Information, Cairo University, Giza, Egypt
{iman.saleh,n.eltazi}@fci-cu.edu.eg

Abstract. The use of social media platforms such as social networks, weblogs and community question answering websites increased largely in the last few years. This increase in usage contributed to the vast explosion in available online content. Some of these platforms use social tagging, tags manually inserted by content authors, as a way to facilitate content description and discovery. In this paper, our goal is to automatically group semantically related tags in order to organize the large amount of tags contributed by various users. Our approach is based on using a topic model to discover topics of documents, and then grouping top tags related to documents assigned to each topic. We perform a set of experiments using different number of topics to provide different levels of details for the generated tag groups. The dataset used in our experiments is extracted from Stack Overflow, a community question answering website used by programming professionals. Tag groups generated by our technique are presented and evaluated.

Keywords: Stack overflow · Tags · Topic modelling · LDA · Organize tags

1 Introduction

The amount of data available online in social media platforms is increasing largely. This large amount of data contains hidden information that need to be extracted. A way of organizing and categorizing this massive amount of data is to use tags inserted by users. However, the number of tags can increase largely over time. It is necessary in such a case to organize tags themselves to help users discover hidden topics in the data. In our work we are interested in Stack Overflow community question answering platform and how to organize its tags.

Stack Overflow is the largest online community question answering website designed to help developers share their programming knowledge. It allows them to post programming related questions and receive answers to them. The website contains[1] around 13.4 million questions tagged with around 48,228 tags. Questions are manually tagged by users and revised by privileged users. Tags in Stack Overflow are used in research studies to predict tags of posts [10,16–18,22,25],

---

[1] All numbers mentioned in the paper date back to March 2017.

© Springer International Publishing AG 2017
M. Kirikova et al. (Eds.): ADBIS 2017, CCIS 767, pp. 235–245, 2017.
DOI: 10.1007/978-3-319-67162-8_23

map tags to Wikipedia concepts [11], suggest and group tag synonyms [3,4], analyze trends of topics [2] and mine challenges encountered by developers [20]. Tags are important because they help categorize posts and hence do proper analysis of Stack Overflow content. The large number of tags in Stack Overflow makes it important to organize them into meaningful groups to be able to browse and analyze Stack Overflow content easily.

In this paper we introduce an automatic method to organize Stack Overflow tags. Our method is based on using Latent Dirichlet Allocation (LDA) topic model [5] to group semantically related tags. We first categorize a sample of posts in Stack Overflow into a number of topics. Each topic is assigned a set of key phrases that serve as a title for the topic using Open Calais tool[2] to provide a meaningful title for each group of words produced by LDA. Then we extract top tags assigned to the posts of each topic and consider them a group of related tags. We do some filtering on groups of tags to remove tags that are common among most of the topics. In order to evaluate how accurate tag groups are, we introduce a method inspired from word intrusion task in [6]: tag intrusion. This method is used to measure how semantically coherent tags are and whether they correspond to natural grouping for humans. We perform a set of experiments using different number of topics and provide proper evaluation of our technique.

The following is a summary of the contributions introduced by this paper:

- An automatic method to group Stack Overflow tags into semantically meaningful topics using LDA topic model (Sect. 3).
- A formal evaluation technique for our method (Sect. 3.4).

2   Related Work

Social tags in social media platforms were utilized in different contexts. [23] consider Twitter hashtags as topics and use them to automatically generate an overall sentiment polarity for a given hashtag. [7] utilize Twitter hashtags as an indicator of events. They use hashtags to discover breaking events. Hashtags are considered topics in Twitter as explained in [13] and used to predict the popularity of a topic. Semantic relationship between Twitter hashtags was explored in [8]. They construct a graph of hashtags and entities drawn from tweets such that edges capture semantic relatedness. A predictive model that incorporates the effect of time on individual hashtag reuse and social hashtag reuse was introduced in [12]. Their goal is to introduce a hashtag recommendation algorithm that considers time dimension in its recommendations. Stack Overflow and Twitter are similar in using tags to define topics. However, Twitter tags are much more diverse and changes frequently overtime unlike Stack Overflow tags which are more stable over time.

Grouping Stack Overflow tags was previously studied in [3]. Their approach was mainly focused on finding tags that are synonyms to each other. They used some strategies to achieve their goal based on studying the set of tag synonyms

---

[2] http://www.opencalais.com/.

that is provided by Stack Overflow. Their work focused only on android related tags. They further extended their work to group tag synonyms using graph clustering techniques [4]. In our work we are interested in similar tags in a broader sense and not only tags that are synonyms to each other. Working on tag synonyms only will result in a large number of tag groups that cannot be browsed easily. [9] used agglomerative clustering technique to group Stack Overflow tags. They only considered tags themselves and did not use Stack Overflow posts in their experiments. We introduce an alternative method that makes use of Stack Overflow posts and proper evaluation of our method. [21] also used k-medoids clustering in order to infer semantically related software tags. They measure the similarity of two tags based on the number of documents tagged by both of them and the similarity of textual content of documents tagged by them. However, the dataset they used is much smaller than our dataset and we believe their approach might not scale well to a large number of tags. [19] used Stack Overflow posts to build a software specific resource similar to WordNet. Their approach is based on the textual content of the posts and not their tags. A large taxonomy of subsumption relations between pairs of Stack Overflow tags was created using machine learning in [26]. They use several features such as lexical tags features and features extracted from tags wiki description. Their work focuses mainly on discovering relationships between tag pairs instead of creating tag groups. Finding trends of topics in Stack Overflow was explored in [2]. They also used LDA to find topics of Stack Overflow posts. Clusters of tags under each topic are manually created and referred to as technologies. Trends of topics and technologies are then extracted over time. In our work we want to automate the process of grouping tags under a certain topic. We also think that tag trends themselves carry important information in addition to topic or technology trends. Our analysis is based on tags themselves and how to compare between them. [20] also used LDA to find topics in Stack Overflow related to different Web API's. Their goal is to find common challenges encountered by developers. They investigate how topics related to different Web API's evolve over time. In our work we are interested in all topics in Stack Overflow rather than specific Web API's. [1] used LDA to mainly analyze problems faced by developers in Stack Overflow. They also provide an analysis of Stack Overflow usage in week days.

3   Grouping Stack Overflow Tags Using LDA

LDA is a generative probabilistic model that is used to model text corpora. It is based on (1) representing documents as a random mixture over k latent topics (2) modeling topics as distributions over words. In this model, topics are groups of semantically related words in a corpus of documents. For example, a group of words like "entity, model, database, query, mongodb" are semantically related and refer to concepts related to database and data management. Also, a document similar to the one shown in Fig. 1 is clearly about git and version control systems. So it belongs to the topic represented by the following group of words: "git, branch, repository, svn, repo, github". We also notice that the document is labeled with the following tags: "git, alias, git-push". Therefore these three tags can be grouped together since they are semantically related and refer to the same topic.

Title: Why does git push origin @ not work?
Post: We can push the head of a branch like below
$ git push origin HEAD
And, we can use @ for alias of HEAD.
$ git show @
Then why does the below command gives me an error?
$ git push origin @
fatal: remote part of refspec is not a valid name in @
Tags: git, alias, git-push

Fig. 1. An example of a stack overflow post

In our work we are interested in creating tag groups that are semantically related in order to organize Stack Overflow massive content. Working towards this end, we train LDA using a subset of Stack Overflow posts, extract tags assigned to each post, and then associate these tags with the most likely topic assigned to the post. Details of our method are explained in the following subsections.

3.1   Training LDA Model

We use MALLET [15] implementation of LDA to model a Stack Overflow posts dump[3]. When we investigated posts we found that some of them are of low quality and create noise in our model. Therefore we filtered data to include only a subset of posts that are assigned a score greater than or equal to 5. The score of a post is assigned by Stack Overflow community users as the difference between the number of upvotes and downvotes.

We split the dataset into 2 folds to make sure our technique is valid for Stack Overflow data. The first fold, DS_1, consists of 463,983 posts while the second, DS_2, contains 360,610 posts. Data in each fold was selected randomly first then filtered based on score. Code snippets were removed and tags of each post were included in training data. Best answer for a post is also included if available. We used Stanford parser [14] to stem data and use only noun words found in a post. LDA has three parameters: k, alpha and beta. The first parameter is the number of topics. The second parameter, alpha, controls the mixture of topics in a document; whether a document is likely to contain a mixture of most topics rather than only specific few topics. The third parameter, beta, controls the mixture of words in a topic; whether a topic is likely to contain most words

---

[3] Dump downloaded in October 2016.

rather than only specific words. We train LDA using 10, 30, 50 and 80 topics and compare between the four values. Alpha value used in our work is 50/k and beta value is 200/V, as recommended in [24], where V is the vocabulary size. Vocabulary size for DS_1 is 184,147, and 143,023 for DS_2. The model is trained using 2000 iterations since this is a reasonable number for the model to converge and find good topics.

3.2 Assigning Titles to Each Topic

Topics generated by LDA are mainly groups of words that are semantically related. To generate topics that are meaningful while browsing tag groups, we need to have a title for each group of words; or a set of phrases that describe the topic represented by those words. In order to generate a title for each topic, we use the top 20 words of each topic as an input to Open Calais API, a tool used for tagging text. For each topic, a set of tags and their relevance to input text are produced (as shown in Table 1). These tags can be used as topics titles. Tags common among all topics such as "computing", "software" and "software engineering" are excluded from the set of title tags.

Table 1. Open calais tags of $topic_{15}$ and $topic_{27}$ and their relevance to top words

Topic index	Topic top 20 words	Open calais tags
15	css, html, div, tag, width, image, browser, style, height, content, javascript, jquery, bootstrap, size, font, svg, background, selector, chrome, border	JavaScript libraries 66%, HTML 66%, Ajax 66%, Bootstrap 66%, Open formats 66%, JQuery 66%, Cascading Style Sheets 66%, HTML element 66%, Web design 66%
27	table, sql, database, column, query, row, mysql, server, record, index, statement, oracle, postgresql, sqlite, procedure, key, insert, transaction, order, clause	Data management 100%, SQL keywords 66%, Database management systems 66%, Database trigger 66%, MySQL 66%, Databases 66%, PostgreSQL 66%, Join 66%, Insert 66%

3.3 Creating Tag Groups Using LDA

LDA model assigns a mixture of topics to each document. For each topic, we collect all posts that most likely belong to it. Then we extract tags assigned to these posts. We refer to the tag group of $topic_i$ produced using an LDA model trained with k topics as TTG_i^k. For each TTG_i^k, we count how many times a tag appears associated with that $topic_i$. Each TTG_i^k is then assigned the top 5% tags according to frequency of occurrence in topic posts. An example of top tags associated with $topic_{15}$ (Table 1) is shown in Table 2.

Table 2. TTG_{15}^{30} top tags and their frequency of occurrence

Tags in TTG_{15}^{30}	Frequency
css	10,462
html	7,292
javascript	3,645
jquery	3,079
css3	2,087
twitter-bootstrap	1,454
html5	1,271

Table 3. $topic_{15}$ top tags that appear in 50% or less of TTG_i^{30}

Tags in UTG_{15}^{30}
css
css3
twitter-bootstrap
html5
svg
google-chrome
css-selectors

However, grouping tags is not a straightforward task. We notice that some tags appear in all topics or most of them. For instance, javascript tag can appear in topics related to database, version control systems, android as well as regular expressions. Therefore, a tag that appears in most or all of the topics cannot be assigned to a specific group of tags since it is shared among topics. Therefore, we create a group of tags called "shared tags". We refer to shared tags group created using an LDA model trained with k topics as STG_k. An STG_k group contains tags that appear in 50% or more of all TTG_i^k. For instance, when k = 10, if a tag_i is found in 5 or more TTG_i^k groups, it is a shared tag. Table 4 shows a sample of shared tags produced by LDA model with k = 10, 30, 50 and 80 and appeared in more than 50% of all TTG_i^k. We also create another group of tags for each topic that contains the remaining tags. These are tags that are either unique to a topic or appear in less than 50% of all TTG_i^k. We refer to the group of unique tags of a $topic_i$ created using an LDA model trained with k topics as UTG_i^k. Table 3 shows tags in UTG_{15}^{30}.

Table 4. Example of shared tags extracted using LDA model trained with k topics

	100% of topics	90% or more and less than 100%
STG_{10}	android, asp.net, c++, c#, cocoa, file, ios, iphone, java, javascript, jquery, json, .net, objective-c, php, python, ruby, ruby-on-rails, string, visual-studio, performance, visual-studio-2010	arrays, angularjs, api, delphi, eclipse, c, css, html, json, node.js, osx, qt, r, unit-testing, vb.net, windows, winforms, wpf, xml
STG_{30}	c#, java, javascript, python, .net	android, c++, ios, php, python, ruby
STG_{50}	c#, java	javascript, .net, php, python
STG_{80}	N/A	c#, java, javascript, python

Table 5. Tag intrusion results for datasets DS_1 and DS_2

	k = 10	k = 30	k = 50	k = 80
DS_1	0.73	0.80	0.81	0.83
DS_2	0.77	0.79	0.76	0.80

3.4 Evaluating Tag Groups

An appropriate quantitative measure is needed to evaluate extracted tag groups. It should be able to measure the coherence of tags in a tag group and whether they are actually semantically related. In order to achieve this goal, we use tag intrusion, a measure inspired from word intrusion task explained in [6]. Word intrusion task measures semantic coherence of top words in an LDA topic. This measure was shown to capture aspects of LDA models that are not captured by other measures of model quality based on held out likelihood. So we believe that a similar measure can be useful in assessing semantic coherence of tag groups. In word intrusion task, top words of a $topic_i$ produced by an LDA model are selected. A word $w_{intruder}$ is then selected randomly from the top words of another $topic_j$ and included together with the set of top words of $topic_i$. $w_{intruder}$ must not be present in the top words of $topic_i$. Annotators are then asked to find the word that does not belong to the group. In our case we call the evaluation task tag intrusion. We aim at finding how cohesive tag groups are, i.e. degree of relatedness between tags in a tag group UTG_i^k. We create the task as follows: given an LDA model trained with k topics, for each UTG_i^k, the top five tags are selected. Then another random UTG_j^k is selected. Next, we randomly choose a tag $tag_{intruder}$ from the top 3 tags in UTG_j^k such that $tag_{intruder} \notin UTG_i^k$. We count how many times the intruder tag was selected correctly by 3 different annotators and use the percent as a measure of tag groups cohesion. Given that tag_{answer} is the intruder tag selected by annotator s and S is the total number of annotators, Eq. 1 shows how we calculate tag intrusion. Table 5 shows the results of tag intrusion using tag groups extracted from LDA trained with k topics equal to 10, 30, 50 and 80.

$$TagIntrusion_k = \frac{(\sum_s(\sum_{i=1}^{k}(tag_{intruder} = tag_{answer})/k))}{S} \tag{1}$$

4 Discussion

The results of tag intrusion (Table 5) show that our method achieves at least 73% score for tag intrusion. We used two different sets in order to validate our results and notice that the difference between scores for both is less than or equal to 5% which indicates our algorithm generalizes well on Stack Overflow data. Also, shared tags and tags under specific topic groups are almost the same for both datasets as shown in Table 6. Tag groups for "Regular Expressions and

Character Encoding" and "Version control software" topics extracted from both datasets are almost the same. The table also shows that shared tags extracted from both datasets are very similar.

Table 6. Tag groups sorted based on frequency of appearance with topic, LDA k = 30

	DS_1	DS_2
Regular expressions and character encoding	regex, string, unicode, utf-8, parsing, perl, encoding, bash, character-encoding, split	regex, string, unicode, utf-8, perl, encoding, bash, character-encoding, split, replace
Version control software	git, svn, github, version-control, mercurial, jenkins, branch, merge, tortoisesvn, tfs	git, svn, github, version-control, mercurial, branch, merge, tortoisesvn, docker, jenkins
Tags in ST_{30} appearing in all topics	javascript, java, c#, python	Java, c#, .net, python

The number of topics does not have a significant effect on tag intrusion score. We think that the choice of the number of topics depends on whether we are interested in high level or low level topics. The lower the number of LDA topics, the more diverse tags will be grouped under a single topic. This is clear in Table 4 were the number of shared tags is inversely proportional to the number of LDA topics. For example, "regex" tag is considered as a shared tag for k = 10. When k is increased to 30, the tag appears as the most frequent tag under a specific topic: "character encoding and regular expressions". Also, as the number of topics increases, some topics are close to each other and its hard to distinguish between them. For instance, when we set k = 80, two topics produced very similar top tags as shown in Table 7. We think that organizing tags into groups is particularly useful to explore tag trends and how they change over time. A user can first choose a topic of interest and then browse tags trends under a topic. This can also facilitate comparing trends of two tags over time. For example, frequent tags include "javascript, java, c#". Then "ios" comes next with lower frequency. Non-frequent tags, compared to the previous tags, include "objective-c", "scala". We find this result some how consistent with the frequency of questions in which these tags appear as shown in Table 8[4].

---

[4] Numbers obtained from http://StackOverflow.com/tags.

Table 7. Similar tag groups produced using LDA trained with k=80

Tag group	Topic titles	Top tags
UTG^{80}_{64}	Smartphones, IOS SDK, Integrated development environments, Xcode, Tablet computers, IPad, Swift, GPS navigation devices, Cocoa	xcode, swift, cocoa, osx, xcode6, cocoa-touch, core-data, ipad, ios8, xcode4
UTG^{80}_{36}	IOS, IOS SDK, TvOS, Xcode, Swift, IPhone	uitableview, swift, cocoa-touch, xcode, ipad, ios8, ios7, cocoa, uiview

Table 8. Frequencies of top shared tags in stack overflow

	javascript	java	c#	ios	objective-c	scala
Freq.	1.3 m	1.2 m	1.0 m	481.3 k	273.9 k	63.1 k

5 Conclusion and Future Work

In this paper we introduced an automatic method to organize Stack Overflow tags. We use LDA to group tags that are semantically similar. We evaluated our method using tag intrusion score; a measure inspired from word intrusion that is used to evaluate LDA. We plan to extend our work by finding better methods to group Stack Overflow tags based on semantic similarity. A comparison between our method and other methods used to organize social tags needs to be performed as well. Finally, we plan to launch a demo to visualize our results.

Acknowledgements. This work is funded by Information Technology Academia Collaboration program organized by Information Technology Industry Development Agency in Cairo, Egypt. We thank Tarek Nabhan, Ahmed Hany and Noura Hassan for their guidance.

References

1. Allamanis, M., Sutton, C.: Why, when, and what: analyzing stack overflow questions by topic, type, and code. In: Proceedings of the 10th Working Conference on Mining Software Repositories, pp. 53–56. IEEE Press (2013)
2. Barua, A., Thomas, S.W., Hassan, A.E.: What are developers talking about? an analysis of topics and trends in stack overflow. Empir. Softw. Eng. **19**(3), 619–654 (2014)
3. Beyer, S., Pinzger, M.: Synonym suggestion for tags on stack overflow. In: Proceedings of the 2015 IEEE 23rd International Conference on Program Comprehension, ICPC 2015, pp. 94–103. IEEE Press (2015)
4. Beyer, S., Pinzger, M.: Grouping android tag synonyms on stack overflow. In: Proceedings of the 13th International Conference on Mining Software Repositories, MSR 2016, pp. 430–440. ACM (2016)
5. Blei, D.M., Ng, A.Y., Jordan, M.I.: Latent dirichlet allocation. J. Mach. Learn. Res. **3**(Jan), 993–1022 (2003)

6. Chang, J., Gerrish, S., Wang, C., Boyd-Graber, J.L., Blei, D.M.: Reading tea leaves: how humans interpret topic models. In: Advances in Neural Information Processing Systems, pp. 288–296. Curran Associates, Inc. (2009)
7. Cui, A., Zhang, M., Liu, Y., Ma, S., Zhang, K.: Discover breaking events with popular hashtags in twitter. In: CIKM (2012)
8. Ferragina, P., Piccinno, F., Santoro, R.: On analyzing hashtags in twitter. In: ICWSM (2015)
9. Gajduk, A., Madjarov, G., Gjorgjevikj, D.: Intelligent tag grouping by using an aglomerative clustering algorithm. In: The 10th Conference for Informatics and Information Technology CIIT (2013)
10. Gruetze, T., Krestel, R., Naumann, F.: Topic shifts in stackoverflow: ask it like socrates. In: Métais, E., Meziane, F., Saraee, M., Sugumaran, V., Vadera, S. (eds.) NLDB 2016. LNCS, vol. 9612, pp. 213–221. Springer, Cham (2016). doi:10.1007/978-3-319-41754-7_18
11. Joorabchi, A., English, M., Mahdi, A.E.: Automatic mapping of user tags to wikipedia concepts: the case of a q&a website-stackoverflow. J. Inf. Sci. **41**, 570–583 (2015)
12. Kowald, D., Pujari, S.C., Lex, E.: Temporal effects on hashtag reuse in twitter: a cognitive-inspired hashtag recommendation approach. In: WWW (2017)
13. Ma, Z., Sun, A., Cong, G.: On predicting the popularity of newly emerging hashtags in twitter. JASIST **64**, 1399–1410 (2013)
14. Manning, C.D., Surdeanu, M., Bauer, J., Finkel, J.R., Bethard, S., McClosky, D.: The stanford coreNLP natural language processing toolkit. In: ACL (System Demonstrations), pp. 55–60 (2014)
15. McCallum, A.K.: Mallet: a machine learning for language toolkit (2002). http://mallet.cs.umass.edu
16. Romero, J.J.F., Guerrero, M.G., Calder, F.: Multi-class multi-tag classifier system for stackoverflow questions. In: 2015 IEEE International Autumn Meeting on Power, Electronics and Computing (ROPEC), pp. 1–6. IEEE (2015)
17. Saha, A.K., Saha, R.K., Schneider, K.A.: A discriminative model approach for suggesting tags automatically for stack overflow questions. In: Proceedings of the 10th Working Conference on Mining Software Repositories, pp. 73–76. IEEE Press (2013)
18. Stanley, C., Byrne, M.D.: Predicting tags for stackoverflow posts. In: 12th International Conference on Cognitive Modelling, pp. 414–419 (2013)
19. Tian, Y., Lo, D., Lawall, J.: Automated construction of a software-specific word similarity database. In: 2014 Software Evolution Week - IEEE Conference on Software Maintenance, Reengineering and Reverse Engineering (CSMR-WCRE), pp. 44–53 (2014)
20. Venkatesh, P.K., Wang, S., Zhang, F., Zou, Y., Hassan, A.E.: What do client developers concern when using web APIs? an empirical study on developer forums and stack overflow. In: IEEE International Conference on Web Services (ICWS), pp. 131–138. IEEE (2016)
21. Wang, S., Lo, D., Jiang, L.: Inferring semantically related software terms and their taxonomy by leveraging collaborative tagging. In: International Conference on Software Maintenance, pp. 604–607. IEEE Computer Society (2012)
22. Wang, S., Lo, D., Vasilescu, B., Serebrenik, A.: EnTagRec: an enhanced tag recommendation system for software information sites. In: Proceedings of the 2014 IEEE International Conference on Software Maintenance and Evolution, pp. 291–300. IEEE Computer Society (2014)

23. Wang, X., Wei, F., Liu, X., Zhou, M., Zhang, M.: Topic sentiment analysis in twitter: a graph-based hashtag sentiment classification approach. In: CIKM (2011)
24. Wood, J.: Source-LDA: Enhancing probabilistic topic models using prior knowledge sources. CoRR abs/1606.00577 (2016)
25. Xia, X., Lo, D., Wang, X., Zhou, B.: Tag recommendation in software information sites. In: Proceedings of the 10th Working Conference on Mining Software Repositories, pp. 287–296. IEEE Press (2013)
26. Zhu, J., Shen, B., Cai, X., Wang, H.: Building a large-scale software programming taxonomy from stackoverflow. In: The 27th International Conference on Software Engineering and Knowledge Engineering, pp. 391–396. KSI Research Inc. and Knowledge Systems Institute Graduate School (2015)

Fuzzy Recommendations
in Marketing Campaigns

S. Podapati[1], L. Lundberg[1], L. Skold[2], O. Rosander[1],
and J. Sidorova[1(✉)]

[1] Department of CS and Engineering,
Blekinge Institute of Technology, Karlskrona, Sweden
julia.a.sidorova@gmail.com
[2] Telenor, Stockholm, Sweden

Abstract. The population in Sweden is growing rapidly due to immigration. In this light, the issue of infrastructure upgrades to provide telecommunication services is of importance. New antennas can be installed at hot spots of user demand, which will require an investment, and/or the clientele expansion can be carried out in a planned manner to promote the exploitation of the infrastructure in the less loaded geographical zones. In this paper, we explore the second alternative. Informally speaking, the term Infrastructure-Stressing describes a user who stays in the zones of high demand, which are prone to produce service failures, if further loaded. We have studied the Infrastructure-Stressing population in the light of their correlation with geo-demographic segments. This is motivated by the fact that specific geo-demographic segments can be targeted via marketing campaigns. Fuzzy logic is applied to create an interface between big data, numeric methods for its processing, and a manager who wants a comprehensible summary.

Keywords: Intelligent data mining · Call detail records · Fuzzy membership function · Geo-demographic segments · Marketing

1 Introduction

In the era of big data a mapping is desired from multitudes of numeric data to a useful summary and insights expressed in a natural language yet with a mathematical precision [1]. Fuzzy logic bridges from mathematics to the way humans reason and the way the human world operates. Clearly, the "class of all real numbers which are much greater than 1," or "the class of beautiful women," or "the class of tall men," do not constitute classes or sets in the usual mathematical sense of these terms. Yet, "the fact remains that such imprecisely defined notions play an important role in human thinking, particularly in the domains of decision-making, abstraction and communication of information" [2]. Few works exist in business intelligence that use fuzzy logic due to certain inherent difficulties of creating such applications, and yet; despite them, such applications are possible and very useful, e.g. the reader can be referred to a review [3]. The challenges include the following. Firstly, not every problem permits trial and error calibration of threshold values. Secondly, the operators, membership

© Springer International Publishing AG 2017
M. Kirikova et al. (Eds.): ADBIS 2017, CCIS 767, pp. 246–256, 2017.
DOI: 10.1007/978-3-319-67162-8_24

functions and inference methods need to have tangible meanings, which can be very context-dependent. Thirdly, fuzzy theory is a borderline discipline with psycholinguistics, which is less objective than formal sciences (such as logic or set theory) and may require yet unavailable knowledge about human cognition. The notion of fuzzieness has distinct understandings and there are important consequences of those discrepancies, e.g. [4] is a review of theoretical models and their empirical validations. The main two types of fuzzy technology are fuzzy knowledge based systems [3] and fuzzy clustering [5]. Our idea is neither of the two, and it aims to implement the above mentioned insight by Zadeh about completing a useful summary from multitudes of data. Fuzzy logic enables us to formulate a natural language interface between big data, numeric analytics, and a manager, hiding the compexity of data and methods and providing him/her with a comprehensible summary. We summarize data using linguistic hedges (very, rather, highly) and formulate queries such as *"Tell me which neighbourhoods are safe to target, if I want more clients but my infrastructure is highly loaded"*. *"Tell me, whether the infrastructure is rather loaded or highly loaded in the region."* Our specific application is targeting different user segments to fill in the spare capacity of the network in a network-friendly manner. In [6], the notion of *Infrastructure-Stressing* (IS) Client was proposed together with the method to reveal such clients from the customer base. Informally, IS clients use the infrastructure in a stressing manner, such as always staying in the zones of high demand, where the antennas are prone to service failures, if further loaded. Being IS is not only a function of the user's qualities, but also of the infrastructure, and of the relative mobility of the rest of the population.

For marketing campaigns geodemographic segmentations (like ACORN or MOSAIC) are used, since it is known how the segments can be targeted to achieve the desired goal, as for example, the promotion of a new mobile service in certain neigbourhoods. The client's home address determines the geodemographic category. People of similar social status and lifestyle tend to live close [7, 8]. Geodemographic segmentation provides a *collective view point*, where the client is seen as a representative of the population who live nearby. However, in recent research, it has been shown that the problem of resource allocation in the zones with nearly overloaded and underloaded antennas is better handled relying on *individual modelling* based on the client's historical trajectories [9]. The authors completed a user segmentation based on clustering of user trajectories and it was demonstrated that network planning is more effective, if trajectory-based segments are used instead of geo-demographic segments. Our aim is to explore the ways to connect the individual trajectory-based view on IS customers and the geodemographic view in order to devise analytics capable to complete the efficient analysis based on the individual view point and yet be useful in marketing campaigns in which geodemographic groups are targeted. As a practical conclusion, we have compiled a ranked list of the segments according to their propensity to contain IS clients and crafted two queries:

1. Which segments contain a low or moderate number of IS clients? (target them, while the infrastructure is still rather underloaded)
2. Which segment is highly devoid of IS clients? (target them, when the customer base becomes mature and the infrastructure becomes increasingly loaded).

The simulation of the resulting fuzzy recommendations guarantees the absence of false negatives, such as, concluding that certain segments are safe to hire from, but in fact that would lead to a service failure at some place in the network.

The rest of the paper is organised as follows. Section 2 describes the data set. In Sect. 3 the proposed methodology is explained. In Sect. 4, the experiments are reported, and finally the conclusions are drawn and discussion is held in Sect. 5.

2 Data Set

The study has been conducted on anonymized geospatial and geodemographic data provided by a Scandinavian telecommunication operator. The data consist of CDRs (Call Detail Records) containing historical location data and calls made during one week in a mid-size region in Sweden with more than one thousand radio cells. Several cells can be located on the same antenna. The cell density varies in different areas and is higher in city centers, compared to rural areas. The locations of 27010 clients are registered together with which cell serves the client. The location is registered every five minutes. During the periods when the client does not generate any traffic, she does not make any impact on the infrastructure and such periods of inactivity are not included in the resource allocation analysis. Every client in the database is labeled with her MOSAIC segment. The fields of the database used in this study are:

- the cells IDs with the information about which users it served at different time points,
- the location coordinates of the cells,
- the time stamps of every event (5 min resolution),
- the MOSAIC geodemographic segment for each client, and
- the Telenor geodemographic segment for each client.

There are 14 MOSAIC segments present in the database; for their detailed description the reader is referred to [15]. The six in-house Telenor segments were developed by Telenor in collaboration with InsightOne, and, to our best knowledge, though not conceptually different from MOSAIC, they are especially crafted for telecommunication businesses.

3   A Link Between IS and Geodemographic Segments

3.1   Notation and Definitions of Fuzzy Logic

Definition (in the style of [2]). A fuzzy set A in X is characterized by a membership function $f_A(x)$, which associates with each point in X a real number in the interval $[0, 1]$, with the value of $f_A(x)$ at x representing the "grade of membership" of x in A. For the opposite quality: $f_{notA}(x) = 1 - f_A(x)$.

Fuzzy membership scores reflect the varying degree to which different cases belong to a set. Under the six value fuzzy set, there are six degrees of membership 1: fully in, $[0.9-1)$: mostly but not fully in, $[0.6-0.9)$: more or less in, $[0.4-0.6)$: more or less out, $[0.1-0.4)$: mostly but not fully out, $[0-0.1)$: fully out. For a comprehensive guide of good practices in fuzzy logic analysis in social sciences the reader is referred to, for example, [10].

Linguistic Hedges:

- *Rather* will be added to a quality A, if the square root of its membership function $f_A(x)^{1/2}$ is close to 1.
- *Very* will be added to a quality A, if the square of its membership function $f_A(x)^2$ is close to 1.
- *Extremely* will be added to a quality A, if $f_A(x)^3$ is close to 1.

The principles for calculating the values of hedged membership functions, for example $f_{veryA}(x) = f_A(x)^2$, are described in [14]. Then, given the new membership function, the same principle applies: the closer to 1, the higher is the degree of membership.

3.2   Query Formulation

To keep the formulations and questions naturally sounding, the word infrastructure-friendly (IF) is used. The quality IF is defined as the opposite to IS: $f_{IF}(segment_i) = 1 - f_{IS}(segment_i)$, for some segment i. As mentioned above, within the same geodemographic segment, the clients differ with respect to the degree of being IS. When the infrastructure is not overloaded, that is, the recent historical load is still significantly smaller than the capacity, then virtually any client is welcome. As the infrastructure becomes more loaded, the operator wants to be more discriminative. We define being "loaded" for an antenna as a fuzzy variable:

$$f_{loaded}(antenna\,j) = max_{all\,t}\{load(j,t) \times capacity(antenna\,j)^{-1}\}.$$

This quality is measured in man units. Being loaded for infrastructure is defined as:

$$f_{loaded}(infrastructure) = max_{all\,antennas\,j}\{f_{loaded}(antenna\,j)\}.$$

Since being loaded is a dangerous quality, we set the strength of the system to be equal to the strength of its weakest component, and for this reason the equation above we use the max operator.

Queries:

1. Which segments to target, provided that *rather* IF users are acceptable clientele?
2. Which segments to target, provided that only *very* IF are wanted?

Depending on the load, there are different rankings of segments. If initially some segments were in the same tier, for example, "very IF segments", some of them fall out of the tier, as the hedge operator is applied and the value of the membership function is squared (for "extremely IF"). The context, when to apply Query 1 or 2, becomes clarified via calculating f_{loaded} (infrastructure) and checking the applicability of different hedges. The method to obtain fuzzy heuristics is summarized to the sequence of the following steps.

Step 1: The IS clients in the customer base are revealed with the method [6] (the algorithm is reproduced as function *reveal_IS* clients in Algorithm 1), and each client is labeled with the IS/notIS descriptor.

Step 2: The propensity of a segment to contain IS clients is defined as the frequency of IS clients among its members and it is calculated from the data:

$$f_{IS}(segment_i) = frequency_{IS}(segment_i)$$

For linguistic convenience the term Infrastructure-Friendly (IF) is introduced and is set to be opposite to IS:

$$f_{IF}(segment_i) = 1 - f_{IS}(segment_i)$$

Step 3: The ranking of segments is carried out with respect to their IF quality and the hedged values of the mebership function are calculated: for all segments $i, f_{rather\ IF}(segment_i), f_{very\ IF}(segment_i), and\ f_{extremely\ IF}(segment_i)$. Given a hedge, which also codes the severity of the context, the segments fall into the different tiers (corresponding to one of the six fuzzy values): "fully in", "mostly but not fully in", "more or less in", and so on.

Step 4: Locally for the region under analysis, the infrastructure is assessed as *loaded, very loaded,* or *extremely loaded,* and thus the severity of the context is assessed. A ranking from Step 3 corresponding to a particular hedge is selected (as a leap of faith further verified in the next section).

The above is depicted as a flow chart in Fig. 1, and formalized as Algorithm 1. The reasoning behind combinatorial optimization is discussed in detail in [11]. The reasoning behind the function *label_ISclients* is discussed in detail in [6].

Algorithm 1: computation of the fuzzy recommendation heuristic.

Variables:
- clientSet: set of with IDs of clients;
- I: the set with geodemographic segments {$segment_1$, ..., $segment_k$};
- D: the mobility data for a region that for each user contain client's ID, client's geodemographic segment, time stamps when the client generated traffic, and which antenna served the client.
- S_i: the number of subscribers that belong to a geodemographic segment i;
- $\Sigma_{all\ i}\ S_{i,t,j}$: the footprint, i.e. the number of subscribers that belong to a geodemographic segment i, at time moment t, who are registered with a particular cell j;
- C_j: the capacity of cell j in terms of how many persons it can safely handle simultaneously;
- x: the vector with the scaling coefficients for the geodemographic segments or other groups such as IS clients;
- x_{IS}: the coeffecient for the IS segment from the vector x;
- $N_{t,j}=$ number of users at cell j at time t;.

Input: data set D: <$user_{ID}$, time stamp t, cell j>.

```
label_ISclients;
for i in I{
        ratherIF[i] = false
        veryIF[i] = false
   extremelyIF[i] = false
        degreeIS = frequency(userID_IS,I)
        degreeIF = 1- degreeIS
        if (degreeIF^{1/2} ≥ 0.9) then ratherIF[i]=true
   if (degreeIF^2 ≥ 0.9) then veryIF[i]=true
   if (degreeIF^3 ≥ 0.9) then extremelyIF[i]=true
}
```

function label_ISclients{

[I. Characterize each user with respect to her relative mobility.]
```
   for each user_ID {
      trajectory_ID = cell_{t1}, ..., cell_{t2016};
      relativeTrajectory_ID = N_{t1,j}, ..., N_{t2016,j};
      sortedTrajectory_ID = sort_{decreas_or.}(relativeTrajectory_ID);
      topHotSpots_ID = Σ_{k=1..100(5%)}sortedTrajectory_ID[k];
```

```
    userTopHotSpots = <userID, topHotSpotsID>
    }
rankedUserList = sortdecreasing_or(rankedUserList)

[II. Initialization.]

        xstressing = 0;
        setStressingUsers = ∅.
```

[III. Reveal the infrastructure-stressing clients.]

```
While (xstressing = 0) do {
    tentativeStressingUsers = head1%(rankedUserList);
    setFriendlyUsers = bottom1%(rankedUserList);
    otherUsers = rankedUserList - tentativeSetStressingUsers -
    setFriendlyUsers;

    [Confirm the tentative labeling via combinatorial optimization.]
    I = {stressing, medium, friendly};
        {xstressing, xmedium, xfriendly} = combinatorial_optimization(I,D);

    IF (xstressing = 0), THEN {
    [take the field userID from tentativeStressingUsers]
    tentativeSetStressingUsers = tentativeStressingUsers< userID >;
    setStressingUsers = setStressingUsers +
    tentativeSetStressingUsers<UserID>

    D = D - Dstressing
} [end of while]

for id in <userIDs> do {
    if (id ∈ setStressingUsers) then label(id,"IS")
    else label(id,"notIS")
} [end loop on id in <userIDs>]
} [end reveal_ISclients]
```

function combinatorial_optimization(I,D){
```
        solve
        Maximize        Σi∈{IF,other,IS} Si xi,
            subject to:

            for all j,t, Σi∈{IF,other,IS} Si,t,j xi ≤ Cj
```
} returns {xIF, xother, xIS}.

Output: array ratherIF[], veryIF[], extremelyIF[].

3.3   Query Simulation

In the above, when deciding which context should be applied, we relied on an intuitive rule: If the infrastructure is *<hedge>* loaded, then *<hedge>* IS segments are suitable to hire clients from. For example, in the case of a *rather* loaded infrastructure, *rather* IS segments are suitable targets. Given the expected success of the campaign, e.g. the campaign can attract 300 new clients or 1500 clients, it is possible to simulate the impact of the expected result on the infrastructure. A warning is thrown, if some antenna is overloaded, i.e. when the expected footprint by the segment violates a restriction for some segment i, at some antenna j, some time moment t:

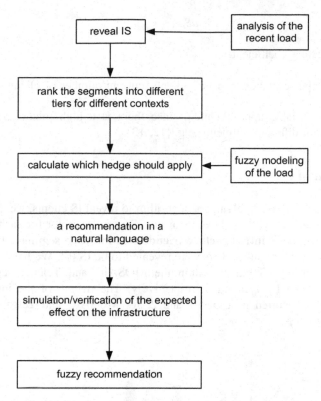

Fig. 1. The flow chart for the calculation of fuzzy recommendation for a marketing campaign.

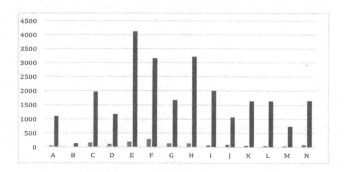

Fig. 2. The number of IS clients in different MOSAIC categories.

$$aS_{i,j,t} \leq C_j,$$

where α is a scaling coefficient:

$$a = \text{ expected number of new clients'(current number of clients)}^{-1}.$$

This is a justifiable approximation, since there is a high predictability in user trajectories within different segments, e.g. [12, 13].

4 Experiment

1. **Reveal the IS clients.** Applying the algorithm to reveal IS clients, we have added a field to the data set with the label IS or not IS as a descriptor for each client.
2. **Calculate degree of infrastructure-friendliness for each segment.** In the whole customer base, 7% of subscribers were revealed to be IS [6]. We have obtained the distribution of the IS clients within the MOSAIC and Telenor segments and depicted them in Figs. 2 and 3, respectively. The degree of the infrastructure-friendliness is reported in Tables 1 and 2, for MOSAIC and Telenor segments, respectively.

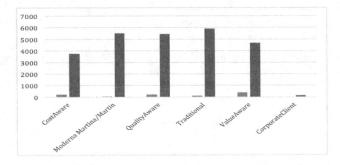

Fig. 3. The number of IS clients in different Telenor segments.

3. **Reasoning behind the queries.** Tables 1 and 2 simulate the reasoning behind the query results for different contexts (codified via a hedge) for the MOSAIC and Telenor segments, respectively. Each of the *14* MOSAIC classes qualifies as *rather IF*, which are those with $f_{IF}(i)^{1/2} > 0.9$. Once the customer base becomes larger and the spare capacity diminishes, only *very IF* will be wanted, which are those with $f_{IF}(i)^2 > 0.9$. Out of those, only 9 segments qualify as *very* IF and five segments qualify as *extremely* IF $(f_{IF}(i)^3 > 0.9)$. The customer population was subjected to the same analysis with respect to Telenor segmentation. As follows from Table 2, each of the six Telenor segments is rather friendly, and there are four and three very and extremely IF segments, respectively.

Table 1. The reasoning behind the query results for the MOSAIC segments.

Segment	fIF(i)	fIF(i)$^{1/2}$	rather IF?	fIF(i)^2	very IF?	fIF(i)^3	extremely IF?
A	0.96	0.97	yes	0.92	yes	0.88	no
B	0.98	0.98	yes	0.96	yes	0.94	yes
C	0.93	0.96	yes	0.86	no	0.79	no
D	0.92	0.95	yes	0.84	no	0.77	no
E	0.96	0.97	yes	0.92	yes	0.88	no
F	0.92	0.95	yes	0.86	no	0.79	no
G	0.93	0.96	yes	0.86	no	0.79	no
H	0.96	0.97	yes	0.92	yes	0.88	no
I	0.97	0.98	yes	0.94	yes	0.91	yes
J	0.92	0.95	yes	0.86	no	0.79	no
K	0.97	0.98	yes	0.94	yes	0.91	yes
L	0.98	0.98	yes	0.96	yes	0.94	yes
M	0.96	0.97	yes	0.92	yes	0.88	no
N	0.95	0.97	yes	0.9	yes	0.85	no

Table 2. The reasoning behind the query results for the Telenor segments.

Segment	fIF(i)	fIF(i)$^{1/2}$	rather IF?	fIF(i)^2	very IF?	fIF(i)^3	extremely IF?
CA	0.94	0.97	yes	0.88	no	0.82	no
MM	0.99	0.89	yes	0.98	yes	0.97	yes
QA	0.96	0.92	yes	0.92	yes	0.88	no
T	0.98	0.87	yes	0.96	yes	0.94	yes
CC	0.92	0.8	yes	0.86	no	0.79	no
VA	0.97	0.91	yes	0.94	yes	0.91	yes

5   Results

When it comes to designing strategies of accommodating many more clients, being IS-prone for a segment is an important quality. We have studied the correlation between IS users and the geo-demographic segments, motivated by the fact that we can target the geo-demographic segments (MOSAIC and Telenor) in marketing campaigns. For different contexts, we have completed candidate rankings of geo-demographic segments, and, given the absence of other preferences, the top-tier segments are preferable. Which ranking out of several candidate ones is taken depends on the hedge calculated for the intensiveness of infrastructure exploitation. The simulation of the expected effect guarantees no false negatives, such as saying that certain segments are safe to hire from, but in fact that would lead to a service failure at some place and time in the network. For the implementation, please check https://sourceforge.net/projects/telenor-user-mobility/?source=navbar.

References

1. Zadeh, L.: Fuzzy logic and beyond - a new look. In: Zadeh, L., King-Sun, F., Konichi, T. (eds.) Fuzzy Sets and Their Applications to Cognitive and Decision Processes: Proceedings of the US-Japan Seminar on Fuzzy Sets and Their Applications, Held at University of California, Berkeley, California, 1–4 July 2014. Academic Press (2014)
2. Zadeh, L.A.: Fuzzy sets. Information and control **8**(3), 338–353 (1965)
3. Meyer, A., Zimmermann, H.J.: Applications of fuzzy technology in business intelligence. Int. J. Comput. Commun. Control **6**(3), 428–441 (2011)
4. Bilgiç, T., Türksen, I.B.: Measurement of membership functions: theoretical and empirical work. Fundam. fuzzy sets **1**, 195–232 (2000)
5. Tettamanzi, A., Carlesi, M., Pannese, L., Santalmasi, M.: Business intelligence for strategic marketing: predictive modelling of customer behaviour using fuzzy logic and evolutionary algorithms. In: Applications of Evolutionary Computing, pp. 233–240 (2007)
6. Sidorova, J., Skold, L., Lundberg, L.: (A) Revealing Infrastructure-Stressing Clients in the Customer Base of a Scandinavian Operator using Telenor Mobility Data and HPI Future SoC Lab Hardware Resources. Hasso Plattner Institute. A Technical report. https://www.researchgate.net/publication/312153463_Optimizing_the_Utilization_in_Cellular_Networks_using_Telenor_Mobility_Data_and_HPI_Future_SoC_Lab_Hardware_Resources
7. Haenlein, M., Kaplan, A.M.: Unprofitable customers and their management. Bus. Horiz. **52**(1), 89–97 (2009)
8. Debenham, J., Clarke, G., Stillwell, J.: Extending geodemographic classification: a new regional prototype. Environ. Planning A **35**(6), 1025–1050 (2003)
9. Sagar, S., Skold, L., Lundberg L., Sidorova, J.: Trajectory segmentation for a recommendation module of a customer relationship management system. In: The 2017 International Symposium on Advances in Smart Big Data Processing (in press). https://www.researchgate.net/publication/316657841_Trajectory_Segmentation_for_a_Recommendation_Module_of_a_Customer_Relationship_Management_System
10. Ragin, C.C., Rihoux, B.: Qualitative Comparative Analysis Using Fuzzy Sets (fsQCA) (2009)
11. Sidorova, J., Skold, L., Rosander, O., Lundberg L.: Discovering insights in telecommunication business from an interplay of geospatial and geo-demographic factors. In: The 1st International Workshop on Data Science: Methodologies and Use-Cases (DaS 2017) at 21st European Conference on Advances in Databases and Information Systems (ADBIS 2017). LNCS, 28–30 August 2017, Larnaca, Cyprus (2017, in press)
12. Song, C., Qu, Z., Blumm, N., Barabási, A.L.: Limits of predictability in human mobility. Science **327**(5968), 1018–1021 (2010)
13. Lu, X., Wetter, E., Bharti, N., Tatem, A.J., Bengtsson, L.: Approaching the limit of predictability in human mobility. Sci. Rep. **3** (2013)
14. Zadeh, L.A.: A fuzzy-set-theoretic interpretation of linguistic hedges (1972)
15. InsightOne MOSAIC lifestyle classification for Sweden, http://insightone.se/en/mosaic-lifestyle/. Accessed 15 Apr 2017

Efficient Data Management for Putting Forward Data Centric Sciences

Genoveva Vargas-Solar[(✉)]

Univ. Grenoble Alpes, CNRS, Grenoble INP, LIG-LAFMIA, 38000 Grenoble, France
genoveva.vargas@imag.fr
http://www.vargas-solar.com

Abstract. The novel and multidisciplinary data centric and scientific movement promises new and not yet imagined applications that rely on massive amounts of evolving data that need to be cleaned, integrated, and analysed for modelling, prediction, and critical decision making purposes. This paper explores the key challenges and opportunities for data management in this new scientific context, and discusses how data management can best contribute to data centric sciences applications through clever data science strategies.

1 Introduction

Data centric sciences are leading to different trendy applications for smart cities, homes and energy; urban computing, monitoring and student assessment from kinder garten to university, automatic health control and monitoring, finances self-management. These applications still provide partial solutions to multi-variable and multi-facet problems related to the understanding of complex systems. For example, the personalization of drugs for better attacking diseases, understanding social behaviours of individuals and societies, modelling and reproducing human skills, and processes like creativity, artistic creation, neuronal behaviour and prediction of natural changes and phenomena. The backbone of these problems are data collections that must be harvested and they must be representative enough and with specific properties so that they can serve as raw material to processing and analysis algorithms. These processes and algorithms can reproduce and understand such phenomena and systems, analyse them, and deduce valid conclusions.

Thus, it is necessary to deal with data collections characterized by their volume, production rate (velocity), variety, multiple veracity, and value. Besides these properties, data collections are also determined by the type of content they group and the conditions in which they are consumed. Paul McFedries in his article Beyond Just Big Data [11] classifies data collections into:

- thick data, which combines both quantitative and qualitative analysis;
- long data, which extends back in time hundreds or thousands of years;
- hot data, which is used constantly, meaning it must be easily and quickly accessible, and

© Springer International Publishing AG 2017
M. Kirikova et al. (Eds.): ADBIS 2017, CCIS 767, pp. 257–266, 2017.
DOI: 10.1007/978-3-319-67162-8_25

- cold data, which is used relatively infrequently, so it can be less readily available.

These types of data collections can be consumed according to different patterns that determine the conditions in which they are accessed:

- shared by several users and accessed in a concurrent manner, with specific isolation or atomicity requirements;
- stored only for archival purposes;
- distributed for promoting parallel data processing and analytics;
- continuous processing for data flows with efficient read/writes.

Cloud and High Performance Computing have evolved to respond to emerging data management and processing challenges introduced by data centric sciences. There is a long path to go through, but the scientific community and industry has started designing the next generation of data management and processing stacks and evolve towards a data science. Data science will deal with the exascales and its associated variety, not in terms of format but in terms of different properties: hot/cold, long/cold, thick data.

On the other hand, the emergence of Big Data some years ago denoted the challenge of dealing with huge collections of heterogeneous data continuously produced and to be exploited through data analytics processes. First approaches have addressed data volume and processing scalability challenges. Solutions have addressed the problem of balancing and scheduling the delivery of computing, storage, memory, and communication resources (bandwidth and reliability) for greedy analytics and mining processes with high in-memory and computing cycles requirements.

Based on these previous results, emerging data centric sciences (e.g., data science, network science, computational science, social electronic sciences, digital humanities) develop methodologies weaving data management, greedy algorithms, and programming models that must be tuned to be deployed in different target computer architectures.

This paper focuses data collections with different uses and access patterns because these properties tend to reach limits of:

- the storage capacity (main memory, cache and disks) required for archiving data collections permanently or during a certain period, and
- the pace (computing speed) in which data must be consumed (harvested, prepared and processed).

These aspects must be addressed when dealing with data centric sciences problems. They are discussed in this paper. Accordingly the remainder of the paper is organized as follows. Section 2 gives an overview of existing approaches addressing data storage infrastructures, data analytics under a data science vision, and infrastructures providing computing capacity for processing data using greedy algorithms. Section 3 profiles data centric sciences and introduces their particular requirements with respect to data as an enabling backbone. It gives the general lines for storing and delivering data and thereby contributing to the efficient execution of data analytics processes running on parallel environments. Finally, Sect. 4 concludes the paper and discusses perspectives.

2 Scaling up Data Processing

The emergence of Big Data has dealt with huge collections of heterogeneous data continuously produced and to be exploited through data analytics processes.

Big Data management and processing are no longer only associated to scientific applications with prediction, analytics requirements, data centric sciences also call for data aware management to address the understanding and automatic control of complex systems, to the decision making in critical and non-critical situations. Given that data centric sciences rely on data collections that stem from complex and diverse gathering processes, two data management issues are key for enabling analysis workflows:

- data storage and distribution, and
- runtime environments that provide enough computing resources for executing greedy data exploration and analytics algorithms.

The next sections discuss these topics.

2.1 Data Storage and Distribution

Data management in many data analytics workflows underlines the need to reduce overhead when data are read, updated, and put into storage supports (memory, cache) as stated by the RUM conjecture proposed in [3]. Several platforms address some aspect of the problem like Big Data stacks [1,6]; data processing environments (*e.g.,* Hadoop, Spark, CaffeonSpark); data stores dealing with the CAP (consistency, availability and partition tolerance) theorem (*e.g.,* NoSQL's); and distributed file systems (*e.g.,* HDFS). The principle is to define API's (application programming interface) to be used by programs to interact with distributed data management, processing, and analytics layers that can cope with distributed and parallel architectures.

Objects and components persistence has been an important issue addressed already by consolidated middleware such as JBOSS and PiJAMA. The new exascale requirements introduced by greedy processes often related to Big Data processing has introduced objects persistence issues again. Particularly for the "passivation" (storage of the execution state and environment of a component) of greedy processes execution in the presence of failures. Instead of loosing costly processes (in time and computing resources) they can be stored and the restarted in without loosing already executed tasks.

In order for exascale and/or Big Data systems to deliver the needed I/O performance, new storage devices such as NVRAM or Storage Class Memories (SCM) need to be included into the storage/memory hierarchy. Given that the nature of these new devices are closer to memory than to storage (low latencies, high bandwidth, and byte-addressable interface) using them as block devices for a file system does not seem to be the best option. DataClay [5] proposes object storage to enable both the programmer, and DataClay, to take full advantage of the coming high-performance and byte-addressable storage devices. Today, given the lack of such devices, DataClay performs a mapping of such abstractions to key-value stores such as Kinetic drives from Seagate[1].

Data structures and associated functions are sometimes more important for some requirements rather than non functional properties like RUM or CAP. Non-relational databases have emerged as solutions when dealing with huge data sets and massive query work load. These systems have been redesigned to achieve scalability and availability at the cost of providing only a reduced set of low-level data management functions, thus forcing the client application to take care of complex logic.

The large spectrum of data persistence and management solutions are adapted for addressing workloads associated with Big Data volumes; and either simple read write operations or with more complex data processing tasks. The challenge today is choosing the right data management combination of tools for variable application requirements and architecture characteristics. Plasticity of solutions is from our point of view the most important property of such tools combination.

Once data has been stored, possibly under a distributed strategy for addressing storage space and data availability, analytics processes can run on top of the whole collection or some sample of it. These processes require very often to be parallelized in order to process the whole data collections in reasonable time. Many data centric science solutions require this type of parallel settings. Thus, there exist different types of parallel runtime environments that enable to deploy analytics processes with different degrees of transparency.

2.2   Parallel Runtime Environments

Today maybe because of the emergence of Big Data and greedy algorithms and applications requiring computing resources, parallel architectures have come back in the arena. There are different kinds of computing, memory, and storage resources providers that adopt their own method for delivering such resources for executing programs. According to [9] there are three categories of resources provision: PaaS frameworks, programming models for computing intensive workloads, and programming models for Big Data.

Platform-as-a-Service (PaaS) layers offer APIs to write applications. For example, in the Microsoft Azure Cloud programming model applications are structured according to roles, which use APIs to communicate (queues) and to access persistent storage (blobs and tables). Microsoft Generic Worker proposes a mechanism to develop a Worker Role that eases the porting of legacy code in the Azure platform [13]. Google App Engine provides libraries to invoke external services and queue units of work (tasks) for execution; furthermore, it

---

[1] http://www.seagate.com.

allows to run applications programmed in the Map-Reduce model. Data transfer and synchronization are handled automatically by the runtime. Environments for computing workload intensive applications use in general the bag of tasks execution model conceiving an application as composed of independent parallel tasks. For example, the Cloud BigJob, Amazon EC2, Eucalyptus, Nimbus Clouds, and ProActive that offer a resource manager developed to mix Cloud and Grid resources [2,12]. Map-Reduce programming is maybe the most prominent programming model for data intensive applications. Map-Reduce based runtime environments provide good performance on cloud architectures above all on data analytics tasks working on large data collections. Microsoft Daytona[2] proposes an iterative Map-Reduce runtime for Windows Azure to support data analytics and machine learning algorithms. Twister [7] is an enhanced Map-Reduce runtime with an extended programming model for iterative Map-Reduce computations. Hadoop [4] is the most popular open source implementation of Map-Reduce on top of the Hadoop Distributed File System (HDFS). The use of Hadoop avoids the lock into a specific platform allowing to execute the same Map-Reduce application on any Hadoop compliant service, as the Amazon Elastic Map-Reduce[3].

For data centric sciences parallel tasks, strategies for dealing with data persistence, efficient read and write in memory operations, data preparation (transformation, cleaning, collocation) must be integrated within the runtime environments. These environments must provide interfaces, either API or annotations to enable programmers to transparently tag their data with their associated properties. It will be up to the runtime environment to provide data management services or integrate existing ones that can ensure efficient data I/O thanks to well adapted operations with underlying processes for providing the right data in the right moment and in the right place.

3 Supporting Data Centric Sciences Experiments

More than ever, researchers in all disciplines find themselves wading through more and more kinds of data. Frequently, there is no standard system for storing, organizing, or analysing this data. Doing so requires a set of skills researchers must largely learn independently. In recent years, data science has emerged as the field that exists at the intersection of math and statistics knowledge, expertise in a science discipline, and so-called hacking skills, or computer programming ability. Yet, data science is more of an emerging interdisciplinary philosophy, a wide-ranging modus operandi that entails a cultural shift in the academic community. The term means something different to every data scientist, and in a time when all researchers create, contribute to, and share information that describes how we live and interact with our surroundings in unprecedented detail, all researchers are data scientists. The philosophy is promising and opens new

---

[2] http://research.microsoft.com/en-us/projects/daytona/.
[3] Amazon elastic Map-Reduce. http://aws.amazon.com/documentation/elasticmap reduce/.

possibilities to develop methodologies that take advantage of information and communication technologies that find themselves new challenges and scientific opportunities in other sciences included in the data science umbrella.

The rapid evolution of computer architectures delivering an increasing amount of resources have opened new opportunities for data centric sciences. As said before, these sciences address complex problems modelled by several related variables that interact as complex systems with plenty of constraints (climate change, traffic control, social behavior in crowds, simulation of biological processes and social phenomena, prediction of diseases). For addressing these kind problems, data centric sciences perform experiments that weave data management, with greedy artificial intelligence and data mining algorithms and computing architectures services.

We consider that efficient data management is the backbone of these experiments. For us, it is no longer pertinent to reason with respect to a set of computing, storage, and memory resources. Instead, it is necessary to conceive algorithms and processes considering an unlimited set of resources usable via a *pay as U go* model, energy consumption or services reputation and provenance models. Rather than designing processes and algorithms considering as threshold the resources availability, computing service providers (e.g. cloud providers) change this vision and rather take into consideration the economic cost of the processes vs. the resources they consume.

It is necessary to develop novel strategies and tools for storing and delivering terabytes of on-line data collections consisting of billions of records, optimizing the consumption of resources while reducing the overhead of exploiting them by parallel programs.

3.1 Data Analytics for Data Science

Methods for querying and mining Big Data are fundamentally different from traditional statistical analysis on small samples. Big Data are often noisy, dynamic, heterogeneous, inter-related, and untrustworthy. Nevertheless, even noisy Big Data could be more valuable than tiny samples. Indeed, general statistics obtained from frequent patterns and correlation analysis usually overpower individual fluctuations and often disclose more reliable hidden patterns and knowledge.

Big Data forces to view data mathematically (e.g., measures, values distribution) first and establish a context for it later. For instance, how can researchers use statistical tools and computer technologies to identify meaningful patterns of information? How shall significant data correlations be interpreted? What is the role of traditional forms of scientific theorizing and analytic models in assessing data? What you really want to be doing is looking at the whole data set in ways that tell you things and answers questions that you are not asking. All these questions call for well-adapted infrastructures that can efficiently organize data, evaluate and optimize queries, and execute algorithms that require important computing and memory resources.

Big Data has enabled the next generation of interactive data analysis with real-time answers. In the future, queries towards Big Data will be automatically generated for content creation on websites, to populate hot-lists or recommendations, and to provide an ad-hoc analysis of data sets to decide whether to keep or to discard them [8]. Scaling complex query processing techniques to terabytes while enabling interactive response times is a major open research problem today.

Analytical pipelines can often involve multiple steps, with built-in assumptions. By studying how best to capture, store, and query provenance, it is possible to create an infrastructure to interpret analytical results and to repeat the analysis with different assumptions, parameters, or data sets. Frequently, it is data exploration and visualization that allow Big Data to unleash its true impact. Visualization can help to produce and comprehend insights from Big Data. Visual.ly, Tableau, Vizify, D3.js, R, are simple and powerful tools for quickly discovering new things in increasingly large datasets.

In parallel to data science addressing (Big) Data under different perspectives, emerges computational science uses advanced computing capabilities to:

- understand and solve complex problems fusing numerical and non-numerical algorithms and modelling and simulation tools;
- computer and information science that develop advanced hardware, software, networking, and data management systems needed to solve computationally demanding problems; and
- computing infrastructure that supports science and engineering problem solving.

In practice, it is the application of computer simulation and other forms of computation from numerical analysis and theoretical computer science to solve problems in various scientific disciplines, for example the one reports in [10] about the application of Computational Science in Archaeology. A collection of problems and solutions in computational science can be found in [14]. Computational science techniques are being applied to perform digital humanities research that are exported for experiment reproducibility. For example, the action Museum 2.0 that explores ways that web 2.0 philosophies can be applied in museum design. Museums share their data collections for digital social computational scientist to perform research and visualize them.

3.2   Data Collections Sharding and Storage for Data Centric Sciences

Data sharding has its origins in centralized systems that had to partition files, either because the file was too big for one disk, or because the file access rate could not be supported by a single disk. Relational distributed databases use data sharding when they place relation fragments at different network sites. Data sharding allows parallel database systems to exploit the I/O bandwidth of multiple disks by reading and writing them in parallel. Relational DBMS implement strategies for sharding data (i.e., tuples): round robin seems appropriate for processes accessing the relation by sequentially scanning all of it on each query, hash-partitioning seems suited for sequential and associative access to data avoiding the overhead of starting queries on multiple disks.

While sharding is a simple concept that is easy to implement, it raises several physical database design issues. Each data collection must have a sharding strategy and a set of disk fragments. Increasing the degree of sharding usually reduces the response time for an individual query and increases the overall throughput of the system. In parallel DBMS for sequential scans, the response time decreases because more processors and disks are used to execute the query; for associative scans, the response time improves because fewer tuples are stored at each node, and hence the size of the index that must be searched decreases.

The NoSQL momentum introduces new datasets storage possibilities. Graph, key-value, multidimensional records stores can provide storage solutions with efficient read/write operations possibilities. Sharding can be also guided by the structure of data, and it can be coped to ad-hoc stores that can ensure persistency and retrieval. This implies also a tuning effort of different NoSQL stores that should then provide efficient that reads/writes with specific degrees of availability, fault tolerance, and consistency. NoSQL stores rely on automatic replication mechanisms. Reads and data querying implemented at the application level are then executed by applying Map-Reduce execution models. Sharded data collections are thus the key to parallel executions.

The NoSQL approach leads to performant ad-hoc per problem solutions. The DBMS approaches, in contrast, promote one-fits all solutions where sharding strategies can be tuned. What is true, it that the pertinence of sharding strategies depends on analysis and processing requirements and available storage and memory resources. In both cases, data collections sharding requires effort, expertise, and a lot of testing for finding the best balance to achieve good data processing and analysis performance. We believe that sharding data collections must be based on clever data organizations and indexing on file systems, data stores, caches, or memories to reduce effort and ensure performant exploitation (retrieval, aggregation, analysis).

Therefore, automatic and elastic data collections sharding toolsare necessary to parametrize data access and exploitation by parallel programs willing to be performant and scale-up in different target architectures:

- Delivering of datasets along distributed process units avoiding bottlenecks.
  - Different data structures used to reduce the overhead associated to read, updates and persistence requests (RUM conjecture) on different target architectures.
  - New strategies and algorithms to enable the transfer of large data sets between distributed processing units. We will explore techniques such as data streaming and process transfer instead of bulk data transfer, as potential solutions.
  - Ensure properties like availability, consistency, and partition tolerance (CAP theorem) that can be reinforced in an adaptable and dynamic manner.

– Assessing and predicting storage (disk, cache, memory) resources for optimally delivering data and scaling up parallel processing.

With the new computing, storage, and memory requirements stemming from data centric science problems, the use of cluster oriented architectures providing such resources has increased and is somehow democratized particularly with the emergence of the cloud. Data management requirements have to be revisited they vary from simple storage coupled with read/write requirements with reasonable data processing performance needs to real critical applications dealing with huge data collections to be processed by greedy algorithms. In such settings it is possible to exploit parallelism for processing data by increasing availability and storage reliability thanks to duplication and collocation.

4 Conclusion and Perspectives

Digging into data today requires data science and data engineering methodologies that can apply information and communication technologies in the best way to address the challenges introduced by the characteristics of data collections (big, continuous, multi-form, and multimedia with different veracity degrees) and that require computing resources in order to be exploited. Yet, data collections are not only a digital artifact, they represent content determined by the conditions in which it has been authored, produced, collected, and digitalized, by its provenance, by the intention behind its exploitation, and by the conditions and rules in which it can be reused. These characteristics have to be exhibited and be accessible to (social) scientists wishing to use them as raw material to perform research and analysis.

It is no longer possible to practice digital social sciences without being supported by data science and engineering to avoid empirical use of technology to maintain and curate data collections. The way algorithms and technology are used has to avoid technological bias pollution of results. If some data processing and cleaning is necessary because of mathematical or technical reasons, this has to be known by the scientist performing the analysis, and it has to be reported in the results obtained. The mathematical and technical conditions in which analysis and visualisation of results are done must be considered in the interpretation of those results. This will ensure the credibility of the experiment performed on digital data using ICT and it will provide an objective perception of the results and interpretations. Data science and engineering do not have sense without having concrete problems with explicit requirements, rules and expectations. The challenge is to have tools that can change their preferences towards data management and analytics and provide elastic strategies for implementing these operations. Such strategies should evolve as data acquire different structures and semantics as a result of the data processing operations applied on them.

Acknowledgement. This work has been partially funded by the project MULTI-POINT, the cooperation contract Clouding Things and the COST EU Actions KEYSTONE.

References

1. Alexandrov, A., Bergmann, R., Ewen, S., Freytag, J.C., Hueske, F., Heise, A., Kao, O., Leich, M., Leser, U., Markl, V., et al.: The stratosphere platform for big data analytics. VLDB J. **23**(6), 939–964 (2014)
2. Amedro, B., Baude, F., Caromel, D., Delbé, C., Filali, I., Huet, F., Mathias, E., Smirnov, O.: An efficient framework for running applications on clusters, grids, and clouds. In: Antonopoulos, N., Gillam, L. (eds.) Cloud Computing, pp. 163–178. Springer, London (2010)
3. Athanassoulis, M., Kester, M., Maas, L., Stoica, R., Idreos, S., Ailamaki, A., Callaghan, M.: Designing access methods: The rum conjecture. In: International Conference on Extending Database Technology (EDBT) (2016)
4. Borthakur, D.: The hadoop distributed file system: Architecture and design. Hadoop Proj. Website **11**(2007), 21 (2007)
5. Cortes, T., Queralt, A., Martí, J., Labarta, J.: DataClay: Towards Usable and Shareable Storage Big Data and Extreme-Scale Computing (BDEC), White paper, pp. 1–3. http://www.exascale.org/bdec/sites/www.exascale.org.bdec/files/whitepapers/dataClay%20at%20BDEC%20Barcelona.pdf
6. Franklin, M.: The berkeley data analytics stack: Present and future. In: 2013 IEEE International Conference on Big Data, pp. 2–3. IEEE (2013)
7. Gunarathne, T., Zhang, B., Tak-Lon, W., Qiu, J.: Scalable parallel computing on clouds using twister4azure iterative mapreduce. Futur. Gener. Comput. Syst. **29**(4), 1035–1048 (2013)
8. Idreos, S., Alagiannis, I., Johnson, R., Ailamaki, A.: Here are my data files. here are my queries. where are my results? In: Proceedings of 5th Biennial Conference on Innovative Data Systems Research, number EPFL-CONF-161489 (2011)
9. Lordan, F., Tejedor, E., Ejarque, J., Rafanell, R., Alvarez, J., Marozzo, F., Lezzi, D., Sirvent, R., Talia, D., Badia, R.M.: Servicess: An interoperable programming framework for the cloud. J. Grid Comput. **12**(1), 67–91 (2014)
10. Marwick, B.: Computational reproducibility in archaeological research: basic principles and a case study of their implementation. J. Archaeol. Method Theor. **24**, 1–27 (2016)
11. McFedries, P.: Beyond just big data, We're all data geeks now. IEEE Spectr. **53**(8), 29 (2015). http://spectrum.ieee.org/computing/software/
12. Peng, J., Zhang, X., Lei, Z., Zhang, B., Zhang, W., Li, Q.: Comparison of several cloud computing platforms. In: 2009 Second International Symposium on Information Science and Engineering (ISISE), pp. 23–27. IEEE (2009)
13. Simmhan, Y., Van Ingen, C., Subramanian, G., Li, J.: Bridging the gap between desktop and the cloud for escience applications. In: 2010 IEEE 3rd International Conference on Cloud Computing (CLOUD), pp. 474–481. IEEE (2010)
14. Steeb, W.H., Hardy, Y., Hardy, A., Stoop, R.: Problems and solutions in scientific computing with C++ and java simulations world scientific publishing (2004)

Towards a Multi-way Similarity Join Operator

Mikhail Galkin[1,2,3](✉), Maria-Esther Vidal[2], and Sören Auer[1,2]

[1] Enterprise Information Systems (EIS), University of Bonn, Bonn, Germany
{galkin,auer}@cs.uni-bonn.de
[2] Fraunhofer Institute for Intelligent Analysis and Information Systems (IAIS),
Sankt Augustin, Germany
vidal@cs.uni-bonn.de
[3] ITMO University, Saint Petersburg, Russia

Abstract. Increasing volumes of data consumed and managed by enterprises demand effective and efficient data integration approaches. Additionally, the amount and variety of data sources impose further challenges for query engines. However, the majority of existing query engines rely on binary join-based query planners and execution methods with complexity that depends on the number of involved data sources. Moreover, traditional binary join operators are not able to distinguish between similar and different tuples, treating every incoming tuple as an independent object. Thus, if tuples are represented differently but refer to the same real-world entity, they are still considered as non-related objects. We propose MSimJoin, an approach towards a multi-way similarity join operator. MSimJoin accepts more than two inputs and is able to identify duplicates that correspond to similar entities from incoming tuples using Semantic Web technologies. Therefore, MSimJoin allows for the reduction of both the height of tree query plans and duplicated results.

Keywords: Semantic data management · Semantic Web · Join operators

1 Introduction

Data integration problems have tackled by the research community for many years. Growing volumes of data and the need for deeper insights have fostered new data integration paradigms. In enterprises, data processing pipelines might involve dozens of data sources which comprise a significant burden on the query processing engines as their query planning mechanisms construct tree-like plans relying on binary join operators. Additionally, an abundance of existing data sources often contains references to the same real-world entities expressed in different formats, terms, schemata, or *RDF vocabularies*. Such a heterogeneity poses a substantial challenge for query processing as few of common join techniques tackle entities equivalence unless such entities are encoded in exactly the same way with matching variables names and respective values.

The increasing adoption of Resource Description Framework (RDF), Linked Open Data, and Semantic Web technologies within enterprises in various

© Springer International Publishing AG 2017
M. Kirikova et al. (Eds.): ADBIS 2017, CCIS 767, pp. 267–274, 2017.
DOI: 10.1007/978-3-319-67162-8_26

domains [10] provides evidences of the maturity of the technology stack, and how this is used to tackle large-scale heterogeneous data integration problems. Furthermore, Semantic Web technologies offer a set of tools to bridge the gap between large-scale heterogeneous data and unified query processing. Specifically, RDF[1] is able to serve as a *lingua franca* model that uniformly expresses a plethora of domain-specific schemata. SPARQL[2] is a query language to extract data from RDF data sources.

In this article, we present an approach that contributes to both outlined problems, i.e., an idea towards a multi-way similarity join operator MSimJoin that is capable of processing more than two input sources and tackle semantic equivalence of the input entities (i.e., that refer to the same real-world object). A join with an arbitrary arity allows for simplify query plans whereas semantic similarity mechanisms deal with data redundancy, thus increasing overall answer completeness. The proposed architecture describes how such components comprise a physical operator and specify inputs and outputs of the operator.

The remainder of this article is structured as follows: Sect. 2 motivates the need in multi-way similarity join operators on an illustrative example. Section 3 gives an overview of the proposed join algorithm. Section 4 analyzes state of the art in federated query processing and similarity joins. Finally, Sect. 5 draws our conclusions and outlines the directions of our future work.

2   Motivating Example

Figure 1 illustrates the intuition behind the idea of a multi-way join plan. Given a complex SPARQL query (cf. Fig. 1a) that consists of ten triple patterns and only one data source is able to answer a particular triple pattern, i.e., the query environment is federated. The task of answering the query presents severe difficulties for existing federated query engines that employ only binary join plans. For instance, some variants of bushy plans (cf. Fig. 1b) and (cf. Fig. 1c) have to perform nine joins that is computationally expensive. On the other hand, employing a multi-way join operator, i.e., 5-way as presented in Fig. 1d, allows for a significant simplification of a query plan. That is, only three joins have to be performed, two multi-way joins that integrate the results of five respective triple patterns each (t1–t5 that share a join variable ?s1 and t6–t10 that share ?a) and one binary join to produce a final answer of t1 triple pattern, i.e., a join between ?s1 and ?a.

Moreover, usually data sources are not aligned among each other and therefore contain replicated and redundant data. For instance, Fig. 2 illustrates the case when three data sources that answer t8, t9, t10 of a query in Fig. 1a return entities, namely <CCR> and <Creedence_Clearwater_Revival>, that refer to the same real-world object. Traditional join operators consider tuples with such entities as completely different and thus, given that ?a is a join variable, do not yield

---

[1] https://www.w3.org/TR/2014/REC-rdf11-concepts-20140225/.

[2] https://www.w3.org/TR/sparql11-query/.

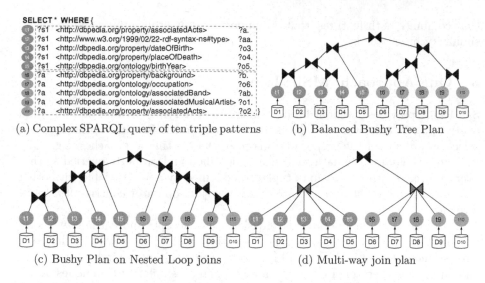

(a) Complex SPARQL query of ten triple patterns

(b) Balanced Bushy Tree Plan

(c) Bushy Plan on Nested Loop joins

(d) Multi-way join plan

Fig. 1. Motivating example. (a) A complex SPARQL query. (b) An example of a bushy tree plan. (c) An example of a nested loop joins. (d) Multi-way plan.

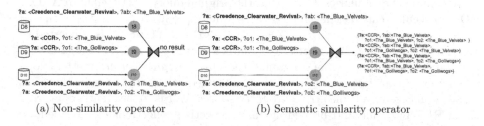

(a) Non-similarity operator

(b) Semantic similarity operator

Fig. 2. Motivating Example (continued). (a) A non-similarity join operator is not able to identify semantically equivalent entities and therefore produces zero results. (b) A semantic similarity operator matches `<CCR>` with `<Creedence_Clearwater_Revival>` and produces four tuples.

any output as `<CCR>` and `<Creedence_Clearwater_Revival>` do no match syntactically. With the growing number of accessible data sources it is hardly ever possible to ensure that only unique entities arrive to the engine.

We call entities *semantically equivalent* (or *equivalent*) if they refer to the same real-world entity but are represented differently. For instance, both notions `<CCR>` and `<Creedence_Clearwater_Revival>` refer to one music band, and thus they are *semantically equivalent*. Furthermore, the task of the join operator when executing against a federation of sources is to tackle equivalent entities and produce valid results. The example of such an operator is presented in Fig. 2b. The operator is able to match values of the syntactically different join variable ?a and yields four results (obtained as a Cartesian product of non-join variable values ?ab, ?o1, ?o2) whereas a non-similarity operator is not able to yield any result (cf. Fig. 2a). This article presents an approach towards such a join operator that combines both features, that is, support for numerous inputs and semantic similarity mechanisms.

3   Our Approach: MSimJoin

SPARQL provides standardized means for querying federations of data sources [3]. In this work the problems of query rewriting and optimal source selection are out of the scope, we thus resort to existing approaches, e.g., [15]. Instead, we assume the data sources are identified and properly queried so that every source returns a stream of tuples to be joined by a join variable with other streams. Therefore, the input state of a n-way join operator consists of n input streams.

A multi-way join operator is envisioned to accepts more than two such streams and produce a join only if an instantiation of a join variable is shared in tuples among all sources. Figure 3 illustrates the intuition behind the proposed similarity join operator. In order to apply semantic similarity techniques, input tuples are converted to the format of *RDF molecules* [5], a semantic data structure that allows for additional annotation of variables and their values in a tuple using common vocabularies. Given that the original query is written in SPARQL, RDF molecules can be constructed using query predicates, e.g., the returned tuple from the data source ?a: <CCR>, ?o1: <The_Golliwogs> is converted to an RDF triple <CCR> dbp:associatedMusicalArtist <The_Golliwogs>.

The operator employs and extends the classical *probe-insert* methodology to support an arbitrary arity and store intermediate results. For instance, as depicted in Fig. 3a, a tuple arrives from the source A. It is subsequently converted to an RDF molecule and probed against respective collections of intermediate results. To produce results as soon as they are ready, a molecule is probed against intermediate results in a specific order, i.e., at first, against a collection $BCDE$ that might contain join results arrived before from B, C, D, and E sources, respectively, i.e., of $n-1$ length. If the tuple from A and a tuple from $BCDE$ share the same join variable instantiation, then the join is found and the result is put into the output collection $ABCDE$. If the join condition is not met, the molecule is probed against collections of $n-2$ length that do not contain tuples from A, namely BCD, BCE, BDE, CDE. Contrary, if the join condition is satisfied, then a new intermediate result is inserted into a respective collection, i.e., $ABCD$, $ABCE$, $ABDE$, $ACDE$. Otherwise, the molecule is probed against collections of $n-3$ length that combine intermediate results from two sources according to Fig. 3. Finally, if all intermediate collections did not satisfy a join condition, the molecule is compared against main collections that accumulate direct RDF molecules from sources B, C, D, E. Any yielded intermediate results are inserted into AB, AC, AD, AE, respectively. The pipeline for a tuple arrived from the source A is depicted in Fig. 3a. A similar pipeline for a tuple arrived from B is depicted in Fig. 3b. The main difference for the source B is in the new probing collections that do not contain B and inserts into respective collections with B, e.g., the intermediate join among B and CDE is inserted into $BCDE$.

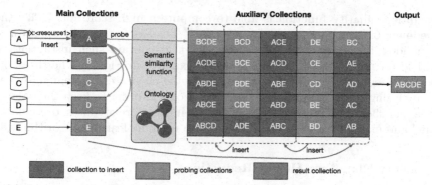

(a) A `resource1` arrives from A. The tuple is probed against the auxiliary tables that do not contain A of length four (BCDE), three (BCD, BCE, etc), two (DE, CE, CD, BE, etc), and against the main collections (B, C, D, E).

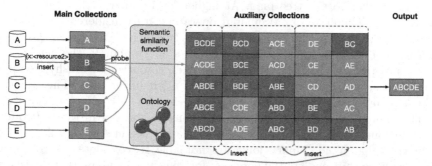

(b) A `resource2` arrives from B. The tuple is probed against the auxiliary tables that do not contain B of length four (ACDE), three (CDE, ADE, etc), two (DE, CE, CD, AE, etc), and against the main collections (A, C, D, E).

Fig. 3. Multi-way similarity join intuition. A resource tuple arrives from one of the source datasets. Auxiliary collections contain intermediate join result obtained from respective sources. The incoming the tuple is semantically probed against a set of auxiliary and main collections. The completed joinable result is put in the output collection (ABCDE). New intermediate results are inserted into a respective collection, e.g., a join between A and BC is stored in ABC.

Moreover, a distinctive feature of the approach consists in the probing mechanism. While traditional join operators search for a full syntactic matching of a join variable instantiation, a similarity join operator performs a semantic comparison of the probed RDF molecules. The operator resorts to semantic similarity functions, for example, *GADES* [13] that are able to deduce a numerical value of relatedness of two molecules given an ontology. An ontology provides additional knowledge and facts, logical axioms, class hierarchies necessary for computation of a similarity score. If the resulting similarity value exceeds a certain threshold then the molecules are considered similar and the join operator processes the given join variable instantiations as the same. For instance, a semantic similarity function is able to compute a high score for <CCR> and <Creedence_Clearwater_Revival> resources. Therefore, the join operator is able to identify a join among the molecules that contain such instantiations and yield a complete result as aimed in the example in Fig. 2b. The threshold value for a semantic similarity function might be chosen either manually, i.e., after analyzing the distribution of similarity scores in order to keep the balance between number of possible candidates and their overall relatedness, or might be learned automatically if the function resorts to machine learning algorithms [8].

3.1   Query Plan Tree Height Reduction

Managing arbitrary arity allows the query optimizer to identify tree plans of lower height. This property of MSimJoin-based plans is illustrated in Fig. 1. MSimJoin joins triple patterns that share the same join variable and pushes the result further to the tree plan. At higher levels binary joins still have to be performed. Therefore, Y MSimJoin operators are considered as Y inputs for binary joins which leads to the following tree height reduction. Given Y MSimJoin operators of arity in range $[n; k]$ in a query plan P, the height of P converges to $\lceil log_2(Y) + 1 \rceil$ in the best case, i.e., comprising a bushy tree plan of MSimJoins, and to Y in the worst case of a left-linear plan. The resulting height h is thus lying in the range $\lceil log_2(Y) + 1 \rceil \leq h \leq Y$. Note that arity of MSimJoin operators does not affect the resulting height. For instance, two five-way MSimJoins in Fig. 1d comprise a bushy plan of height $h = \lceil log_2 2 + 1 \rceil = 2$.

On the other hand, binary joins are not able to complete the join of Y $[n; k]$ sources within one step. Therefore, for binary join plans there exists an additional sub-tree that increases the tree plan height, i.e., the sub-tree of $[n; k]$ sources that share the same join variable. In the best case of a bushy tree plan (cf. Fig. 1b) the additional height h_a converges to $h_a = \lceil log_2(n) \rceil$. In the worst case of a left linear plan the additional height converges to $h_a = k - 1$ as depicted in Fig. 1c. Having in mind that the first layer when computing MSimJoin tree height is already included in h_a the overall height of a binary plan tree equals to $h = \lceil log_2(n) \rceil + \lceil log_2(Y) \rceil$ in the best case and $h = k + Y - 2$ in the worst case when both additional and main components comprise a left linear join plan. For instance, a bushy tree-like variant of a plan in Fig. 1b has a total height of $\lceil log_2 5 \rceil + \lceil log_2 2 \rceil = 4$, whereas the height of another bushy tree-like option with a left linear additional component in Fig. 1c equals to $h = 5 - 1 + \lceil log_2 2 \rceil = 5$.

Clearly, application of MSimJoin is able to reduce the query plan tree height.

4   Related Work

Existing federated query engines [1,2,11,14] that process SPARQL queries employ various binary join algorithms and construct query plans complying with the binary joins. *ANAPSID* [2] is an adaptive query engine that reacts on possible data source delays or blocks; it is able to adjust query execution schedulers accordingly. *ANAPSID* employs a range of binary join operators and defines two own operators, namely *agjoin* and *adjoin*. *FedX* [11] collects metadata about the given source sending auxiliary SPARQL queries. *FedX* resorts to nested loop binary joins as well. *TPF* [14] introduces an original web access interface based on direct triple patterns requests and build query plans based on nested loop binary joins. *nLDE* [1] extends a TPF query mechanism involving a binary *gjoin* operator in the query processing. All the approaches employ binary join operators, whereas our approach supports a multi-way setup with numerous inputs.

On the other hand, applications of semantic similarity functions in SPARQL query processing received little attention. Instead, there exist techniques that apply string similarity algorithms [4,6,16], set similarity algorithms [7,9], and graph similarity techniques [12,17]. However, such algorithms have never been applied to SPARQL. In contrast, our approach resorts to semantic similarity mechanisms in order to deduce relatedness between join variable instantiations, and thus achieving higher query completeness.

5   Conclusion

We presented an approach towards a multi-way similarity join operator, MSimJoin, that tackles data duplication and redundancy by applying semantic similarity mechanisms. The operator is able to both support numerous input streams and compute similarity scores among received resources. The future work aims at implementing a physical multi-way semantic similarity join operator and evaluating it against state of the art benchmarks and real RDF datasets.

References

1. Acosta, M., Vidal, M.-E.: Networks of linked data eddies: an adaptive web query processing engine for RDF data. In: Arenas, M., et al. (eds.) ISWC 2015. LNCS, vol. 9366, pp. 111–127. Springer, Cham (2015). doi:10.1007/978-3-319-25007-6_7
2. Acosta, M., Vidal, M.-E., Lampo, T., Castillo, J., Ruckhaus, E.: ANAPSID: an adaptive query processing engine for sparql endpoints. In: Aroyo, L., Welty, C., Alani, H., Taylor, J., Bernstein, A., Kagal, L., Noy, N., Blomqvist, E. (eds.) ISWC 2011. LNCS, vol. 7031, pp. 18–34. Springer, Heidelberg (2011). doi:10.1007/978-3-642-25073-6_2
3. Buil-Aranda, C., Arenas, M., Corcho, O., Polleres, A.: Federating queries in SPARQL1.1: syntax, semantics and evaluation. Web Semant. Sci. Serv. Agents World Wide Web **18**, 1–17 (2013)
4. Feng, J., Wang, J., Li, G.: Trie-join: a trie-based method for efficient string similarity joins. VLDB J. **21**(4), 437–461 (2012)
5. Fernández, J.D., Llaves, A., Corcho, O.: Efficient RDF interchange (ERI) format for RDF data streams. In: Mika, P., et al. (eds.) ISWC 2014. LNCS, vol. 8797, pp. 244–259. Springer, Cham (2014). doi:10.1007/978-3-319-11915-1_16
6. Li, G., Deng, D., Wang, J., Feng, J.: Pass-join: a partition-based method for similarity joins. PVLDB **5**(3), 253–264 (2011)

7. Mann, W., Augsten, N., Bouros, P.: An empirical evaluation of set similarity join techniques. PVLDB **9**(9), 636–647 (2016)
8. Morales, C., Collarana, D., Vidal, M.-E., Auer, S.: MateTee: a semantic similarity metric based on translation embeddings for knowledge graphs. In: Cabot, J., Virgilio, R., Torlone, R. (eds.) ICWE 2017. LNCS, vol. 10360, pp. 246–263. Springer, Cham (2017). doi:10.1007/978-3-319-60131-1_14
9. Ribeiro, L.A., Cuzzocrea, A., Bezerra, K.A.A., Nascimento, B.H.B.: Incorporating clustering into set similarity join algorithms: the *SjClust* framework. In: Hartmann, S., Ma, H. (eds.) DEXA 2016. LNCS, vol. 9827, pp. 185–204. Springer, Cham (2016). doi:10.1007/978-3-319-44403-1_12
10. Schmachtenberg, M., Bizer, C., Paulheim, H.: Adoption of the linked data best practices in different topical domains. In: Mika, P., et al. (eds.) ISWC 2014. LNCS, vol. 8796, pp. 245–260. Springer, Cham (2014). doi:10.1007/978-3-319-11964-9_16
11. Schwarte, A., Haase, P., Hose, K., Schenkel, R., Schmidt, M.: FedX: optimization techniques for federated query processing on linked data. In: Aroyo, L., Welty, C., Alani, H., Taylor, J., Bernstein, A., Kagal, L., Noy, N., Blomqvist, E. (eds.) ISWC 2011. LNCS, vol. 7031, pp. 601–616. Springer, Heidelberg (2011). doi:10.1007/978-3-642-25073-6_38
12. Shang, Z., Liu, Y., Li, G., Feng, J.: K-join: knowledge-aware similarity join. IEEE Trans. Knowl. Data Eng. **28**(12), 3293–3308 (2016)
13. Traverso, I., Vidal, M.-E., Kämpgen, B., Sure-Vetter, Y.: Gades: a graph-based semantic similarity measure. In: SEMANTiCS, pp. 101–104. ACM (2016)
14. Verborgh, R., Sande, M.V., Hartig, O., Herwegen, J.V., Vocht, L.D., Meester, B.D., Haesendonck, G., Colpaert, P.: Triple pattern fragments: a low-cost knowledge graph interface for the web. J. Web Sem. **37–38**, 184–206 (2016)
15. Vidal, M.-E., Castillo, S., Acosta, M., Montoya, G., Palma, G.: On the selection of SPARQL endpoints to efficiently execute federated SPARQL queries. In: Hameurlain, A., Küng, J., Wagner, R. (eds.) Transactions on Large-Scale Data- and Knowledge-Centered Systems XXV. LNCS, vol. 9620, pp. 109–149. Springer, Heidelberg (2016). doi:10.1007/978-3-662-49534-6_4
16. Wandelt, S., Deng, D., Gerdjikov, S., Mishra, S., Mitankin, P., Patil, M., Siragusa, E., Tiskin, A., Wang, W., Wang, J., Leser, U.: State-of-the-art in string similarity search and join. SIGMOD Rec. **43**(1), 64–76 (2014)
17. Wang, Y., Wang, H., Li, J., Gao, H.: Efficient graph similarity join for information integration on graphs. Front. Comput. Sci. **10**(2), 317–329 (2016)

Workload-Independent Data-Driven Vertical Partitioning

Nikita Bobrov[1], George Chernishev[1,2(✉)], and Boris Novikov[1,2]

[1] Saint Petersburg University, Saint Petersburg, Russia
nikita.v.bobrov@gmail.com, chernishev@gmail.com, borisnov@acm.org
[2] JetBrains Research, Prague, Czech Republic
http://www.math.spbu.ru/user/chernishev/

Abstract. Vertical partitioning is a well-explored area of automatic physical database design. The classic approach is as follows: derive an optimal vertical partitioning scheme for a given database and a workload. The workload describes queries, their frequencies, and involved attributes.

In this paper we identify a novel class of vertical partitioning algorithms. The algorithms of this class do not rely on knowledge of the workload, but instead use data properties that are contained in the workload itself. We propose such algorithm that uses a logical scheme represented by functional dependencies, which are derived from stored data. In order to discover functional dependencies we use TANE — a popular functional dependency extraction algorithm. We evaluate our algorithm using an industrial DBMS (PostgreSQL) on number of workloads. We compare the performance of an unpartitioned configuration with partitions produced by our algorithm and several state-of-the-art workload-aware algorithms.

Keywords: Physical design · Vertical partitioning · Functional dependency

1 Introduction

Vertical partitioning is a technique which aims to improve a database performance by reducing the amount of data to be read from disk. This problem was of interest to database community since the earliest days. Despite being old, the problem of automatic selection of a vertical partitioning scheme is not yet solved. A database administrator still plays the critical role in the selection of this type of data layout.

However, a significant amount of experience was accumulated, a large number of approaches was developed. In overall there were dozens of studies published. This number of studies is justified by the fact that an optimal solution is an NP-hard problem for many different formulations of the vertical partitioning

This work is partially supported by Russian Foundation for Basic Research grant 16-57-48001.

M. Kirikova et al. (Eds.): ADBIS 2017, CCIS 767, pp. 275–284, 2017.
DOI: 10.1007/978-3-319-67162-8_27

problem [2,20,29]. The number of different vertical partitioning schemes for a single table is equal to the Bell number of attributes [15]. Thus, the algorithms proposed in these studies are approximate.

All studies use one and the same problem formulation: for a given database and a given workload find the vertical partitioning scheme that optimizes some performance metric. The metric is usually represented by a response time or a throughput and the database is represented by a single table or a group of tables. The workload is the information regarding queries, their frequencies, involved attributes, arrival patterns, and so on.

However, there are scenarios when vertical partitioning is required, but queries are not known, or known but can not be used for tuning. These scenarios lead to a novel class of vertical partitioning algorithms, that has the following properties:

Workload-independence states that the algorithm does not rely on the knowledge of a workload. There are several cases [13] when a workload is not available or hard to identify, for example proxy database caching [22]. Another example is the case when there are errors in physical parameter estimations that make the knowledge of the workload useless.

Data-driven property states that the algorithm pays special attention to the stored data. The data contains semantic information that can be exploited for vertical partitioning scheme selection. The semantic information is largely ignored in the existing works on automatic physical database tuning. The only semantic information that is commonly employed is the one discovered and used by a query optimizer during "what-if" invocations. However, it is employed only during configuration assessment and the algorithms do not make use of it directly.

The goal of this study is to construct a workload-independent data-driven vertical partitioning algorithm. The core idea of our approach is to exploit the logical scheme of a database by extracting the functional dependencies from database tables and use them to perform partitioning.

In our previous study [4] we have evaluated TANE — an algorithm for functional dependency extraction. Unlike other evaluations we studied precision-related metrics of this algorithm. We have demonstrated algorithm suitability for the construction of the workload-independent data-driven advisor — the results were surprisingly good for the approximate algorithm. In this paper we continue our work and use TANE for a functional dependency extraction stage of our partitioning algorithm.

The contributions of this paper are the following:

- We identify a novel class of vertical partitioning algorithms that does not include a workload as its component. We justify its existence by describing several use-cases when such an algorithm will be useful.
- We develop a vertical partitioning algorithm which uses functional dependencies embedded in data to produce a partitioning scheme.
- We perform an evaluation of our approach by comparing our algorithm not only with unpartitioned configurations, but also with several state-of-the-art workload-aware algorithms.

2   Related Work

There are dozens of vertical partitioning studies. All of them followed the same approach: for a given database and a workload derive an optimal vertical partitioning scheme.

One can classify these works into two types [7,8]:

1. In heuristic approach [6,17,20,24] some procedure is used to construct the partitioning scheme. This procedure does not use any cost-function, instead it uses some assumptions regarding the solution. This approach was popular in the 80s-90s, then it was abandoned in favor of the cost-based one due to the low quality of solutions. However, recently appeared a number of promising applications [12,13], like an on-line vertical partitioning [19,28], where a database adapts its vertical partitioning schema on the fly.
2. In cost-based approach [9,15,21,25,26] a cost function and an enumeration algorithm are devised and used to construct a partitioning scheme. The cost function uses various workload and hardware parameters to estimate benefits of the candidate partitioning scheme. Currently, this is the most popular approach due to quality and reliability of results. The most active subarea is the vertical partitioning for in-memory systems [14,16].

Several studies [1,23] also adopt mixed approach, where poor configurations are pruned by heuristics followed by a cost-based enumeration. Initially, the popularity of heuristic and mixed studies was justified by the tiny computational resources available in the 80s. Nevertheless, some of the recent studies [1] still employ it.

We do not discuss these approaches in detail due to space constraints, we refer an interested reader to surveys [7,8].

Another physical design aspect that needs to be mentioned is usage of logical scheme information. Currently, this aspect is largely ignored by the physical design research community. There is only a handful of studies which employ it [5,10,26,27,30] (see a more detailed survey in [4]). At the same time, exploitation of the information hidden in data may allow to significantly improve quality of the obtained configurations.

Finally, a recent study [18] presents a comparison of several classic vertical partitioning algorithms. Authors conclude with the following — "HillClimb is the best algorithm for disk-based systems". In our paper we rely on findings of this comparison: we use HillClimb [16] and Navathe's algorithms [23] for comparison of our algorithm with the workload-aware approach.

3   Heuristic Affinity-Based Vertical Partitioning

One of the most popular heuristic approaches is the affinity-based one [7], which is as follows:

1. Construct an Attribute Usage Matrix (AUM) using a workload. The AUM is a matrix which records usage of each attribute by each query. It stores zero on (i, j) position if the attribute i is not used in the query j and one otherwise.

2. Transform the AUM into Attribute Affinity Matrix (AAM), a symmetric square matrix. The elements of this matrix denote the degree of closeness between two attributes. The transformation algorithm is specific to each study.
3. Generate a vertical partitioning scheme from the AAM using a matrix clustering [6,17,23], graph-based approaches [20,24], and others [7].

This approach was used not only for vertical partitioning in relational databases, but also for various physical design problems [3,11]. In this study we use mined functional dependencies inside the affinity-based approach.

4   Partitioning Algorithm

The algorithm is presented in the Listing 1. Its input is the AAM built using the functional dependencies extracted from the source data (a single table). To extract functional dependencies we use the TANE algorithm, which returns a set of functional dependencies in the form of $LHS \rightarrow RHS$ (left and right hand sides). TANE presents output in such a way that RHS consists only of one attribute and LHS may contain several attributes. The algorithm is controlled using the $FdDepth$ parameter which specifies the maximum number of attributes that are allowed to appear in the LHS.

In order to construct $F(N, N)$ — the AAM, firstly, functional dependencies are extracted. Next, for each pair of attributes (i, j) the number of functional dependencies such that $i, \ldots \rightarrow j$ is calculated.

The lines 4–9 are used to detect attributes which can be considered key candidates. These attributes would be duplicated in each A_k and used to reconstruct the original relation. In order to detect such attributes the algorithm scans the whole column and checks whether it contains all zeroes. In other words it selects all attributes which do not appear in RHS in the found dependencies.

Then, we begin to construct partitions in the loop (lines 12–29), until all attributes are used. On each iteration we start with the highest affinity (the most "frequent") among all unused attributes, i.e. $max(F(N, N))$. We put in the $MaxRelatioIndex$ RHS of this affinity. Then we look at $Top(FdDepth + 1)$ attributes in a sorted $MaxRelatioIndex$ column and add them to $Relation_r$. Consider this set of attributes as a tree, where $MaxRelatioIndex$ is the root, and $MostFrequent$ are its children.

Next, for each of this child (lines 17–23), the same search and addition of potential children is conducted, but only for $Top(2)$ attributes. We also record all $NonFrequent$ attributes, which were not added to the relation, but their ancestors are in it. After all children of the current $MaxRelatioIndex$ were considered and the partition is almost constructed, we check whether the $Top(1)$ of each $NonFrequent$ attribute is in the relation (lines 25–27), and if this is true — we add $NonFrequent$ to the relation.

Algorithm 1. FD-based algorithm

Data: Affinity matrix $F(N, N)$, where N – number of attributes in relation R;
$F(i, j)$ – frequency of attribute i in functional dependencies with i in
LHS and j in RHS, e.g. $i, \ldots \to j$

Result: Set of partitions $A_1, \ldots, A_n$: $\bigcup\limits_{1 \leq k \leq n} A_k = R$ and $\forall i, j A_i \bigcap A_j = \varnothing$

```
 1 begin
 2 │   UsedAtts ⟵ {}, CandidateKeyAtts ⟵ {};
 3 │   Key ⟵ True;
 4 │   for i = 1 : N do
 5 │   │   for j = 1 : N do
 6 │   │   │   if F(j, i)! = 0 then
 7 │   │   │   │   Key ⟵ False;
 8 │   │   │   │   break;
   │   │   │   end
   │   │   end
 9 │   │   if(Key) then CandidateKeyAtts ⟵ CandidateKeyAtts ∪ {i};
   │   end
10 │   UsedAtts ⟵ AllRelationAtts \ CandidateKeyAtts;
11 │   r ⟵ 0;
12 │   while SizeOf(UsedAtts) < N do
13 │   │   MaxRelatioIndex ⟵ max(F(N, N));
14 │   │   MostFrequent ⟵ Top(FdDepth + 1) of MaxRelatioIndex column in
   │   │   F(N, N);
15 │   │   Relation_r ⟵ MostFrequent ∪ MaxRelatioIndex;
16 │   │   NonFrequent ⟵ {};
17 │   │   for i ∈ MostFrequent do
18 │   │   │   Nearest_i ⟵ Top(2) of i column in F(N, N);
19 │   │   │   for j ∈ Nearest_i do
20 │   │   │   │   if j ∈ Relation_r then
21 │   │   │   │   │   Relation_r ⟵ Relation_r ∪ {j};
22 │   │   │   │   │   UsedAtts ⟵ UsedAtts ∪ {j};
   │   │   │   │   else
23 │   │   │   │   │   NonFrequent ⟵ NonFrequent ∪ {j};
   │   │   │   │   end
   │   │   │   end
   │   │   end
24 │   │   for k ∈ NonFrequent do
25 │   │   │   if Top(1)ofF(N, k) ∈ Relation_r then
26 │   │   │   │   Relation_r ⟵ Relation_r ∪ {k};
   │   │   │   end
   │   │   end
27 │   │   Print(Relation_r);
28 │   │   r ⟵ r + 1;
   │   end
   end
```

---

5   Evaluation

In order to evaluate our approach we implemented the proposed algorithm and used the PostgreSQL DBMS to evaluate the performance of partitioning schemes. In our experiments an existing[1] implementation of the TANE algorithm was used.

5.1   Experimental Setup and Evaluation Procedure

In our experiments the following hardware and software configuration was used: Intel(R) Core(TM) i5-4670K CPU 3.40 GHz (4 CPUs) RAM 16 GB, Ubuntu Linux 16.04 (64 bit), PostgreSQL 9.5.6.

To prepare the experiments we had to select benchmarks which provide not only data, but also queries. The following two benchmarks were used:

1. TPC-H — lineorder (primary) and lineitem, orders, customer, nation; synthetic; 150K rows; 6 queries.
2. IOWA Liquor[2]: single table; real; 2.7 million rows; 22 attributes; 6 queries.

TPC-H is a well-known benchmark which offers both data and queries. However, it already possess a well-designed data scheme that does not need vertical partitioning. The queries in this benchmark also fit this scheme and, thus, it may be impossible to achieve any improvement at all.

In order to address this problem we have also selected the IOWA Liquor benchmark. Unfortunately, this benchmark contains only data, but not queries. Thus, we had to collect and adapt queries employed by users of the "bigquery" subreddit[3]. However, we were not able to translate all the queries to PostgreSQL. The resulting queries and their descriptions are presented on the website[4].

The evaluation procedure was performed as follows:

1. Run our partitioning algorithm and populate the obtained vertical partitioning scheme with data. Note, that we do not use queries on this stage.
2. Using populated scheme run benchmark queries and record execution times.
3. Compare these times with:
   (a) the ones obtained on unpartitioned configuration;
   (b) the ones obtained on configurations generated by the workload-aware algorithms (HillClimb and Navathe's);

Both experiments were conducted with the *FdDepth* parameter set to 4, an empirically chosen value.

---

[1] TANE implementation. http://www.cs.helsinki.fi/research/fdk/datamining/tane/.

[2] Iowa Liquor, https://data.iowa.gov/Economy/Iowa-Liquor-Sales/m3tr-qhgy.

[3] https://www.reddit.com/r/bigquery/comments/37fcm6/iowa_liquor_sales_dataset_879mb_3million_rows/?st=j3ppu30u\&sh=35bdeeb2.

[4] http://www.math.spbu.ru/user/chernishev/papers/iowa_queries.txt.

5.2 Results

Our first series of experiments used the IOWA Liquor dataset. The results are presented in the Table 1 and the Fig. 1. Note the logarithmic axis in the Fig. 1. We also used the bold font in our tables to indicate the best algorithm for every query. The table presents query run times and the overall workload run time measured in milliseconds. The column "Improvement" shows the relative improvement of the partitioning scheme for the whole workload compared to the baseline — the "Original" configuration.

There are six queries in this benchmark, in our evaluation each query was benchmarked individually. From this experiment you can see that for every query any partitioned configuration performs better than the unpartitioned one.

Another observation is the following: the cost-based workload-aware algorithms perform better than our algorithm. This is not surprising, since these algorithms employ a workload and a cost model. Though, it is interesting to note that the workload-aware algorithms outperform our algorithm 1.5–2 times. The outcome of this experiment also corroborates Jindal et al. results [18]: Hill-Climb algorithm is superior to Navathe's.

The second series of experiments involved the TPC-H benchmark. You can see the results in the Table 2 and the Fig. 2. Here we also use the logarithmic axis.

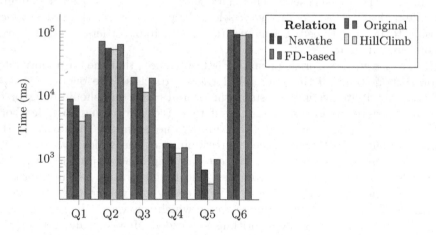

Fig. 1. Query time execution, Iowa Liquor

Table 1. IOWA results

Relation	Q1	Q2	Q3	Q4	Q5	Q6	Overall	Run Time	Improvement
Original	8325	69544	18568	1658	1100	105168	204363	—	—
Navathe's	6574	53669	12681	1635	635	90285	165479	30671	19.0%
HillClimb	**3729**	**51550**	**10736**	**1167**	**382**	**87356**	154920	47	24.2%
FD-based	4721	61996	18118	1436	930	90297	177498	21	13.1%

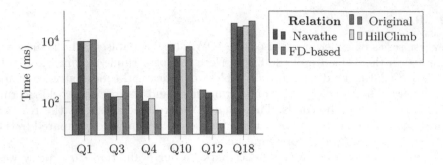

Fig. 2. Query time execution, TPC-H

Table 2. TPC-H results

Relation	Q1	Q3	Q4	Q10	Q12	Q18	Overall	Run Time	Improvement
Original	**414**	182	320	7123	231	35885	44155	—	—
Navathe's	9248	**141**	101	**2993**	185	**27232**	39900	44157	9.3%
HillClimb	9289	142	122	3004	51	29643	42251	128	4.3%
FD-based	10614	323	**51**	6125	**18**	42192	59323	190	-34.3%

The TPC-H is a synthetic benchmark with several of its queries spanning multiple tables. In this experiment we select the primary table (lineorder) which is going to be partitioned. Secondary tables (lineitem, orders, customer, and nation) are left intact.

This series yields surprising results. First of all, note that the Q1 performance dropped more than 20 times for all of the approaches. The queries Q3, Q4, and Q12 contribute negligible costs in the overall result and should be ignored. The workload-aware algorithms managed to reduce processing costs for both Q10 and Q18 — the most resource-consuming queries in the workload. At the same time our algorithm produced partitioning scheme which was only 14% better than the original one, while the cost-based approaches produced the 57% improvement. The query Q18 produced a discouraging result for our approach, it increased the processing times for the most costly query almost 7%. In overall, our algorithm produced partitioning scheme which was worse than the original one, while the workload-aware algorithms succeeded. However, another surprise was the superiority of Navathe's algorithm over the HillClimb: it achieved twice the improvement than the HillClimb.

We have also evaluated the time it takes to produce a solution by each of the considered algorithms. It is presented in the "Run Time" column in Tables 1 and 2. Here, we can see that the HillClimb and our algorithms are approximately equal. Moreover, this time is negligible compared to the run times of the workload. On the other hand, Navathe's algorithm exhibits significant run times for both benchmarks.

6   Conclusions

In this paper we have introduced a notion of the workload-independent data-driven vertical partitioning. We have identified the new class of vertical partitioning algorithms, that possess these properties. We have outlined scenarios when such algorithms can be of use. Next, we have described an algorithm belonging to this class. It employs functional dependencies mined from the data to construct a partitioning scheme. Finally, we have performed an evaluation using the PostgreSQL DBMS for both real and synthetic benchmarks.

Our algorithm succeeded in generating beneficial partitioning scheme for the benchmark that had no predefined data schema. As expected the algorithm produces solutions of slightly worse quality than that of the workload-aware ones. However, it has much wider application area: it can operate in some scenarios when workload-aware algorithms can not, i.e. when no workload knowledge is available.

Acknowledgments. We would like to thank anonymous reviewers for their valuable comments on this work. This work is partially supported by Russian Foundation for Basic Research grant 16-57-48001.

References

1. Agrawal, S., Narasayya, V., Yang, B.: Integrating vertical and horizontal partitioning into automated physical database design. In: SIGMOD 2004, pp. 359–370. ACM, 2004

2. Apers, P.M.G.: Data allocation in distributed database systems. ACM Trans. Database Syst. (TODS) **13**(3), 263–304 (1988)

3. Bellatreche, L., Benkrid, S.: A joint design approach of partitioning and allocation in parallel data warehouses. In: Pedersen, T.B., Mohania, M.K., Tjoa, A.M. (eds.) DaWaK 2009. LNCS, vol. 5691, pp. 99–110. Springer, Heidelberg (2009). doi:10. 1007/978-3-642-03730-6_9

4. Bobrov, N., Chernishev, G., Grigoriev, D., Novikov, B.: An evaluation of TANE algorithm for functional dependency detection. In: Ouhammou, Y., et al. (eds.) MEDI 2017. LNCS, vol. 10563, pp. 208–222. Springer International Publishing, Cham (2017). doi:10.1007/978-3-319-66854-3_16

5. Boehm, A.M., Seipel, D., Sickmann, A., Wetzka, M.: Squash: a tool for analyzing, tuning and refactoring relational database applications. In: Seipel, D., Hanus, M., Wolf, A. (eds.) INAP/WLP -2007. LNCS (LNAI), vol. 5437, pp. 82–98. Springer, Heidelberg (2009). doi:10.1007/978-3-642-00675-3_6

6. Cheng, C.-H.: A branch and bound clustering algorithm. IEEE Trans. Syst. Man Cybern. **25**, 895–898 (1995)

7. Chernishev, G.: A survey of dbms physical design approaches. SPIIRAS Proceedings **24**, 222–276 (2013)

8. Chernishev, G.: The design of an adaptive column-store system. J. Big Data **4**(5), 25 (2017)

9. Cornell, D., Yu, P.: An effective approach to vertical partitioning for physical design of relational databases. IEEE Trans. SE **16**, 248–258 (1990)

10. De Marchi, F., Lopes, S., Petit, J.-M., Toumani, F.: Analysis of existing databases at the logical level: the DBA companion project. SIGMOD Rec. **32**, 47–52 (2003)
11. Fung, C.-W., Karlapalem, K., Li, Q.: Cost-driven vertical class partitioning for methods in object oriented databases. VLDB J. **12**, 187–210 (2003)
12. Galaktionov, V., Chernishev, G., Novikov, B., Grigoriev, D.: Matrix clustering algorithms for vertical partitioning problem: an initial performance study. In: DAMDID/RCDL 2016, Russia, pp. 24–31 (2016)
13. Galaktionov, V., Chernishev, G., Smirnov, K., Novikov, B., Grigoriev, D.A.: A study of several matrix-clustering vertical partitioning algorithms in a disk-based environment. In: Kalinichenko, L., Kuznetsov, S.O., Manolopoulos, Y. (eds.) DAMDID/RCDL 2016. CCIS, vol. 706, pp. 163–177. Springer, Cham (2017). doi:10.1007/978-3-319-57135-5_12
14. Grund, M., Krüger, J., Plattner, H., Zeier, A., Cudre-Mauroux, P., Madden, S.: HYRISE: a main memory hybrid storage engine. Proc. VLDB Endow. **4**, 105–116 (2010)
15. Hammer, M., Niamir, B.: A heuristic approach to attribute partitioning. In: SIGMOD 1979, pp. 93–101 (1979)
16. Hankins, R.A., Patel, J.M.: Data morphing: an adaptive, cache-conscious storage technique. In: VLDB 2003, pp. 417–428 (2003)
17. Hoffer, J.A., Severance, D.G.: The use of cluster analysis in physical data base design. In: VLDB 1975, pp. 69–86 (1975)
18. Jindal, A., Palatinus, E., Pavlov, V., Dittrich, J.: A comparison of knives for bread slicing. Proc. VLDB Endow. **6**, 361–372 (2013)
19. Li, L., Gruenwald, L.: SMOPD: a vertical database partitioning system with a fully automatic online approach. In: IDEAS 2013, pp. 168–173 (2013)
20. Lin, X., Orlowska, M., Zhang, Y.: A graph based cluster approach for vertical partitioning in database design. Data Knowl. Eng. **11**, 151–169 (1993)
21. Ma, H., Schewe, K.-D. Kirchberg, M.: A heuristic approach to fragmentation incorporating query information. In: Databases and Information Systems IV - Selected Papers from the Seventh International Baltic Conference, DB&IS 2006, Vilnius, Lithuania, 3–6 July 2006. Frontiers in Artificial Intelligence and Applications, vol. 155. IOS Press (2006). ISBN 978-1-58603-715-4
22. Malik, T., Wang, X., Burns, R., Dash, D., Ailamaki, A.: Automated physical design in database caches. In: ICDEW 2008, pp. 27–34 (2008)
23. Navathe, S., Ceri, S., Wiederhold, G., Dou, J.: Vertical partitioning algorithms for database design. ACM Trans. Database Syst. **9**, 680–710 (1984)
24. Navathe, S., Karlapalem, K., Ra, M.: A mixed fragmentation methodology for initial distributed database design. J. Comput. Softw. Eng. **3**(4) (1995)
25. Pai-Cheng, C.: A transaction-oriented approach to attribute partitioning. Inf. Syst. **17**, 329–342 (1992)
26. Papadomanolakis, S., Ailamaki, A.: AutoPart: automating schema design for large scientific databases using data partitioning. In: SSDBM 2004, pp. 383–392 (2004)
27. Qian, L., LeFevre, K., Jagadish, H.V.: CRIUS: user-friendly database design. Proc. VLDB Endow. **4**, 81–92 (2010)
28. Rodríguez, L., Li, X.: A dynamic vertical partitioning approach for distributed database system. In: SMC 2011, pp. 1853–1858 (2011)
29. Sacca, D., Wiederhold, G.: Database partitioning in a cluster of processors. ACM Trans. Database Syst. **10**, 29–56 (1985)
30. Wiese, D., Rabinovitch, G., Reichert, M., Arenswald, S.: Autonomic tuning expert: A framework for best-practice oriented autonomic database tuning. In: CASCON 2008, pp. 327–341 (2008)

Can SQ and EQ Values and Their Difference Indicate Programming Aptitude to Reduce Dropout Rate?

Juris Borzovs[✉], Natalija Kozmina[✉], Laila Niedrite[✉],
Darja Solodovnikova[✉], Uldis Straujums[✉], Janis Zuters[✉], and Atis Klavins

Faculty of Computing, University of Latvia, Raina blvd. 19, Riga, Latvia
{juris.borzovs,natalija.kozmina,laila.niedrite,
darja.solodovnikova,uldis.straujums,janis.zuters}@lu.lv

Abstract. A crucial problem that we are currently facing at the Faculty of Computing of the University of Latvia is that during the first study semester on average 30% of the first-year students drop out, whereas after the first year of studies the number of dropouts increases up to nearly 50%. Thus, our overall goal is to determine in advance applicants that most likely will not finish the first study year successfully. A hypothesis formulated in another research study was that programming aptitude could be predicted based on the results of two personality self-report questionnaires – Systemizing Quotient (SQ) and Empathy Quotient (EQ) – taken by students. The difference between the SQ and EQ scores had a strong correlation with grades received for programming test. We reproduced the circumstances of mentioned empirical study with our first-year students using similar tests to calculate SQ and EQ, and semester grades in introductory programming course as a quantitative measure to evaluate programming ability. In this paper, we elaborate on the empirical setting, measures, and estimation methods of our study, which produced the results that made us call the stated hypothesis into question and disprove it.

Keywords: Programming aptitude · Systemizing quotient · Empathy quotient · Correlation · Dropout rate

1 Introduction and Motivating Example

A notable number of students leave the universities in their first years of study, especially at the engineering and computer science programs. This has been observed in higher education institutions in many countries for a long period of time, and remains an acute issue also in recent years. In last years, there is a persistent trend that up to half of the students drop out during their first year of studies at the Faculty of Computing at the University of Latvia. In its turn, it served us as an impulse to search for a method on how to determine in advance applicants that will not finish the first study year successfully. This problem

M. Kirikova et al. (Eds.): ADBIS 2017, CCIS 767, pp. 285–293, 2017.
DOI: 10.1007/978-3-319-67162-8_28

becomes especially important during the programs' evaluations, because the high dropout rate is often associated with a low study quality, meanwhile, the experts evaluating the study programs ask questions on what has been done to reduce the attrition. However, the attrition causes found by other researchers [1,4,11,14] are very different depending on the country and often are not connected with the quality of study programs. Also, attrition causes vary for economic reasons, e.g. high study fee and living expenses for students till psychological reasons such as adaptation problems to the university environment or lack of information about the chosen study program when the students face with subjects that they are not interested in.

The Faculty of Computing at the University of Latvia offers first level higher education Professional study program and Bachelor's study program. We accept applications of 260 new students annually, but during the first study semester on average 30% of them dróp out. After the first year of studies the number of dropouts increases up to nearly 50%. Therefore, to investigate the dropout causes, we have performed a number of studies, including the analysis of students' data from admission of the year 2013, to find what characterizes the dropouts [7]. Several factors were studied: high school grades, intermediate grades for core courses at the university level, and prior knowledge of programming. The conclusions of the study were that none of the studied factors is determinant to identify the students who will drop out. The results also showed that many of the dropouts did not really begin their studies. Thus, we concluded that planned activities should be performed before the admission, for instance, providing more information about the contents of the program or about the profession of the programmer for prospective students. There should be offered a possibility to evaluate students' potential to study computer science.

Our second study [6] was performed to systematize the existing approaches and find out the most promising ones that provide methods of how to determine the programming aptitude of the prospective students. Searching in Google Scholar by 'computer programming aptitude test' search string, at least 60 publications for the period from 1960 to 2014 were obtained. Unfortunately, in neither of them including the latest publications an acceptable solution was found.

All existing approaches to determine computer programming aptitude could be divided into two categories. The first ones are based on the psychological tests [5,20]. The rest are based on solving specifically designed non-programming tasks [9,13,17]. In practice, hybrid tests that contain elements from both groups of tests are often employed [18]. According to our conclusions, a summary of the most promising psychological and problem solving tests should be offered to the prospective students for self-evaluation purposes [5,18,20].

We have decided to take a closer look at the most promising psychological tests found during our previous study, namely, the tests in Wray's research [20]. We performed case studies using Wray's approach to compare acquired results with the original study before preparing a self-test for programming aptitude evaluation for all prospective students. Our first attempt was reflected in a student's Master thesis [12] where schoolchildren were interviewed and the results of their tests were analysed. However, the results were in contradiction with [20], so we decided to repeatedly apply the tests for the first-year students in the year 2016. The results of the case study are presented in this paper.

The rest of the paper is organized as follows: Sect. 2 gives an overview of the related work, our experimental study and its results are presented in Sect. 3, and Sect. 4 finalizes the paper with conclusions and future work.

2   Related Work

Bishop-Clark and Wheeler [5] have based their study on Myers-Briggs test available online at [8] to determine which types of college students have better results in the programming introductory course. Myers-Briggs test evaluates 4 aspects of personality such as (1) general attitude: Extraverted vs. Introverted; (2) perception function: Sensing vs. Intuition; (3) judging function: Thinking vs. Feeling; (4) perception–judging domination: Judging vs. Perceiving. The study [5] concludes that "sensing students performed better on programming assignments than intuitive students, and that judging students achieved higher programming averages than perceptive students". Nevertheless, our goal is to determine the potential dropouts, so these results cannot be used directly for this purpose.

The paper [20] discusses an experiment carried out among 19 students who took a programming module of BSc(hons) course in Telecommunications Systems Engineering. The experiment tested a hypothesis that programming aptitude could be predicted based on the results of different tests taken by students. Students who took part in the experiment completed a programming test after finishing a programming module and after 5 months were invited to take 4 tests that could potentially predict programming aptitude.

The best correlation between test results and programming ability determined by the programming test was obtained for the difference between the scores of two personality self-report questionnaires Systemizing Quotient (SQ) [2] and Empathy Quotient (EQ) [3] produced by the Autism Research Group at Cambridge. SQ measures the interest of a person in systems of objects and EQ determines the ability of a person to interact with other people. The difference SQ–EQ is suggested to have a link with autistic traits in a person.

Both questionnaires are composed of 60 questions, 40 out of them are used to determine empathizing or systemizing ability and 20 filler questions are not scored and added to distract respondent from paying too much attention to empathizing or systemizing. Each question is expressed in a form of a statement and has 4 alternatives that a respondent can choose: "strongly agree", "slightly agree", "slightly disagree", and "strongly disagree". A score 1 or 2 is assigned depending on the chosen alternative and the sum of scores calculated out of all questions is in range from 0 to 80.

The results of the study [20] revealed that each quotient separately had a low correlation with the programming test, but their difference had a strong correlation, therefore, the author concluded that these tests could be effectively used to predict programming aptitude of students before teaching a programming course.

3   An Empirical Study

The goal of our empirical study was to reproduce the method applied in [20] to either confirm or disprove the assumption stated in the paper. More precisely, we wanted to check if there is a strong correlation between combination of SQ and EQ values (i.e. SQ−EQ) as well as a moderate correlation between SQ and EQ individually and students' grades for the test in programming. In our case, though the methodology of the experimental study is the same, the context and population sample differs from that in [20].

3.1   Settings of the Experimentation

Population sample for this study consists of 73 students (29 female and 44 male students) who were randomly selected of out of 260 first-year students of the Faculty of Computing at the University of Latvia. The students were offered to fill in two personality questionnaires − SQ [16] and EQ [10] surveys produced by the Autism Research Group at Cambridge and available online − to assess systemizing and empathizing respectively. Both tests were translated into Latvian language by the author of [12] and were available online during the time period from September 23rd to October 9th of 2016 via Google Docs.

Since the population of our study includes many females, we used a revised version of the Systemizing Quotient [19] with 75 questions in our experimental study, where items from the original SQ questionnaire with 60 questions that is considered male-oriented were supplemented by additional questions that are more relevant to females in general population to avoid bias of study results. The EQ test contains 60 questions, 40 of which are rated with points, whereas 20 remaining questions are not since they are meant for distraction. Each of the test questions has 4 different outcomes − "strongly agree", "slightly agree", "slightly disagree", and "strongly disagree" − which correspond to 2, 1, or 0 points depending on the question. Keys for SQ and EQ tests are available online [10,16]. We created a mapping between students' answers and key values, and calculated the total sum of all points for SQ and EQ tests.

We have taken the final grade for "Programming I" course that the students received at the end of fall semester (January, 2017) as a quantitative measure to evaluate programming ability. Course grade is the sum of multiple grades that include at-home practical tasks (40% of the grade), 7–10 min long in-class written tests (20% of the grade), a compulsory written exam (30% of the grade), and non-obligatory tasks with higher complexity (10% of the grade).

The distribution of final grades for all 73 students is given in Fig. 1. Shapiro-Wilks normality test returns a p-value of 1.038e-06 meaning that grade data is not normally distributed; the dataset can be classified as multi-model distribution with 3 peaks − $x = 0, 4, 7$. The most frequent grade is 7, which was received by 14 students. Also, there were 12 students who hadn't accomplished any task and, thus, didn't pass the course. We did not exclude such students' data from our dataset to check whether there is a correlation between those who failed and do not have appropriate programming skills and their corresponding SQ, EQ,

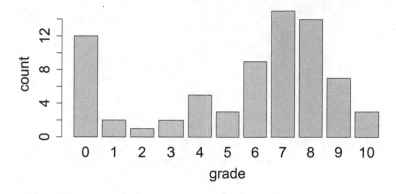

Fig. 1. Distribution of the final grades of 73 students.

SQ–EQ values. In addition, grade values are ordinal data, which implies that to compute correlation coefficient we should use Spearman's rank (ρ).

3.2   Data Analysis

A preliminary step of this research was a study on exploring correlation between SQ, EQ, SQ–EQ values and semester grades received by schoolchildren [12]. The whole experimentation procedure was the same as described above. The population samples included 30 pupils of 11th form (14 boys and 16 girls) and 20 pupils of 12th form (11 boys and 9 girls). For pupils of 11th form, SQ correlation was very weak (≤ 0.04 for either boys or girls), EQ correlation was classified as negative and weak (-0.24 for boys and -0.28 for girls), while SQ–EQ correlation was positive and weak (0.31 for boys and 0.37 for girls). For pupils of 12th form, correlation was classified as very weak (between 0.03 and 0.12) for all values and categories of participants excluding SQ correlation values for girls – in this group there was a weak positive correlation (0.24). Thus, the results of the preliminary study seemed to arise a contradiction with the statement put forward in [20] – namely, we didn't spot a tendency of supporting the hypothesis of positive correlation between SQ–EQ values and semester grades. Moreover, the preliminary study served as a motivation for us to conduct another empirical study that would yield valuable results for the University of Latvia by analyzing data of the first-year students of the Faculty of Computing.

The resulting dataset of the current empirical study consisting of student ID, gender, SQ, EQ, SQ–EQ, and grade values was loaded into RStudio [15] for further statistical analysis. We formulated null hypothesis that we would like to verify in term of our empirical study as *H0: There is no relationship between SQ, EQ, or SQ–EQ, and the final grade for "Programming I" course.*

We built multiple boxplots with final grade data as independent variable (x) and SQ, EQ, SQ–EQ as dependent variable (y). Acquired boxplots for SQ, EQ, and SQ–EQ are presented in Figs. 2, 3, and 4 respectively. Each figure includes 3 plots that contain as an input final grade data: (a) for all students, (b) for male students only, and (c) for female students only.

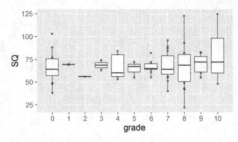

(a) SQ values for all students

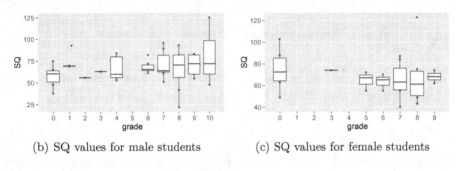

(b) SQ values for male students   (c) SQ values for female students

Fig. 2. Boxplots representing students' SQ values plotted against the final grade.

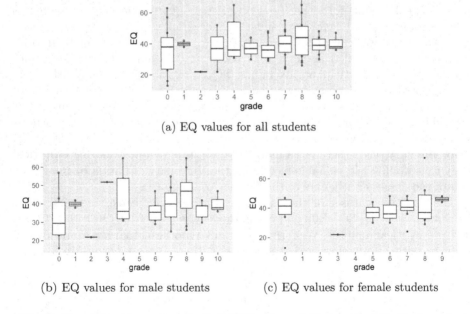

(a) EQ values for all students

(b) EQ values for male students   (c) EQ values for female students

Fig. 3. Boxplots representing students' EQ values plotted against the final grade.

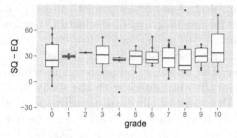

(a) SQ−EQ values for all students

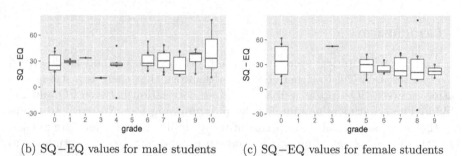

(b) SQ−EQ values for male students   (c) SQ−EQ values for female students

Fig. 4. Boxplots representing students' SQ−EQ values plotted against the final grade.

There is a summary of values of Spearman's rank correlation coefficient (ρ) calculated for each case in Table 1. Similarly as in boxplots Figs. 2, 3, and 4, we split the data into 3 groups: all students, only male, and only female, with n indicating the number of students in the corresponding group. We performed an evaluation for all students regardless of the grade as well as for a subset of students who received a final grade in course (*"Grades > 0"*) to check if there would be a significant difference in ρ values.

We can notice that almost all ρ values in Table 1 − except for the ones in bold font − indicate a very weak negative or positive correlation between SQ, EQ, or SQ−EQ, and the final grade. There is a weak positive SQ correlation of 0.231

Table 1. Values of Spearman's ρ between SQ, EQ, and SQ−EQ, and all grades or grades higher than 0 grouped by all students, female, and male.

		All Grades			*Grades > 0*			
	n	$\rho(SQ)$	$\rho(EQ)$	$\rho(SQ-EQ)$	n	$\rho(SQ)$	$\rho(EQ)$	$\rho(SQ-EQ)$
All students	73	0.072	0.152	−0.044	61	0.142	0.069	−0.021
		(p = 0.546)	(p = 0.2)	(p = 0.711)		(p = 0.274)	(p = 0.599)	(p = 0.87)
Only male	44	**0.231**	0.163	0.092	38	0.133	0.052	0.098
		(p = 0.132)	(p = 0.291)	(p = 0.552)		(p = 0.426)	(p = 0.756)	(p = 0.56)
Only female	29	−0.173	0.161	**−0.233**	23	−0.024	**0.32**	**−0.225**
		(p = 0.371)	(p = 0.405)	**(p = 0.225)**		(p = 0.914)	**(p = 0.136)**	**(p = 0.303)**

within the male group of students (*"All grades"* subset), and a weak positive EQ correlation of 0.32 for female group of students (*"Grades > 0"* subset).

In both grade subsets (*"All grades"* and *"Grades > 0"*) there is a weak negative SQ−EQ correlation within female group of students (−0.233 and −0.225 respectively). However, the acquired values have no similarity with the findings from our preliminary study, neither with the study performed by the other author in [20]. Given that none of the corresponding p-values is less than 0.05, the null hypothesis ($H0$) defined in Subsect. 3.2 cannot be rejected, meaning that there is no evidence of correlation between SQ, EQ, or SQ−EQ, and the final grade for "Programming I" course.

4   Conclusion

The goal of our empirical study was to apply the same method as described in Wray's research [20] to either confirm or disprove that there is a strong correlation between SQ−EQ as well as moderate correlation between SQ and EQ individually and student's grade for the programming test. We tried to reproduce the circumstances of the experiment using similar tests to calculate SQ and EQ with the actual set of questions, and we invited our first-year students to take part in the experimentation. The exception was that we used the results of the introductory "Programming I" course instead of the specially developed test with programming tasks.

The acquired values have no similarity with the study performed by the author in [20]. The null hypothesis ($H0$) defined in Subsect. 3.2 cannot be rejected, which implies that there is no proof of correlation between SQ, EQ, or SQ−EQ, and the final grade for "Programming I" course.

The dropout problem still remains a very pressing one. Therefore, we need to bring in new ideas for future research. For instance, the same set of data can be used for a more targeted evaluation − we can filter out the data for students with the final grade 0 and integrate it with the data from other sources to have a set of records on "dropouts", because students with academic debts also received grade 0. Then, we can investigate more precisely the dropouts' data and look for specific causes.

Other research direction could include application and analysis of another group of tests that are mostly oriented on problem solving.

Finally, we hope that for our computing study programs it would be feasible to reach acceptable improvements regarding the dropout rate and find an appropriate tool to predict potential dropouts in the future.

References

1. Araque, F., Roldán, C., Salguero, A.: Factors influencing university drop out rates. Comput. Edu. **53**(3), 563–574 (2009)
2. Baron-Cohen, S., Richler, J., Bisarya, D., Gurunathan, N., Wheelwright, S.: The Systemizing Quotient: an investigation of adults with asperger syndrome or high-functioning autism, and normal sex differences. Philos. Trans. Royal Soc. B: Biol. Sci. **358**(1430), 361–374 (2003)

3. Baron-Cohen, S., Wheelwright, S.: The empathy quotient: an investigation of adults with asperger syndrome or high-functioning autism, and normal sex differences. J. Autism Develop. Disord. **34**(2), 163–175 (2004)
4. Belloc, F., Maruotti, A., Petrella, L.: University drop-out: an Italian experience. High. Edu. **60**(2), 127–138 (2010)
5. Bishop-Clark, C., Wheeler, D.D.: The Myers-Briggs personality type and its relationship to computer programming. J. Res. Comput. Edu. **26**(3), 358–370 (1994)
6. Borzovs, J., Niedrite, L., Solodovnikova, D.: Computer programming aptitude test as a tool for reducing student attrition. Environment. Technology. Resources. In: Proceedings of the International Scientific and Practical Conference, vol. 3, pp. 29–35 (2015)
7. Borzovs, J., Niedrite, L., Solodovnikova, D.: Factors affecting attrition among first year computer science students: the case of University of Latvia. Environment. Technology. Resources. In: Proceedings of the International Scientific and Practical Conference vol. 3, pp. 36–42 (2015)
8. Carl Jung's and Isabel Briggs Myers' typology test. http://www.humanmetrics. com/hr/you/personalitytype.aspx Accessed 09 June 2017
9. Dehnadi, S., Bornat, R.: The camel has two humps (2006). http://www.eis.mdx. ac.uk/research/PhDArea/saeed/paper1.pdf Accessed 09 June 2017
10. Empathy Quotient (EQ) for Adults: Empathy Quotient (EQ-60) - paper version, EQ scoring key (EQ-60). https://www.autismresearchcentre.com/arc_tests Accessed 09 June 2017
11. Grebennikov, L., Shah, M.: Investigating attrition trends in order to improve student retention. Qual. Assur. Edu. **20**(3), 223–236 (2012)
12. Klavins, A.: Evaluation of potential students of computer programming by using relevance between systemizing quotient and empathy quotient (Master thesis in Latvian). https://dspace.lu.lv/dspace/handle/7/33213 Accessed 09 June 2017
13. Lorenzen, T., Chang, H.-L.: Mastermind©: A predictor of computer programming aptitude. SIGCSE Bull. **38**(2), 69–71 (2006)
14. Paura, L., Arhipova, I.: Cause analysis of students' dropout rate in higher education study program. Procedia-Soc. Behav. Sci. **109**, 1282–1286 (2014)
15. RStudio homepage. https://www.rstudio.com Accessed 09 June 2017
16. Systemizing Quotient (SQ) (Adult): Systemizing Quotient (SQ), SQ scoring key. https://www.autismresearchcentre.com/arc_tests Accessed 09 June 2017
17. Tukiainen, M., Mönkkönen, E.: Programming aptitude testing as a prediction of learning to program. In: Proceedings of the 14th Workshop of the Psychology of Programming Interest Group, pp. 45–57 (2002)
18. University of Kent. Computer programming aptitude test. http://www.kent.ac. uk/careers/tests/computer-test.htm Accessed 09 June 2017
19. Wheelwright, S., Baron-Cohen, S., Goldenfeld, N., et al.: Predicting autism Spectrum Quotient (AQ) from the Systemizing Quotient-revised (SQ-R) and Empathy Quotient (EQ). Brain Res. **1079**(1), 47–56 (2006)
20. Wray, S.: SQ minus EQ can predict programming aptitude. In: Proceedings of the 19th Annual Workshop of the Psychology of Programming Interest Group, pp. 243–254 (2007)

Fusion of Clinical Data: A Case Study to Predict the Type of Treatment of Bone Fractures

Anam Haq$^{(\boxtimes)}$ and Szymon Wilk

Poznan University of Technology, Poznan, Poland
anam.haq@put.poznan.pl, szymon.wilk@cs.put.poznan.pl

Abstract. Clinical data is characterized not only by its constantly increasing volume but also by its diversity. Information collected in clinical information systems such as electronic health records is highly heterogeneous and it includes structured laboratory and examination reports, unstructured clinical notes, images, and more often genetic data. This heterogeneity poses a significant challenge when constructing diagnostic and therapeutic decision models that should use data from all available sources to provide a comprehensive support. A possible response to this challenge is offered by the concept of data fusion and associated techniques. In this paper, we briefly describe the foundations of data fusion and present its application in a case study aimed at building a decision model to predict the type of treatment for patients with bone fractures. Specifically, the model should distinguish between patients who should undergo a surgery and those who should be treated non-surgically.

Keywords: Clinical data · Data fusion · Prediction models · Decision support

1 Introduction

Data fusion is generally defined as the combination of data or information that is obtained from multiple sources of diversified format and structure [1]. The concept of data fusion can be clearly explained by considering the example of human brain perception system. In order to perceive its surrounding conditions, the human brain collects information from all its senses i.e., sight, hearing, smell, taste and touch integrate them together along with the results extracted from the previous memory of similar experiences and generate order accordingly. The use of data fusion concepts in human biological system shows the importance of this notion.

Concept of data fusion and its associated techniques found their use not only in health care where they offer a foundation for developing advanced decision support and smart patient monitoring systems [2], but also in other application areas, such as geospatial system, defense systems, and intelligence services (see [3] for a review).

There are two prevalent approaches to data fusion: *combination of data* (COD) and *combination of interpretation* (COI) [4]. Both techniques are explained below:

© Springer International Publishing AG 2017
M. Kirikova et al. (Eds.): ADBIS 2017, CCIS 767, pp. 294–301, 2017.
DOI: 10.1007/978-3-319-67162-8_29

1. Combination of Data (COD): Features from all data sources are first aggregated and the employed to build a single decision model (i.e., to learn a single classifier).
2. Combination of interpretation (COI): Each data source is used to build a separate decision model and then outcomes of individual models are aggregated by a combiner (which can be considered as a meta-decision model or a meta-classifier) to produce a final outcome. Conceptually, COI is similar to constructing ensembles of classifiers (in particular to the stacking scheme) [5].

Unfortunately, both COI and COD have some inherent drawbacks – COD suffers from the curse of dimensionality, and COI is proved to be sub-optimal as it is not able to preserve dependencies between data coming from various sources. To address these shortcomings Lee et al. [6] developed a *general fusion framework* (GFF) where COI and COD are considered as two extremes of a continuous spectrum.

The general schema of GFF is shown in Fig. 1. Selection of transformations applied to specific data sources to bring them to a common knowledge representation is driven by the characteristics of data, e.g., for sparsely packed data the dimensionality reduction schemes like principal component analysis (PCA) should be used. These transformations may also introduce and use source-specific classifiers. If the latter transformation is applied to all sources, then GFF becomes COI. On the other hand, if all applied transformation simply copies data, GFF boils down to COD.

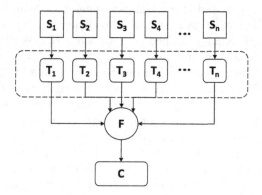

Fig. 1. Schema of GFF (S_1 to S_n are disparate sources, T_1 to T_n are transformations of data sources into a common representation, F is the fused feature space, and C is a classifier based on F) [6].

In this paper we present an on-going clinical case study where a customized data fusion process based on GFF is employed to build a decision model that would predict the type of treatment (surgical vs. non-surgical) for patients with bone fractures. The decision should be based taking into account the general patient state as well as the characteristic of a fracture [7]. Therefore, a therapeutic decision model should be derived from X-ray images and non-image data using data fusion techniques.

The rest of the text is organized as follows. In the next section we present a brief overview on related work on data fusion. Then, in Sect. 3 we introduce our case study and describe its goals, available data sources and the customized data fusion process. Next, in Sect. 4 we give preliminary results of our analysis. Finally, in Sect. 5 we provide conclusions and discuss future work.

2   Related Work

Application of COD and COI approaches are discussed in [5,8–11]. Lanckriet et al. in [8] demonstrated an example of the COD methodology to construct a support vector machine (SVM) to predict the yeast proteins function. Their proposal employs a set of kernels to combine complex protein data, amino acid sequences and gene expressions. K. Kourou et al. in [9] also used the COD approach to develop different classification models using machine learning algorithms that are frequently used in the detection and prognosis of various type of cancers.

Jesneck et al. in [10] described the COI approach that fuses objective findings computed from mammograms, radiologist-interpreted findings and patient history to diagnose breast cancer. A detection theory approach was employed to construct classifiers. Specifically, a separate binary classifier employing likelihood ratio was constructed for each feature (thus there were multiple classifiers for a given data source), and then their outcomes were combined as the joint likelihood ratio of the set of decision variables (capturing outcomes of separate classifiers). Finally, thresholds applied at both levels of classification were optimized using a generic algorithm.

Conceptually, COI is similar to constructing ensembles of classifiers (in particular to the stacking scheme), the concept of ensemble classifiers is explained in [5]. The author describe the approaches of constructing ensemble classifiers, and discussed in details the method for integrating the decisions obtained from each classifier. G. Zorluoglu et al. in [11] created a model for breast cancer diagnosis using decision tree, support vector machine and neural networks along with the ensemble of these three techniques. After comparing the performance of classifiers individually with the ensemble classifier it was concluded that latter performs better in terms of accuracy.

In [4] the author provides a comparison between the COI and COD by using them to develop models for four different kind of bio-medical image processing functions. These functions are segmentation of atlas based images, average image tissue based segmentation, multi-spectral classification and deformation-based group morphometry. The author compared the performance of COI and COD used in construction of models for the above mentioned functions on the bases of their versatility and capability of producing reliable and consistent results.

In the remaining part of the text, we describe an on-going clinical case study where a customized data fusion process based on GFF is employed to build a decision model that would predict the type of treatment (surgical vs. non-surgical) for patients with bone fractures. The decision should be based taking into account a general patient state as well as the characteristic of a fracture [7]. Therefore, a therapeutic decision model should be derived from X-ray images and non-image data using data fusion techniques.

3   Clinical Case Study

3.1   Problem Statement

It is important to distinguish between these patients with bone fractures who require surgery and these who can be managed non-surgically. Several studies have shown that surgery is not needed in every case [12]. Surgical treatments are not only more expensive than non-invasive ones, but they are also more painful. Moreover, there is a group of patients who may be not sufficiently clinically fit for the surgery. Thus, the decision about the type of treatment should be based not only on the characteristic of a fracture, but also on the patient's "fitness", and in order to develop an appropriate therapeutic decision model image and non-image data should be fused together.

3.2   Available Data Sources

In this case study we used a data set provided by the Wielkopolska Center of Telemedicine (https://www.telemedycyna.wlkp.pl/web/guest/home) – a teleconsultation platform for patients with multiple injuries. This data set includes 2030 patients with bone fractures – 1593 (78.5%) underwent a surgery, and the remaining 437 (21.5%) were treated non-surgically. Patients are described using 301 features capturing demographics (e.g., age and gender) results of physical examination and basic laboratory tests (e.g., blood work), and detailed descriptions of injuries. For the sake of simplicity, in the subsequent text we will refer to this data and features as to clinical data and clinical features. Moreover, for almost each patient there is a collection of X-ray images (usually between 2 and 4) of fractured bones.

From this data set we randomly selected 103 patients – 40 non-surgical and 63 surgical cases. For each patient we manually selected a single X-ray image representing a fractured bone at the time when management started.

3.3   Data Fusion Process

The main goal of the presented case study was the application of GFF to build a therapeutic decision model considering both image and non-image features. Specifically, we followed the COD approach to transform heterogeneous data into a single space.

The GFF schema we applied included two data sources S_1 and S_2 corresponding to X-ray images and clinical data respectively. Moreover, transformations T_1 and T_2 applied to S_1 and S_2, respectively, employed diversified techniques of feature extraction/construction and selection. The T_1 transformation was aimed at extracting from an X-ray image a single numerical feature indicating the "severity" of a fracture (such information was not recorded explicitly in the clinical data). It included the following steps (the flow diagram is presented in Fig. 2:

1. Noise removal with a median filter and contrast adjustment,
2. Bone edge detection with the Canny operator and removal of disconnected components,
3. Application of the Hough transform to detect the bone breakage, the process is explained in detail in [13]. The parameter values are set in such a way that the transform produces a single peak in result of minute fractures and multiple peak values for significant fracture bones (mean value of these peak points are considered as feature values),
4. Identifying the location of bone fracture by drawing ellipse around the area.

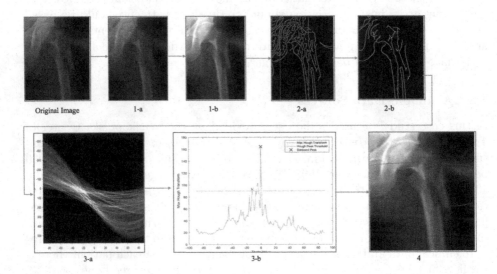

Fig. 2. Results of extracting features from X-ray images: 1(a–b) – Noise removal and contrast enhancement, 2(a–b) – Edge detection and disconnected components removal, 3(a–b) – Hough transformation and peak value extraction, (4) – Identifying bone fracture

The transformation T_2 applied to S_2 was less complex and it involved the following steps:

1. Discretization of numerical features using norms defined by clinical experts,
2. Introduction of additional features capturing information about injuries at a lower granularity level, and
3. Removal of "useless" features (e.g., features with the majority of missing values or extremely low or high variability).

Step (2) may require additional explanation. The originally recorded information about injuries was very detailed (e.g., it indicated a precise location of a fracture) and thus it was impossible to identify any strong patterns in the clinical data. To address this issue we introduced additional variables that captured more general information (e.g., they indicated that a specific bone was broken). Application of T_2 resulted in obtaining 113 features.

Transformation T_1 was implemented in MATLAB, and T_2 was implemented in a custom Java program (discretization and introduction of new features) and in WEKA (feature selection).

Finally, the feature extracted from the X-ray images and 113 features from the clinical data were combined into a fused space that we used to construct and evaluate several possible classifiers (this analysis was also conducted in WEKA). The choice of learning algorithm mostly depends upon the type of dataset. The dataset used in this case study had class information so we have considered supervised learning algorithms for performance evaluation. The clinical data contains the large number of features and values of those features are categorical so we have selected algorithms which performs better when applied over such type of dataset. Specifically, we used a naive Bayes (NB) classifier, a C4.5 decision tree (DT), a random forest (RF) and a support vector machine (SVM) with a linear kernel.

Such selection of classifiers was based on our past experience with analysis of clinical data [14, 15] and on results of other studies related to data fusion [16]. In order to assess the impact of data fusion on the classification performance we considered three versions of each classifier based on the image feature, clinical features and all (fused) features. Finally, classification performance was evaluated using stratified 10-fold cross validation.

4 Results

Set of four classifiers were used, each representing a different learning approach. The use of these classifiers were to compared the performance of the fusion approach on different classification algorithms. Classification performance of specific classifiers is given in Table 1 where we report overall accuracy, accuracy for both decision classes and finally a geometric mean (G-mean) of class-specific accuracies (this measure is less affected by the bias toward any of the classes). Best results for each classifier are marked with bold.

With the results obtained in Table 1 we can derived the following conclusion that:

1. Fusion always improved overall accuracy and most cases G-mean (DT was the only exception).
2. In case of RF fusion improved the accuracy for the non-surgical class while for other classifiers it was more beneficial for the surgical class
3. The overall accuracy of classifiers based on the single image-based feature in most cases was equal or better than the accuracy of classifiers based on clinical features (RF was the only exception). Moreover, the former classifiers turned to be surprisingly accurate in comparison to the classifiers based on fused features.

Table 1. Performance of classifiers for various features (overal = overall accuracy, non-surg, surg = accuracy for the non-surgical and surgical class, respectively)

Classifier	Features	Overal [%]	Non-surg [%]	Surg [%]	G-mean [%]
NB	Image	78.64	**77.50**	79.40	78.44
	Clinical	78.64	62.50	88.90	74.54
	Fused	**83.50**	67.50	**93.70**	**79.53**
SVM	Image	79.61	**80.00**	79.40	79.70
	Clinical	67.96	62.50	71.40	66.81
	Fused	**80.58**	77.50	**82.50**	**80.00**
DT	Image	78.64	**82.50**	76.20	**79.29**
	Clinical	66.20	60.00	69.80	64.71
	Fused	**79.61**	75.00	**82.50**	78.66
RF	Image	70.87	65.00	74.60	69.60
	Clinical	75.72	57.50	**87.30**	70.85
	Fused	**79.61**	**82.76**	78.40	**80.55**

5   Conclusions

In this paper we presented preliminary results from our on-going clinical case study where we applied to combination of data approach to fuse image and clinical data. We checked the performance of for classifiers – a naive Bayes, support vector machine, decision tree and random forest – and in each case we observed the improvement in the overall classification accuracy.

The largest increase was observed for the random forest classifier which is consistent with already published results [16].

This case study is a initial step towards the implementation of a comprehensive framework for clinical data. Such framework that would be able to handle all the challenges associated with it like clinical data, heterogeneity, class imbalance, overlapping of decision boundaries, rare cases, outliers and noise.

In the next steps of our case study we will use more data (especially, we will consider more than one image per patient), introduce additional image-based features extracted with texture analysis, and finally apply the combination of interpretation technique employing specialized ensembles of classifiers.

Acknowledgement. The second author would like to acknowledge support by the Polish National Science Center under Grant No. DEC-2013/11/B/ST6/00963.

References

1. Mitchell, H.B.: Data Fusion: Concepts and Ideas. Springer, Heidelberg (2014). doi:10.1007/978-3-642-27222-6
2. Lahat, D., Adali, T., Jutten, C.: Multimodal data fusion: an overview of methods, challenges, and prospects. Proc. IEEE **103**(9), 1449–1477 (2015). doi:10.1109/JPROC.2015.2460697

3. Castebedo, F.: A review of data fusion techniques. Sci. World J. (2013). doi:10. 1155/2013/704504
4. Rohlfing, T., Pfefferbaum, A., Sullivan, E.V., Maurer, C.R.: Information fusion in biomedical image analysis: combination of data vs. combination of interpretations. In: Christensen, G.E., Sonka, M. (eds.) IPMI 2005. LNCS, vol. 3565, pp. 150–161. Springer, Heidelberg (2005). doi:10.1007/11505730_13
5. Ponti Jr., M.P: Combining classifiers: from the creation of ensembles to the decision fusion. In: 24th SIBGRAPI Conference on Graphics, Patterns and Images Tutorials (2011)
6. Lee, G., Madabhushi, A.: A knowledge representation framework for integration, classification of multi-scale imaging and non-imaging data: preliminary results in predicting prostate cancer recurrence by fusing mass spectrometry and histology. In: International Symposium on Biomedical Imaging: From Nano to Macro. IEEE (2009)
7. Twiss, T.: Nonoperative treatment of proximal humerus fractures. In: Crosby, L.A., Neviaser, R.J. (eds.) Proximal Humerus Fractures. LNCS, pp. 23–41. Springer, Cham (2015). doi:10.1007/978-3-319-08951-5_2
8. Lanckriet, G., Deng, M., Cristianini, N., Jordan, M., Noble, W.: Kernel-based data fusion and its application to protein function prediction in yeast. In: Proceedings of Pacific Symposium on Biocomputing (2004)
9. Kourou, K., Exarchos, T.P., Exarchos, K.P., Karamouzis, M.V., Fotiadis, D.I.: Machine learning applications in cancer prognosis and prediction. Comput. Struct. Biotechnol. J. **13**, 8–17 (2015). doi:10.1109/SIBGRAPI-T.2011.9
10. Jesneck, J., Nolte, L., Baker, J., Floyd, C., Lo, J.: Optimized approach to decision fusion of heterogeneous data for breast cancer diagnosis. Med. Phys. **33**, 2945–2954 (2006). doi:10.1118/1.2208934
11. Zorluoglu, G.M.: Diagnosis of breast cancer using ensemble of data mining classification methods. Int. J. Bioinform Biomed. Eng. **1**(3), 318–322 (2015). doi:10. 5829/idosi.wasj.2014.29.dmsct.4
12. Hossain, M., Neelapala, V., Andrew, J.G.: Results of non-operative treatment following hip fracture compared to surgical intervention. Int. J. Care Inj. **40**, 418–421 (2008)
13. Myint, S., Khaing, A.S., Tun, H.M.: Detecting leg bone fracture in x-ray images. Int. J. Sci. Technol. Res. **5**, 140–144 (2016)
14. Wilk, S., Stefanowski, J., Wojciechowski, S., Farion, K.J., Michalowski, W.: Application of preprocessing methods to imbalanced clinical data: an experimental study. In: Piętka, E., Badura, P., Kawa, J., Wieclawek, W. (eds.) Information Technologies in Medicine. AISC, vol. 471, pp. 503–515. Springer, Cham (2016). doi:10.1007/ 978-3-319-39796-2_41
15. Kubat, M., Matwin, S.: Addresing the curse of imbalanced training sets: one-side selection. In: Proceedings of the 14th International Conference, ICML 1997, pp. 179–186 (1997)
16. Tiwari, P., Viswanath, S., Lee, G., Madabhushi, A.: Multi-model data fusion schemes for integrated classification of imaging and non-imaging biomedical data. In: International Symposium on Biomedical Imaging: From Nano to Macro. IEEE (2011). doi:10.1109/ISBI.2011.5872379

Data Science Techniques for Law and Justice: Current State of Research and Open Problems

Alexandre Quemy[1,2(✉)]

[1] IBM Poland Software Lab, Cracow, Poland
aquemy@pl.ibm.com
[2] Faculty of Computing, Poznań University of Technology, Poznań, Poland

Abstract. By comparing the state of research in Legal Analysis to the needs of legal agents, we extract four fundamental problems and discuss how they are covered by the current best approaches. In particular, we review the recent statistical models, relying on Machine Learning coupled to Natural Language Processing techniques, and the Abstract Argumentation applied to the legal domain before giving some new perspectives of research.

Keywords: Legal analysis · Abstract Argumentation · Case-Based Reasoning

1 Introduction

The legal environment is a *messy concept* [41] that intrinsically poses a certain amount of difficulties to analyze: grey areas of interpretation, many exceptions, non-stationarity, deductive and inductive reasoning, non classical logic, etc.

In this paper, we review the large spectrum of studies on the legal domain to extract a list of major problems to cover: statistical studies, helped by Natural Language Processing techniques, Case-based Reasoning and Abstract Argumentation. A major classification factor appears to be the nature and scope of the information. Some methods will rely on legal knowledge and reasoning techniques while others will exploit the available data: past decisions records, non-legal case features, etc. However, to be broadly adopted by practitioners, a model needs to be explicit and to provide an explanation, meaning it must integrate legal knowledge at some point. This is a challenge, in particular for some Machine Learning techniques such as Random Forests that are well-known to propose non-analytical and hard-to-interpret results.

The organization of the paper is as follow: in Sect. 2, we state the problems to discuss their legitimacy. The next Sections are dedicated to the state of research respectively in statistical models, Case-based Reasoning (CBR) and Abstract Argumentation (AA). Last, we will discuss the current limitations and argue that the fundamental problems did not receive a homogeneous interest from the research community.

© Springer International Publishing AG 2017
M. Kirikova et al. (Eds.): ADBIS 2017, CCIS 767, pp. 302–312, 2017.
DOI: 10.1007/978-3-319-67162-8_30

2   Law and Justice Fundamental Problems

To define the Law and Justice problems and orient the research we need to understand the practitioners. A major debate among the legal community is the interpretation problem, namely *legalism vs realism*, i.e. are the judges objectively applying a method contained in the text (legalism) or do they create their own interpretation (realism). For Frydman [17], the interpretation is an intellectual operation to delimit the extent of the rule in the given context of its application. For Troper [49], the interpretation is iterative because the law is expressed in generic terms and in the interpretation lie some discretionary elements. Despite this genericity, new situations will emerge, creating grey zones of interpretation to be answered by an interpretation [47]. The non-omniscience of judges coupled to the abstraction of the law often offer a room for manoeuvre on the applicability of some legal arguments and the details of sentencing. This can lead to several valid legal justifications up to a certain degree. Finding the best justifications among the possible justifications w.r.t. some criteria is a concrete problem.

For some jurists, the law must be strictly considered as an incitation mechanism to tend to the economic efficiency. For a given legal environment, one might want to control it to reach this goal (e.g. by changing the law). Broadly speaking, studying how it complies with society goals and ideals is a necessity. Due to its decentralized and non-stationary nature some drifts might appear or the current implementation of institutions or policies might not match the theoretical expectations. Some actors cannot directly change the law but they have to take decisions according to the current environment (a company needs to value the risk of its actions w.r.t. the uncertainty and take some decisions). Those observations lead to distinguish four fundamental problems:

- Predicting the outcome of a case given the legal environment. (Prediction)
- Building a legal justification, given some facts, a set of law texts with the jurisprudence and an outcome. (Justification)
- Taking the *best* decisions w.r.t. the legal environment dynamics and some criteria. (Decision)
- Modifying the legal environment dynamics to match some criteria. (Control)

The Prediction problem is challenging, even for the best legal experts: 67.4% and 58% accuracy, respectively for the judges and whole case decision, is observed [42] for the Supreme Court of the United States (SCOTUS). Using crowds, the Fantasy Scotus[1] project reached respectively 85,20% and 84,85% correct predictions. The Justification problem is in a way, an extension of the Prediction one: the reasons of a decision have to be given, including the sentence details. The Justification does not consist in explaining the prediction but finding an explanation based solely on legal factors.

One can illustrate the four problems with the European Union directive to set fines for abuses of dominant position[2]:

---

[1] https://fantasyscotus.lexpredict.com/.
[2] http://eur-lex.europa.eu/legal-content/EN/TXT/?uri=URISERV%3Al26118.

"The level of a fine must be sufficiently high both to punish the firms involved and to deter others from practices that infringe the competition rules. [...] The basic amount is calculated as a percentage of the value of the sales connected with the infringement [...]. The percentage of the value of sales is determined according to the gravity of the infringement (nature, combined market share of all the parties concerned, geographic scope, etc.) and may be as much as 30%."

The Prediction problem consists in determining if the verdict will sanction or not the infringement, possibly using non-legal factors. The Justification problem resides in estimating the fines compatible with the text. It involves estimating the dynamic quantities mentioned such as the combined market share. A possible Control problem would be to determine if in practice the calculation reaches the goals, namely punishing and preventing the other agents to break the law. Conversely, for a company, a Decision problem would be to determine if given the current state of the legal environment, an aggressive takeover would not be perceived as breaking the law and thus leading to undesirable sanctions.

3   Predictive Models

An ideal court, is a court where the judges are perfectly rational, free of all bias and preferences, and omniscient. In this configuration, all judges must reach the same decisions, and two exactly similar cases would result in the same decision. The decisions are not correlated: it is impossible to predict a judgement from an ideal court using the information from the past cases. Of course, in real life there is no such ideal court and statistical methods try to detect hidden patterns between the sequence of decisions to predict future decisions[3].

In [34, 42] a classification tree per Justice is used to predict the votes after being trained on cases described by 6 visible[4] features such as the lower court circuit or the type of petitioner. The model performs significantly better than the experts, while the difference of prediction at the judge level is not significant. An important result is also that the experts were better at drawing good conclusions on the vote of the most extremely ideologically oriented judges[5].

The Stochastic Block Model [19] predicted 77% of individual Justice votes given the votes of the other Justices of the SCOTUS and the history of decisions. Despite the model is not fully predictive and cannot be used to build legal explanation, as it does not use any legal arguments or case-based information, but hidden associations between social actors, it clearly exhibits empirical proofs to support the attitudinalism paradigm[6].

---

[3] One may notice that the less a court is predictable, the closer it is from an ideal court.

[4] That is to say also available to legal experts.

[5] Where the measure is calculated as given by Martin and Quinn [38,39].

[6] On top the predictions, the authors shown the existence of different predictability between judges, implying a difference of attitude toward the law, as well as a decrease in the SCOTUS predictability during some periods or depending on the political party at the presidence.

A general, robust and fully-predictive model has been proposed in [27] using Random Forest [18] and successfully identified 69.7% of the Court decisions over 60 years. The model is built up on more than 300 features, divided into 3 categories: Court and Justice level information, Case information, Historical Justice and Court information. The case features account for 23% of the predictive power while the Court background only for less than 5%. In other words, most of the predictive power holds in the behavioral trend including ideological shifts. More than the exact percentage for this particular model and Court, this comforts the realism paradigm once again. The classifier returns the weights aka the importance of every feature in the prediction. Those weights evolves in time giving additional comprehensive hints on the Court dynamic despite a complex interpretation due to the correlation between the features [48].

Using NLP techniques, the authors of [2] achieve 79% accuracy to predict the decisions of the European Court of Human Rights (ECtHR). They make the hypothesis that the textual content of the European Convention of the Human Rights and the case elements holds hints that will influence the decision of the Judge. They extracted from the case the top 2000 N-grams, calculated their similarity matrix based on cosine measure and partition this matrix using Spectral Clustering to obtain a set of interpretable topics. The binary prediction was made using an SVM with linear kernel. Contrary to the previous studies, they found out the formal facts are the most important predictive factor which tend to favor realism. However, as they used legal documents to extract their features, the non-legal hints are most likely to be less present.

As enlighted by [19,27], one prominent factor in the predictive power of statistical methods is the *ideology* of the judge. To capture this reality, many estimators of the *ideal point*, a latent variable to position the ideology in a continuous space, have been developed. A taxonomy of estimators can be established according to the type of information they rely on: the party affiliation [46], the expert judgements [43,44] or the votes [38,39], possibly including decision-related material [23,30,45]. All of them focus on SCOTUS. The Segal-Cover Score [43,44] is constructed out of editor's assessments published before the nomination of a given judge and whose information are extracted and interpreted manually. The Martin-Quinn Score [33,38,39] uses an Item Response Model [22] with the ideal point modelled as a random walk, thus evolving in time. NLP techniques are used in [30,45] to model the influence of opinion texts (amicus briefs, opinions) on the ideal point of judges while in [23], the ideal point is expressed in the topic space of the legal texts, enabling a more complete *preference profile* per Justice.

4   Case-Based Reasoning

To solve a given problem, a case-based system performs the following cycle [1]:

1. Search for the most related past cases, either by filtering the irrelevant cases or selecting the closest ones depending on a metric and a KNN algorithm.
2. Adapt the best case solution to the new case.

3. Evaluate and revise the proposed solution, including at least why the solution is not satisfying.
4. Integrate the solution to the database.

Among the legal CBRs, CATO [3] is one of the most famous. Used to teach argumentation to students, it consists in 8 basics *reasoning moves* over a database of cases described by manually extracted legal factors. The factors are connected together in a pre-established hierarchy. The links are annotated with a plus (+) or minus (−) to indicate if a factor attacks or defend another one. In a sense, CATO can be seen as an ancestor of Abstract Argumentation.

If CBRs are more efficient and reliable than classical rule-based systems when it comes to law oriented problems [28], many drawbacks subsist: similarity and relevance of precedent cases are dynamic, non-stationary as social and governmental laws evolve [11]. A novel approach to reason by analogy is to learn some rules from a set of similar cases and then to use those rules to reason and predict a new case [25]. The rules represent the prevailing norm in a legal environment at a given moment and thus can evolve while the law corpus remains the same.

5   Abstract Argumentation

Abstract Argumentation (AA) [13] is a hot topic in non-monotonic reasoning, that is to say where there is a need to reason with pieces of information that can turn to be in contradiction with classical logic. The versatility of AA is thus it can be used as a descriptive tool, for instance by including agent's beliefs in order to study and understand a given situation, as well as a normative tool where, given a knowledge base, we want to infer the best action or decision to take [4]. In [10] the author defines the usage of AA with the following three steps:

1. Defining the arguments and the relation(s) between them.
2. Valuating the arguments using their relations, a strength, etc.
3. Selecting some arguments using some criteria (a *semantic*).

From this very succinct summary, one may catch a glimpse on the interest of AA applied to the legal domain, either as a modeling or decision aid tool: the judgment of a case in both Common Law or Civil Law countries follow this exact procedure of collecting the evidence and legal facts, evaluating their suitability in the context of the case, take a decision based on those facts.

In a similar fashion of CATO, an Argumentation Framework (AF) where the arguments are taken manually from a set of cases and the attack deduced from the opinions has been proposed [8,9]. Given a particular case, the arguments that do not apply are removed to obtain a new AF, a subset of the initial one. The key arguments are calculated to determine if the outcome is admissible, i.e. in favor of the plaintiff.

Oren et al. [35,36] instantiate arguments as inference rules and use Subjective Logic [24] to value the strength of the arguments. After giving an algorithm to propagate the strength of the arguments, they define a dialogue game protocol for several agents to argue about the state of the environment. Using Assumption-Based Argumentation [14] , Dung and Thang [16] model a pool of agents arguing about some arguments in front of judges with a final jury taking a decision. If the jury can also introduce new arguments, they are limited to considering the probabilities of causal arguments, while the judges are the only one to determine the admissibility of an argument. Other Assumption-Based Argumentation applications to the the legal domain can be found in [15, 29]. Meanwhile, several quantitative methods to calculate and propagate the strength of arguments have been developed [6, 40] with e.g. Social AA [31] that intends to model debates and decision-making in social networks using a voting system. Future work should consider using them in legal analysis due to good results obtained in systems with similar characteristics.

Recently, CBR based on AA formalism have been proposed. In the previous approaches the arguments are seen as elements of a case while in AA-CBR the arguments are the cases, and the attacks relations between cases. The cases are defined as a set of features and an outcome. In [50], the outcome of a new case is given depending on the *grounded extension* and a justification[7] to support the decision is built using a Dispute Tree [14] while in [5] the outcome is deduced from rules learnt from past cases. In [37], the outcome is given after a deliberation between several agents and a fully adaptive and dynamic approach such that the agents learn from each other and are able to resolve conflict through specific a game protocol.

6   Summary, Limitations, and Future Work

As analyzed in Sects. 2, 3, 4 and 5, the approaches can be broken down into three groups, namely: the statistical models, the case-based reasoning and the abstract argumentation.

Among the **statistical models**, we distinguish two categories. First, some methods use Machine Learning [19, 27, 34, 42] or NLP [2] to predict the outcome of a case given the past cases or votes. Papers in the second category focus on modeling and estimating the ideal point of judges and predict votes solely using this estimation [23, 30, 33, 38, 39, 43–45]. Those methods tackle the prediction problem but cannot handle the justification problem. All models adopt a binary outcome and further work must integrate sentencing elements for more complex and detailed outcomes (Table 1).

Case-base reasoning systems evolved in their formalism from ad-hoc structures to the Abstract Argumentation structure. The most important factor to distinguish between CBR systems is the way they build the justification. In [3, 8, 9] the justification is based on pre-determined legal factor hierarchy and past cases description, while in [5, 37, 50] the factor hierarchy is deduced from past cases relations. Most CBRs neglect the importance of non-legal factors,

---

[7] To be precise, some previous cases and some of their specific features. Thus, this is not a legal justification for Roman Law.

Table 1. Comparison of Law and Justice approaches. The features to compare include: (1) whether the method relies on pre-defined rules or exploit the data (Information), (2) if it uses Legal Knowledge (LK), (3) the capacity to be applied to any case over years (General), (4) to adapt to the environment shifts (Robust), (5) to make a prediction solely based on past information (Fully Pred.), (6) the extra-data used on top of the information about the current case (Extra Data) and (7) the capacity to justify a decision (Just.).

	Information	LK	General	Robust	Fully Pred.	Extra Data	Just.
Pred. Models							
[34,42]	Data-driven	No	No	No	Yes	Past cases	No
[19]	Data-driven	No	Yes	Yes	No	Past votes	No
[27]	Data-driven	No	Yes	Yes	Yes	Past cases	No
[2]	Data-driven	No	Yes	No	Yes	Past cases	No
Ideal Point							
[43,44]	Data-driven	No	Yes	No	Yes	Non-legal	No
[33,38,39]	Data-driven	No	Yes	Yes	Yes	Past votes	No
[30,45]	Data-driven	No	Yes	No	Yes	Amicus	No
[23]	Data-driven	No	Yes	No	Yes	Opinions	No
CBR							
[3]	Rule-based	Yes	No	No	Yes	Past cases & Legal factors	Yes
AA							
[8,9]	Both	Yes	No	No	Yes	Past cases & Legal factors	Yes
[35,36]	Both	Yes	Yes	Yes	Yes	Norm	Yes
[14]	Rule-based	No	Yes	Yes	Yes	-	Yes
[6,40]	Both	No	Yes	No	Yes	-	-
[31]	Rule-based	No	Yes	No	Yes	-	-
AA-CBR							
[50]	Both	No	No	No	Yes	Past cases	Partly
[5]	Both	No	Yes	Yes	Yes	Past cases	Partly
[37]	Both	No	Yes	Yes	Yes	Past cases	Partly

and thus, they implicitly work in an ideal court setting, ruining their capacity to handle the Prediction problem.

Finally, CBRs do not account for the temporal dimension of the legal environment. Future work must focus on integrating *dynamic features* into (AA)-CBR systems. The case features are static and if a feature is shared between cases, e.g. the judge, they may be considered close because they are judged by the same person. However, the preferences of the judges change in time and are influenced by local documents such as case amicus. Thus this cannot directly be used as a feature. Conversely, using the value of the ideal point estimation as a feature is not a good idea: the ideal point strongly depends on many underlying features (the judge, the area of law, other case feature).

In **Abstract Argumentation**, apart from AA-CBR, two kinds of approaches emerge. A *positive* one intends to model real-life decision processes or environment [14,35,36]. A *normative* one tries to elaborate methods to select among the best alternatives and discuss arguments [6,31,40]. If AA is a promising tool to handle the Prediction and Justification problem, the level it operates, mainly at case level, is not directly suitable for the Decision and Control. Despite a rich literature on dynamic in AA [7,12,32] including recent attempts to solve decision process in non-stationnary environments [20,21], as far as we know, there is no attempt to tackle the Control and Decision problem.

A main pitfall of all the approaches is the lack of automation: the Segal-Cover score relies on manually extracted information, the features hierarchy in CATO are manually constructed, etc. Future work should focus on the automation using the recent progresses in NLP and the availability of data. For instance, it is possible to update the Segal-Cover score on a daily basis by extracting the information published over the internet[8].

To conclude, many elements of law look like finance 25 up to 50 years ago [26]. We do believe transferring risk assessment techniques from finance to law is the way to handle all the four problems but requires to corretly quantitatively model the underlying dynamics of the legal environment: a new challenge for data science.

References

1. Aamodt, A., Plaza, E.: Case-based reasoning: foundational issues, methodological variations, and system approaches. AI Commun. **7**(1), 39–59 (1994)
2. Aletras, N., Tsarapatsanis, D., Preoţiuc-Pietro, D., Lampos, V.: Predicting judicial decisions of the European court of human rights: a natural language processing perspective. PeerJ Comput. Sci. **2**, 93–112 (2016)
3. Aleven, V., Ashley, K.D.: Evaluating a learning environment for case-based argumentation skills. In: Proceedings of International Conference on Artificial Intelligence and Law (ICAIL), pp. 170–179 (1997)
4. Amgoud, L.: A unified setting for inference and decision: an argumentation-based approach. In: Proceedings of International Conference on Uncertainty in Artificial Intelligence (UAI), pp. 26–33 (2005)
5. Athakravi, D., Satoh, K., Law, M., Broda, K., Russo, A.: Automated inference of rules with exception from past legal cases using ASP. In: Calimeri, F., Ianni, G., Truszczynski, M. (eds.) LPNMR 2015. LNCS (LNAI), vol. 9345, pp. 83–96. Springer, Cham (2015). doi:10.1007/978-3-319-23264-5_8
6. Baroni, P., Romano, M., Toni, F., Aurisicchio, M., Bertanza, G.: Automatic evaluation of design alternatives with quantitative argumentation. Argum. Comput. **6**(1), 24–49 (2015)
7. Barringer, H., Gabbay, D., Woods, J.: Temporal dynamics of support and attack networks: from argumentation to zoology. In: Hutter, D., Stephan, W. (eds.) Mechanizing Mathematical Reasoning. LNCS, vol. 2605, pp. 59–98. Springer, Heidelberg (2005). doi:10.1007/978-3-540-32254-2_5
8. Bench-Capon, T.J.: Representation of case law as an argumentation framework. In: Proceedings of International Conference on Legal Knowledge and Information Systems (JURIX), pp. 103–112 (2002)

---

[8] e.g.: IBM Watson Services offer query services over hundred of thousands of articles indexed every day.

9. Bench-Capon, T.J.: Try to see it my way: modelling persuasion in legal discourse. Artif. Intell. Law **11**(4), 271–287 (2003)

10. Cayrol, C., Lagasquie-Schiex, M.C.: On the acceptability of arguments in bipolar argumentation frameworks. In: Godo, L. (ed.) ECSQARU 2005. LNCS (LNAI), vol. 3571, pp. 378–389. Springer, Heidelberg (2005). doi:10.1007/11518655_33

11. Delgado, P.: Survey of casebased reasoning as applied to the legal domain (2007, unpublished)

12. Delobelle, J., Haret, A., Konieczny, S., Mailly, J., Rossit, J., Woltran, S.: Merging of abstract argumentation frameworks. In: Proceedings of International Conference on Principles of Knowledge Representation and Reasoning (KR), pp. 33–42 (2016)

13. Dung, P.M.: On the acceptability of arguments and its fundamental role in non-monotonic reasoning, logic programming and n-person games. Artif. Intell. **77**, 321–357 (1995)

14. Dung, P.M., Kowalski, R.A., Toni, F.: Dialectic proof procedures for assumption-based, admissible argumentation. Artif. Intell. **170**(2), 114–159 (2006)

15. Dung, P.M., Thang, P.M.: Towards an argument-based model of legal doctrines in common law of contracts. In: Proceedings of International Conference on Computational Logic in Multi-Agent Systems (CLIMA), pp. 111–126 (2008)

16. Dung, P.M., Thang, P.M.: Towards (probabilistic) argumentation for jury-based dispute resolution. In: Proceedings of International Conference on Computational Models of Argument (COMMA), pp. 171–182 (2010)

17. Frydman, B.: Le sens des lois: histoire de l'interprétation et de la raison juridique. Penser le droit, Bruylant, Bruxelles (2005)

18. Geurts, P., Ernst, D., Wehenkel, L.: Extremely randomized trees. Mach. Learn. **63**(1), 3–42 (2006)

19. Guimerà, R., Sales-Pardo, M.: Justice blocks and predictability of U.S. supreme court votes. PLOS ONE **6**(11), 1–8 (2011)

20. Hadoux, E., Beynier, A., Maudet, N., Weng, P., Hunter, A.: Optimization of probabilistic argumentation with markov decision models. In: Proceedings of International Joint Conference on Artificial Intelligence (IJCAI), pp. 2004–2010 (2015)

21. Hadoux, E., Beynier, A., Weng, P.: Sequential decision-making under non-stationary environments via sequential change-point detection. In: Workshop on Learning over Multiple Contexts (LMCE) at ECML-PKDD (2014)

22. Hambleton, R.: Fundamentals of Item Response Theory. Measurement Methods for the Social Science. SAGE Publications, Newbury Park (1991)

23. Islam, M.R., Hossain, K., Krishnan, S., Ramakrishnan, N.: Inferring multi-dimensional ideal points for US supreme court justices. In: Proceedings of International Conference on Artificial Intelligence (AAAI), pp. 4–12 (2016)

24. Jøsang, A.: Subjective Logic: A Formalism for Reasoning Under Uncertainty. Artificial Intelligence: Foundations, Theory, and Algorithms. Springer, Switzerland (2016)

25. Kannai, R., Schild, U.J., Zeleznikow, J.: There is more to legal reasoning with analogies than case based reasoning, but what? In: Dershowitz, N., Nissan, E. (eds.) Language, Culture, Computation. *Computing of the Humanities, Law, and Narratives*. LNCS, vol. 8002, pp. 440–451. Springer, Heidelberg (2014). doi:10.1007/978-3-642-45324-3_15

26. Katz, D.M., Bommarito, M.J.: Fin(legal)tech - law's future from finance's past (talk) (2017). https://speakerdeck.com/danielkatz/fin-legal-tech-laws-future-from-finances-past-professors-daniel-martin-katz-plus-michael-j-bommarito. Accessed 21 Mar 2017

27. Katz, D.M., Bommarito, M.J., Blackman, J.: Predicting the behavior of the supreme court of the united states: a general approach. SSRN Electron. J. (2014)
28. Kowalski, A.: Case-based reasoning and the deep structure approach to knowledge representation. In: Proceedings of International Conference on Artificial Intelligence and Law (ICAIL), pp. 21–30 (1991)
29. Kowalski, R.A., Toni, F.: Abstract argumentation. Artif. Intell. Law 4(3), 275–296 (1996)
30. Lauderdale, B.E., Clark, T.S.: Scaling politically meaningful dimensions using texts and votes. Am. J. Polit. Sci. 58(3), 754–771 (2014)
31. Leite, J., Martins, J.: Social abstract argumentation. In: Proceedings of International Joint Conference on Artificial Intelligence (IJCAI), pp. 2287–2292 (2011)
32. Mailly, J.: Dynamic of argumentation frameworks. In: Proceedings of International Joint Conference on Artificial Intelligence (IJCAI), pp. 3233–3234 (2013)
33. Martin, A.D., Quinn, K.M., Epstein, L.: The median justice on the united states supreme court. N.C. Law Rev. 83, 1275–1322 (2004)
34. Martin, A.D., Quinn, K.M., Kim, P.T., Ruger, T.W.: Competing approaches to predicting supreme court decision making. Perspect. Polit. 2, 761–767 (2004)
35. Modgil, S., Faci, N., Meneguzzi, F., Oren, N., Miles, S., Luck, M.: A framework for monitoring agent-based normative systems. In: Proceedings of International Conference on Autonomous Agents and Multiagent Systems (AAMAS), pp. 153–160 (2009)
36. Nirn, O.: An Argumentation Framework Supporting Evidential Reasoning with Applications to Contract Monitoring. Ph.D. thesis, University of Aberdeen (2007)
37. Ontañón, S., Plaza, E.: An argumentation-based framework for deliberation in multi-agent systems. In: Rahwan, I., Parsons, S., Reed, C. (eds.) ArgMAS 2007. LNCS (LNAI), vol. 4946, pp. 178–196. Springer, Heidelberg (2008). doi:10.1007/978-3-540-78915-4_12
38. Quinn, K.M., Martin, A.D.: Dynamic ideal point estimation via markov chain Monte Carlo for the U.S. supreme court, 1953–1999. Polit. Anal. 10(2), 134–153 (2002)
39. Quinn, K.M., Park, J.H., Martin, A.D.: Improving judicial ideal point estimates with a more realistic model of opinion content (2006, unpublished)
40. Rago, A., Toni, F., Aurisicchio, M., Baroni, P.: Discontinuity-free decision support with quantitative argumentation debates. In: Proceedings of International Conference on Principles of Knowledge Representation and Reasoning (KR), pp. 63–73 (2016)
41. Rissland, E.L.: AI and similarity. IEEE Intell. Syst. 21(3), 39–49 (2006)
42. Ruger, T.W., Kim, P.T., Martin, A.D., Quinn, K.M.: The supreme court forecasting project: legal and political science approaches to predicting supreme court decisionmaking. Columbia Law Rev. 104(4), 1150–1210 (2004)
43. Segal, J.A., Cover, A.D.: Ideological values and the votes of U.S. supreme court justices. Am. Polit. Sci. Rev. 83(2), 557–565 (1989)
44. Segal, J.A., Epstein, L., Cameron, C.M., Spaeth, H.J.: Ideological values and the votes of U.S. supreme court justices revisited. J. Polit. 57(3), 812–823 (1995)
45. Sim, Y., Routledge, B.R., Smith, N.A.: The utility of text: the case of amicus briefs and the supreme court. Comput. Res. Repos. (2014)
46. Spitzer, M.L., Cohen, L.: Solving the chevron puzzle. J. Law Contemp. Probl. 57, 65–110 (1994)
47. Sunstein, C.R.: How law constructs preferences. Georget. Law J. 86, 2637–2652 (1998)

48. Tolosi, L., Lengauer, T.: Classification with correlated features: unreliability of feature ranking and solutions. Bioinformatics **27**(14), 1986–1994 (2011)
49. Troper, M.: La théorie du droit, le droit, l'état. In: Léviathan. Presses universitaires de France (2001)
50. Čyras, K., Satoh, K., Toni, F.: Abstract argumentation for case-based reasoning. In: Proceedings of International Conference on Principles of Knowledge Representation and Reasoning (KR), pp. 549–552 (2016)

Using Data Analytics for Continuous Improvement of CRM Processes: Case of Financial Institution

Pāvels Gončarovs$^{(\boxtimes)}$ and Jānis Grabis$^{(\boxtimes)}$

Information Technology Institute, Riga Technical University,
Riga LV-1658, Latvia
pavels.goncarovs@gmail.com, grabis@rtu.lv

Abstract. Data analytics capabilities integrated with Customer Relationship Management Systems play an important role to enable customer-centric sales activities at financial institutions. This paper reports a case study on developing a data mining model to identify the Next Best Offer (NBO) for selling financial products to bank's customers. The case study emphasizes importance of collaboration among data scientists and business representatives in iterative refinement of the prediction models. It has been shown that the iterative refinement and combination of various modeling techniques lead to accuracy improvement by 30% and facilitates acceptance of the modeling results.

Keywords: Data analytics · Next Best Offer · Analytical CRM · Data mining process · Combination of association · Classification and clustering

1 Introduction

The ability to access, analyze, and manage vast volumes of data while rapidly evolving the Information Architecture has long been a goal at many Financial institutions [12]. Many have long standing data warehouses and have used analytics tools. Predictive analytics and forecasting models in a Big Data environment enable institutions to make right investment decisions for higher institutional impact.

Financial institutions have found that variety of inbound channels (web, call center, ATM, branch) is increasing in recent years, while traditional outbound channels (cold calling, direct mail and messages) are increasingly challenged [8]. Banks with a list of 20 products to sell to every customer are overloaded with information and questions about the products to sell. Their customers expect personal advice from their advisers more than ever before because they know the banks have data about them. In the past, analysts examined data acquisition extensively, but now research has moved to data analysis [7]. It is this analysis and how the data is subsequently used that makes it so meaningful. Data analysis needs to be used to form responses to real time shifts in customer actions and behavior. It must then help facilitate a comprehensive analysis of the relationship between customers, products, pricing, promotions, and sales.

Data analytics provide an opportunity to transform from a product-centric focus to a more customer-centric view. Making relevant product offers is a key to building

© Springer International Publishing AG 2017
M. Kirikova et al. (Eds.): ADBIS 2017, CCIS 767, pp. 313–323, 2017.
DOI: 10.1007/978-3-319-67162-8_31

succcssful customer relationships. Data analytics, supported by CRM, can be used throughout the organization, from forecasting customer behavior and purchasing patterns to identifying trends in sales activities. Banks improve profitability and loyalty by determining the NBO for every customer interaction. The challenge is to market to the right customer at the right time with the right offering at the right place. That can be achieved by using increasingly granular data, from detailed demographics and psychographics to customers' transactional data to create highly customized offers These are called "next best offers" or NBO [8, 13]. The NBO model is a different approach, representing a single, continuous campaign or marketing program, that selects the optimal product's offer, or decision, at a time. NBO data analysis has been often used as a background application for customer cross-selling.

In this case study, we investigate a new customer up-sell approach based on using data analytics in integration with CRM processes. The case study focuses on iterative development of the analytical model. The model adaptively profiles user's behavior from their transactional records. Customer segmentation and new direct offers can be based on user profiles. Predictive models at the foundation of NBO technologies create recommendations that anticipate what a customer wants even before the customer fully realizes the need. We show how data mining techniques can help in Bank marketing projects. Moreover, we also show some interesting observations, and thus may motivate new research and development.

The rest of the paper is structured as follows. Section 2 introduces key concepts of the propose research. Section 3 reports the case study. Section 4 concludes.

2   Foundations

In applications, a data mining process can be broken into six phases: business understanding, data understanding, data preparation, modeling, evaluation and deployment (Fig. 1), as defined by the CRISP-DM [11]. This process is followed to identify NBO.

2.1   Next Best Offer as a Part of CRM Processes

An integrated approach to NBO management requires a broad business perspective – not just implementing another software package. Typically, the NBO initiative involves integration with the following infrastructure and tools: (1) analytical CRM (Customer information storage and business rules and decision automation engine. Predictive models can be integrated with a business rules engine which drives the workflow.); (2) predictive analysis, data mining, and statistical modeling tools; and (3) visualization tool. (BI). Enterprise CRM supports all aspects of customer's life cycle (Fig. 2). Analytical CRM deals with the analysis of customer data for strategic or tactical purposes to enhance both customer and firm value. The realized outcomes from analytical CRM depend on the quality of the underlying data, the sophistication and skill of analytical methods and their application. Customer data are the lifeblood of CRM, so firms need to build a knowledge management strategy that will support the collection, analysis [9].

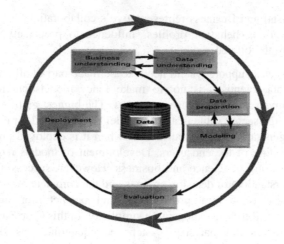

Fig. 1. CRISP-DM (Cross Industry Standard Process for Data Mining) [11]

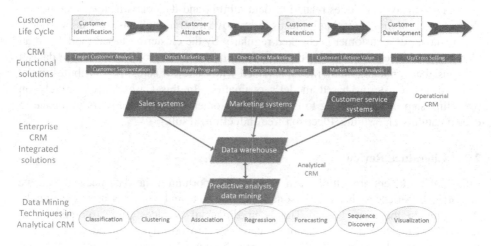

Fig. 2. CRM supports customer life cycle

Analytical CRM framework distinguishes between strategic, analytical, and operational aspects of CRM. Strategic CRM refers to the managerial decision-making processes involved with defining and building a customer-oriented business strategy, business processes and culture, and requisite supporting technology models. This strategic model encompasses a variety of intelligent and creative executive decisions, as captured at the center of the CRM framework.

2.2   Research Questions

The two research concerns addressed in this paper are:

1. IT (data scientists) and Business representative's collaboration;
2. To identify customer behavior profiles' influence to propensity prediction that customer to buy the offered product.

The intelligent sales support system needs to be made part of all the sales activity. The job of IT (data scientists) is not to make fancy models but to facilitate the decision-making process, and help business users. The biggest gratification for job in data science is through seeing value that your model has generated for the business. The models will only generate value for businesses when it is deployed properly with the complete understanding of the end users. Development of models requires predominantly the understanding of science and business. However, successful NBO deployment requires understanding of decision engines and the complete business story. Most importantly it requires creativity to provide right and succinct information to the decision maker to answer all their questions with simplicity. In this Case Study, Bank relies on data mining process and that step-by-step six sequential steps to determine the optimal IT (data scientists) and Business representative's collaboration strategy.

Creating the NBO strategy is not an exact science, so we must experiment and test. We survey many techniques related to data mining and data classification techniques. We select clustering algorithm k-means to improve the training phase of Classification. The behavior of a customer is checked regularly by the clustering-based model and an alert is raised when the behavior deviates from its behavior profile. The source dataset for Classification data mining model is added with behavior categories attributes in the result of the clustering-based model's evaluation. In this Case Study, we rely on modeling step of data mining to determine customer's behavior clusters influence to classification's algorithm's (decision tree induction) result.

2.3   Literature Review

Financial marketers are sitting on a wealth of opportunity; the data needed to drive customer experiences that are personalized, relevant, and seamless across channels. According to the 2016 Digital Trends in Financial Services study, 62 percent of respondents indicate a single customer view is a top priority in the advancement of digital maturity. [7] Ideally, CRM is "a cross-functional process for achieving a continuing dialogue with customers, across all of their contact and access points, with personalized treatment of the most valuable customers, to increase customer retention and the effectiveness of marketing initiatives" [9] Role of Analytical CRM continuously increase in enterprise. Analytical CRM is the use of data to develop relationship strategies.

Creating NBOs is an inexact but constantly improving science. Like any science, it requires experimentation. Some offers will work better than others; companies must measure the performance of each and apply the resulting lessons. It would be hard for any company to incorporate every possible customer, product, and context variable into an NBO model [8].

Historically, business intelligence and data warehouses have been associated with back office employees. Over time, knowledge workers evolved to demand richer, more diverse insights. Pervasive BI is the ability to deliver integrated right-time DW information to all users – including front-line employees, suppliers, customers, and business partners. As usage matured, requirements to include predictive analytics, event-driven alerts, and operational decision support have become the norm [2].

Methods for querying and mining Big Data are fundamentally different from traditional statistical analysis on small samples. Data Mining requires integrated, cleaned, trustworthy, and efficiently accessible data, declarative query and mining interfaces, scalable mining algorithms, and big-data computing environments. The following types of data modeling exist: Association, Classification, Clustering, Forecasting, Regression, Sequence Discovery and Visualization [10].

Clustering algorithms have been studied extensively in the last ten years, with many traditional clustering techniques successfully applied. [4, 5] Clustering techniques has usually been applied as a first step in the data mining process to analyze hidden structures and reveal interesting patterns in the data [6]. The K-Means algorithm is well known for its efficiency in clustering large data sets [1].

Data classification means categorization of data into different category according to rules. Existing research performed over the classification algorithm learns from the training set and builds a model and that is used to classify new objects. Decision tree is useful because construction of classifiers does not require any domain knowledge. It can handle high dimensional data. The learning and classification steps of decision tree induction are simple and fast. Their representation of acquired knowledge in tree form is easy to assimilate by users. Decision tree classifiers have good accuracy [1].

3 Case Study

Data mining is integrated as a part of the CRM systems at a Latvian bank. The integration aim is to improve efficiency of CRM activities and that of cross-sell and up-sell activities in particular. This case study investigates an iterative elaboration of data mining capabilities as a result of collaboration among business actors and IT specialist.

3.1 Business Understanding

As indicated above, the NBO analysis is an important part of the cross-sell and up-sell activities. From the business perspective, in order to identify the NBO for a customer, the bank focuses on four primary questions:

1. How risky is the customer if he takes up the proposed product?
2. What is the propensity of the customer to take up this product?
3. How profitable is the customer expected to be if he takes up the new proposed product?
4. What is the utility of the value propositions of the new product to the customer?

The first three points focus on the profitability for the bank and the last point focuses on the use of this product for the customer. Both profitability and customer centricity need to be balanced. Therefore, all four questions should be considered before making an offer to the customer. An effective NBO action can increase both customer value and customer satisfaction by presenting relevant treatments during each customer interaction. Banks convert a customer from potential attrition to a new cross-sell lead by using propensity of a customer to buy new products to make cross-sell offers. Propensity prediction is based on past trends of product take up by various segments of customers. The advantage of using propensity of a customer to take up a product for targeting is that the success rate is expected to be higher.

The direct product offer for every customer depends on propensity of that customer to buy the new product. NBO is identified following the process shown in Fig. 3.

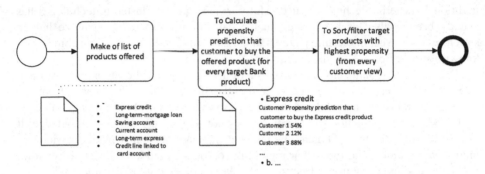

Fig. 3. Identification of the NBO process from the business perspective

1. To Make a target list of Bank products (example list)
2. To Sort/filter target products with highest propensity (from every customer view)
3. To Calculate propensity prediction that customer to buy the offered product (for every target Bank product)

The products with the highest propensity of the customer to take up this product are offered to them.

3.2   Data Understanding

To create an effective NBO, detailed data about customers, bank's offerings and purchasing context must be collected and integrated. The main goal of the Data Understanding step is to answer such questions:

1. What data is available for the task?
2. Is this data relevant?
3. Is additional relevant data available?
4. How much historical data is available?

Demographics, socioeconomic, or geographic characteristics of the customers are the traditionally and widely used variables for customers analysis. Customer intelligence data mining models may be the most powerful and simplest technique for generating knowledge from CRM data, however, this approach does not consider the customer behavior data [14]. By filtering and extracting the necessary data from the data warehouses it is possible to develop the behavioural-based customer intelligence data mining models. In behavioural-based analysis, customer data can be classified by Recency, Frequency and Monetary variables. Recency shows the length of time since the latest purchase. Frequency is the number of purchases in a period and Monetary indicates the total amount of spending in a period.

In this case, data profiling is used to identify data quality problems and to select relevant data for the data mining model. It is a specific kind of data analysis used to discover and characterize important features of data sets. Profiling and other forms of assessment will identify unexpected conditions in the data. A data quality issue is a condition of data that is an obstacle to a data consumer's use of that data. In our approach to resolve a problem means to find a solution and to implement that solution in data source systems.

Talend Data Preparation platform was used to discover data and select relevant data for the data mining proposal. Smart guides and visual tools help anyone quickly understand data attributes and quality status. As an example of data profiling, Fig. 4. shows customers age data profile.

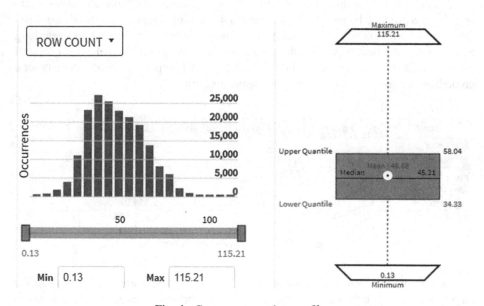

Fig. 4. Customers age data profile

3.3   Data Preparation

The good data preparation is a key to producing valid and reliable models. Data preparation includes table, record and attribute selection, data transformation and cleaning. The following data preparation tasks we performed in this case: missing values replacement, unified date format, data normalization, converting nominal to numeric, discretization of numeric data, data validation and statistics and balancing data.

Balancing data is one of the most significant data preparation activities in our case. Taking the simple route, a modeling system might collapse to always predicting the majority class (is **not** interested in express credit), and thereby claim a performance of 90% (as a measure of simple, unweighted accuracy). But in our case, correctly predicting an example of the minority class (is **interested in** express credit), is more important than accidentally misclassifying an example of the majority class. An approach that has worked for us is balanced stratified sampling.

3.4 Modeling

Predictive models discover patterns by processing complex datasets. There are different types of models depending on their purpose and type. In the NBO identification case, association, classification and clustering methods are used. These methods are used in three sequential steps: (1) to group customers by behavior data (Clustering) (2) to calculate propensity of the customer to take up this product (Decision Tree Induction) and (3) to make bank's products preference recommendations (Association Rule Mining).

Customer behavior data, like accounts balance, payments counts and sums, were selected and grouped by months for last year. The training dataset consists of customers who have already taken a target bank product and their behavior data for one year long period before the deal was made are selected. For each customers' behavior group of metrics, 6–12 clusters were created (Fig. 5). This model outputs the cluster centers for a predefined number of clusters using K-means algorithm.

Row ID	Spark Lines	D APGROZIJUMS_D_01MM	D APGROZIJUMS_D_02MM
cluster_0		621.702	646.041
cluster_1		1,444.157	1,581.008
cluster_2		2,400.603	2,728.428
cluster_3		1,403.654	1,410.476
cluster_4		6,281.945	5,354.079
cluster_5		1,761.82	1,976.595

Fig. 5. Customers behavior clusters

Clusters with a significant proportion of customers who currently have taken the target product, define the target customer's behavior model and indicate a right moment to make the target product offer. The source dataset with behavior categories attributes is added to the data mining model at the end of the first step.

In the second step, we deploy a decision tree because of its nice tree visualization and highlighting property. For algorithm's training reason, we use the 70% partition of the data for training and the small remaining amount (30%) for evaluation. To train a decision tree, we specify the column with the class values to learn (Customer sign the target product agreement), an information (quality) measure, a pruning strategy (if any), the depth of the tree through the number of records per node (20 records), and the split strategies for nominal and numerical values. At the end of the training phase, the decision tree's model shows the decision path through the tree to reach leaves with signing the target product agreement and not signing the target product agreement customers.

In the third step, Data mining workflow takes Customers product data from the CRM system and uses an association predictive analytics technique to make bank's products preference recommendations for each customer. The model uses the Apriori algorithm (Agrawal et al. 1993). Frequent item set mining and association rule induction [Agrawal et al. 1993, 1994] are powerful methods for so-called market basket analysis. With the induction of frequent item sets and association rules one tries to find sets of products that are frequently bought together, so that from the presence of certain products in a customer portfolio one can infer (with a high probability) that certain other products are present. Such information, expressed in the form of rules, can be used to increase the number of bank products sold. In addition, the workflow creates recommendations report that shows the top products and the other bank products associated with each in the form "Others who have X also have....".

Decision tree's model result and association rule's recommendations both are playing a very important role in the NBO. In our case, the recommended product with the highest propensity of the customer to take up this product is offered to them.

3.5 Evaluation

When we trained a model, it is important to check what if the model has not learned anything useful? We need to evaluate it before running it for real on real data. For the evaluation, we use that 30% of data we have kept aside and not used in the training phase, to check model quality. Evaluation process applies the model to all data rows one by one and produces the likelihood that that customer has of signing given his/her contract and operational data (P (Interested in target product = 0/1)). Depending on the value of such probability, a predicted class will be assigned to the data row (Prediction (Interested in target product) = 0/1). The number of times that the predicted class coincides with the original (Interested in target product) class is the basis for any measure for the model quality as it is calculated by the Scoring process.

IT (data scientists) and Business representative's collaboration

The Data mining models will only generate value for businesses when they are deployed properly with the complete understanding of the end users. This is by far the most difficult part of a data scientist's job – developing models in contrast is much

Table 1. NBO pilot project's allocated resources and model accuracy by project iteration

Iteration nr.	Spending time	Final model accuracy
1	120 h	58%
2	+20 h	59%
3	+20 h	66%
4	+25 h	72%
5	+10 h	72%
6	+20 h	73%
7	+20 h	75%

simpler. It is necessary strong communication between IT (data scientist) and business representatives in model evaluation step and in all data mining projects steps.

As shown in Table 1. Seven iterations ware made for NBO pilot project. (Express credit product analysis). It is necessary to have strong cyclical, interactive communication between IT (data scientist) and business representatives.

3.6 Deployment

The last step in the data mining process is to deploy the models to a production environment. Deployment is important because it makes the models or their results available to users so that you can perform any of the following tasks:

1. Use the models to create predictions and make business decisions.
2. Embed data mining functionality directly into an application.
3. Create reports that let users request predictions and view trends.

In that time, we can just provide or business users with predictions reports what they can use in their cross-sell activities. But in future all NBO results will be embedded into Business intelligence application as new separated data mart.

4 Summary and Conclusion

The paper reported on application of data mining technologies at the bank to identify the NBO to customers. The main emphasis was on collaboration among data analysis and business owners. It has been shown that the iterative approach is necessary to improve prediction accuracy and to attain useful results for the business. The prediction accuracy was improved by 30% as the result of seven refinement iterations.

The main observations are that the combination of Association, Classification and Clustering types of data modeling are useful approach to develop the NBO solution, data profiling filters out unessential data, data balancing is required during the data preparation and strong cycling, interactive communication between IT (data scientist) and business representatives is necessary in data mining projects.

References

1. Patel, B.N., Prajapati, S.G., Lakhtaria, K.I.: Bonfring Int. J. Data Min. Efficient classification of data using decision tree **2**, 6 (2012)
2. Markarian, J., Brobst, S., Bedell, J.: Critical success factors deploying pervasive BI. Informatica Teradata MicroStrategy **1007**, 3 (2007). EB-5408
3. Pegasystems Inc.: Next-Best-Action Marketing: A Customer Centric Approach (2012)
4. Wang, K., Zhao, Q., Lu, J., Yu, T.: K-profiles: a nonlinear clustering method for pattern detection in high dimensional data. Biomed. Res. Int. **2015**, 918954 (2015)
5. Jiang, D., Tang, C., Zhang, A.: Cluster analysis for gene expression data: a survey. IEEE Trans. Knowl. Data Eng. **16**(11), 1370–1386 (2004)
6. Hastie, T., Tibshirani, R., Friedman, J.H.: The Elements of Statistical Learning: Data Mining, Inference, and Prediction. Springer, New York (2009)

7. Digital Trends in the Financial Services and Insurance Sector (2016)
8. Sanjiv, K.R.: Next Best Action: The One-To-One Future. www.wipro.com
9. Tanner Jr., J.F., Ahearne, M., Leigh, T.W., Mason, C.H., Moncrief, W.C.: CRM in sales-intensive organizations a review and future directions. J. Pers. Sell. Sales Manag. **25**, 169–180 (2005)
10. Najafabadi, M.M., Villanustre, F., Khoshgoftaar, T.M., Seliya, N., Wald, R., Muharemagic, E.: Deep learning applications and challenges in big data analytics. J. Big Data **2**(1), 1–21 (2015)
11. Chapman, P., Clinton, J., Kerber, R., Khabaza, T., Reinartz, T., Shearer, C., Wirth, R.: CRISPDM 1.0 step-by-step data mining guide. Technical report, CRISP-DM (2000)
12. Sivarajah, U., Kamal, M.M., Irani, Z., Weerakkody, V.: Critical analysis of Big Data challenges and analytical methods. J. Bus. Res. **70**, 263–286 (2016)
13. Croft, J.: Advanced next best offer marketing using predictive analytics. Appl. Mark. Anal. **1**(4), 363–376 (2014). AUTUMN/FALL 2015
14. Sasikala, D., Kalaiselvi, S.: Data Mining for Business Intelligence in CRM System. Sri Vasavi College, (sfw), Bharathiar University, Erode-638 316, India (2016)

Towards a Data Science Environment for Modeling Business Ecosystems: The Connected Mobility Case

Adrian Hernandez-Mendez[(⊠)], Anne Faber, and Florian Matthes

Department of Informatics, Chair of Software Engineering
for Business Information Systems, Technical University of Munich,
Boltzmannstr. 3, 85748 Garching Bei München, Germany
{adrian.hernandez,anne.faber,matthes}@tum.de

Abstract. Modeling the enterprise within its environment is constantly gaining more relevance in research and practice. This holistic view can be formalized as business ecosystem models (BEM). However, the BEM management process is a challenging task, which involves several stakeholders in a data-intensive process. In this paper, we adapted an existing data-science approach and introduced a tool called Business Ecosystem Explorer (BEEx) for supporting the BEM management process in the context of connected mobility. The BEEx provides role-based user interfaces, which empower end-users to manage the BEM. The evaluation of the BEEx was conducted with twelve companies, indicating the usefulness of the BEEx when fulfilling BE related tasks.

Keywords: Data science tool · View-driven approach · Business ecosystems · Visualizations · Connected mobility

1 Introduction

As companies must develop agile mechanisms to adapt to ever-changing conditions within their business ecosystem, business ecosystem models (BEM) have gained relevance [1]. Applying BEM can support enterprises to model themselves within their business ecosystems and thus applying an outside-in approach of enterprise modeling.

In addition to suppliers, customers, business partners, and competitors, a business ecosystem comprises innovative start-ups and companies for potential future collaboration as well. To model a company's ecosystem, it is thus important to continuously analyze the market by scanning news feeds, social media, start-up databases, companies web pages, etc. Besides the variety of data sources, the business ecosystem affects various stakeholders of different business units (e.g. the legal, market analysis, customer department, etc.). Thus, modeling business ecosystem requires the collaboration between different stakeholders through a data-intensive process. Overall, as knowledge gained from the business

© Springer International Publishing AG 2017
M. Kirikova et al. (Eds.): ADBIS 2017, CCIS 767, pp. 324–330, 2017.
DOI: 10.1007/978-3-319-67162-8_32

ecosystem modeling affects the strategic development of an enterprise, the modeling process must conclude by providing relevant information to decision-makers creating an added value.

To address these challenges, visual decision support and visual analytic systems and approaches have been proposed and evaluated (cf., [2–5]), ensuring that visualizing BEM supports stakeholders and decision-makers applying the "wide lens" [6] and making informed decisions. However, these approaches don't address new challenges that have emerged in the domain of information systems to reduce the time to execute the business ecosystem modeling process. First, empowering end-users (i.e., users without programming skills) to adapt not only the BEM but also the visualizations [7]. Second, reducing the complexity of the User Interface (UI) for such systems, which requires UIs with a minimal feature-set and an optimal layout based on the roles of the system [8]. In this paper, we address these challenges by introducing a tool consisting of multiple UIs tailoring different stakeholders' needs to understand the business ecosystems. We refer to such a visual analytic system as a "Business Ecosystem Explorer" (BEEx). For the development of the BEEx, we followed Hevner's Design Science approach [9].

The here presented prototype is applied to the context of connected mobility (CM). CM describes the interconnectedness between means of transportation, especially cars, traffic systems and infrastructure [10] due to advancing digitization of mobility, and exhibits a currently emerging business ecosystem as actors with technology-related business models, such as Google and Apple, challenge the established players [11,12].

The main contribution of this paper is adapting a data science approach to address variety and velocity of BEM's data, through the implementation of a prototype tool, in the context of the project TUM Living Lab Connected Mobility (TUM LLCM)[1], which empowers end-users to manage BEM by providing role-based UIs.

2   The Hybrid Wiki Approach

The development of an information system that provides a collaborative environment supporting the evolution of data models (e.g., BEM and visualizations) at runtime by different end-users (i.e., users without programming knowledge or skills) was addressed in [21] by introducing the Hybrid Wiki approach. This approach has been used in different use cases and domains such as Enterprise Architecture Management [13] and Collaborative Product Development [14]. To instantiate the BEM we used the updated Hybrid Wiki metamodel as presented in [15].

The Hybrid Wiki metamodel contains the following model building blocks: Workspace, Entity, EntityType, Attribute, and AttributeDefinition. These concepts structure the model inside a Workspace and capture its current snapshot in a data-driven process (i.e., bottom-up process). An entity contains a collection of attributes, and the attributes are stored as a key-value pair. The attributes

---

[1] http://tum-llcm.de/en/.

have a name and can store multiple values of different types, for example, strings or references to other Entities. The user can create an attribute at run-time to capture structured information about an entity. An EntityType allows users to refer to a collection of similar entities, e.g., organizations, persons, etc. The EntityType consists of multiple AttributeDefinitions, which in turn contain multiple data validators.

3 Modeling the Connected Mobility Bussines Ecosystems

For modeling the connected mobility business ecosystem, we adapted the four step *data science approach* proposed in [16] and encoded the data in each step using the Hybrid Wiki metamodel. The connected mobility BEM is comprised of three EntityTypes: organizations, relations, and visualizations, which are enhanced with data in the process:

First, the industry structure is determined using trade publications. Categories, of both companies and their interrelations, are identified and each stored as one AttributeType. For companies, these are automotive OEMS, parts supplier, public transportation, technology companies, platform and connectivity provider, new competitors of affected industries, and public institution. As relation types, we identified cooperation, (partially) ownership, funding, negotiation, and supply. In the next step, relevant companies, describing attributes and their relations to other entities of the ecosystem are collected. AttributeDefinitions such as company name, abbreviation, logo, URL, short description, headquarter, CEO, category, and legal form are gathered for each identified company using companies' web pages, newsfeed, social media, etc. Third, the BEM is represented using an explicit visual model language. Each visualization has two elements: the first element is the link between the data model and the visualizations, stored in an AttributedDefintion for each visualization. The second element is the specification of the visualizations, which is described using the visual encodings of the visual grammar Vega[2], as presented and described in [17].

Finally, the collected information are visualized, analyzed, and interpreted. Figure 1 shows the visualizations implemented in the connected mobility BEEx. The visualizations can support stakeholders analyzing the ecosystem. For example, the modified ego-network can be used to understand which types of companies are necessary to provide a mobility service, by visualizing the mobility services in the center and the mobility service providers – offering the mobility service solutions with a variety of different relations – on the outside circle. Additionally, this visualization allows the interpretation which companies already address the growing demand for mobility as a service providing mobility services, thereby acting innovatively.

4 The Connected Mobility Business Ecosystem Explorer

The development and maintaining of an information system that provides a collaborative environment supporting the evolution of both the model and its

---

[2] https://vega.github.io/vega/.

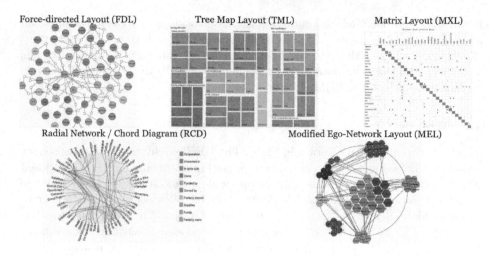

Fig. 1. BEEx Visualizations used in the connected mobility BEM

instances at runtime by stakeholders and ecosystem experts (i.e., the Hybrid Wiki approach) can be considered a difficult task. This process involves nontrivial design decisions and a reference architecture of the connected mobility BEEx.

4.1 Design Considerations

Four main requirements drive the design considerations. First, the BEEx must support semi-structured and non-structured data (i.e., data variety). Second, the end-users should be able to modify the BEM and visualizations at run-time (i.e., without the need to recompile the system). Third, the BEEx must provide role-based UIs. Finally, the BEEx must be supported by WEB-based technologies. In this paper, we describe the decisions we made regarding the following concerns: development process, architecture style, and visualizations technology

We followed a model-driven development (MDD) process, which is used on the Hybrid Wiki approach. This decision ensures that the system can handle different types of data, thereby the CM BEEx can address the data variety challenge in the associated scenario. However, this approach increases the complexity of the system [15]. Therefore, we select the resource-oriented architecture (ROA) as architecture style. The ROA enables the communication between the different components using defined resources encoded in JSON format, which increases the performance in the communication of WEB-based systems (i.e. using the HTTP protocol) [18]. Thereby, the complexity of the system that supports the Hybrid Wiki approach can be encapsulated. Hence, each component in the system utilizes the optimal resources/features that guarantee the role-based UI.

The visualizations technology selected was based on the visual grammar Vega. Instead of using an imperative approach (e.g. D3JS[3]), we select a

---

[3] https://d3js.org.

declarative approach, which guarantees the description of the visualization using a common visual grammar. This decision ensures that end-users can create visualizations, and bind them to the current BEM at run-time. Additionally, this approach addresses the data velocity challenge since it allows the end-users to continuously adapt the BEM and the visualizations.

4.2   Architecture

The BEEx architecture is shown in Fig. 2. The BEEx architecture ensures a clear separation between execution and development environments, and it is composed of four main components. The first component is the BEEx Modeler, which is the core component of the modeling environment, allowing the BE Modeler to create/update/delete the BEM and the visualizations. The next component is the BEEx Client, which is the core component of the execution environment, and it allows the BE User to interact with all the visualizations, providing only read access to the Models Repository through the Query Service. The Model Repository component stores the BEM and the visualizations in JSON format, which are exposes using a RESTful service. Finally, the Query Service component is responsible for extracting the information from the Models Repository, which is required in each environment using the stated security constraints.

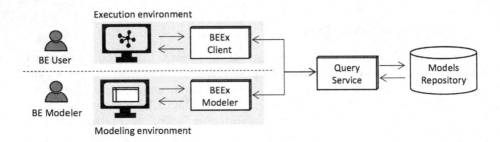

Fig. 2. BEEx architecture with main components.

4.3   First Evaluation of the BEEx Prototype

For the evaluation process, the BEEx is hosted in a TUM University server under the following addresses, the execution environment https://ecosystem-explorer. in.tum.de, and the modeling environment https://vmmatthes17.in.tum.de.

Initially, we conducted nine interviews in semi-structured form with nine different companies within two months, following [19]. We aimed at a balance between receiving some quantifiable evaluation but also enabling interviewees to vary the depth of answer depending on own capability and willingness [20]. The focus of these interviews was to receive feedback regarding the CM BEM. A sample of attributes was presented, as well as some types of relations and a visualization of the combination of entities through relations in the form of a force-directed layout. All interviewees stated that the BEM supported them in mastering ecosystem related tasks and that their knowledge of the connected mobility ecosystem was increased. Some suggested immediately additional scenarios, for example, the patent management in the pharmaceutical industry.

We used the interviews' results to update the connected mobility BEEx. As a next step, we conducted three in-depth interviews with additional three companies. To obtain a wider range of opinions, we selected three companies of different fields of activity. Namely, an automotive OEM, a publicly funded non-research institution and a software company with main business area addressing connected mobility, all actively modeling their business ecosystem. All stated their perceived limitations with the current in use tools. Additionally, two companies confirmed that different stakeholders within in the enterprise have different viewpoints towards the enterprise's business ecosystem. As a limiting factor, it should be noticed that the presented BEEx prototype required further developed within the conduction of these in-depth interviews. Nevertheless, all companies agreed that the prototype foster the understanding of the connected mobility ecosystem and two continued that it would be interesting to use such a tool within in their enterprise to collaboratively manage the business ecosystem evolution.

5   Conclusions

In this paper, we adapted an existing data-science approach and introduced a tool for supporting the BEM process. The design decisions and the architecture ensure that end-users can address variety and velocity of BEM's data in the context of the connected mobility business ecosystem. The BEEx provides role-based UIs, which empowering end-users to manage the BEMs. The first BEEx evaluation indicates that the proposed approach and tool conforms to the enterprise tasks for BEM. However, further evaluation is required to evaluate the potential of the BEEx on supporting different stakeholders in completing different BEM related tasks.

References

1. Bosch, J.: Speed, data, and ecosystems: The future of software engineering. IEEE Softw. **33**(1), 82–88 (2016)
2. Park, H., Bellamy, M.A., Basole, R.C.: Visual analytics for supply network management: System design and evaluation. Decis. Support Syst. **91**, 89–102 (2016)
3. Huhtamaki, J., Rubens, N.: Exploring innovation ecosystems as networks: Four european cases. In: Proceedings of the Annual Hawaii International Conference on System Sciences March 2016, pp. 4505–4514 (2016)
4. Basole, R.C., Huhtamäki, J., Still, K., Russell, M.G.: Visual decision support for business ecosystem analysis. Expert Syst. Appl. **65**, 271–282 (2016)
5. Evans, P.C., Basole, R.C.: Revealing the api ecosystem and enterprise strategy via visual analytics. Commun. ACM **59**(2), 26–28 (2016)
6. Adner, R.: The Wide Lens: What Successful Innovators See That Others Miss. Portfolio Penguin, New York (2013)

7. Satyanarayan, A., Heer, J.: Lyra: An interactive visualization design environment. Comput. Graphics Forum **33**(3), 351–360 (2014)
8. Akiki, P.A., Bandara, A.K., Yu, Y.: Rbuis: Simplifying enterprise application user interfaces through engineering role-based adaptive behavior. In: Proceedings of the 5th ACM SIGCHI Symposium on Engineering Interactive Computing Systems, EICS 2013, pp. 3–12. ACM, NY, USA (2013)
9. Hevner, A.R.: A three cycle view of design science research. Scand. J. Inf. Syst. **19**(2), 4 (2007)
10. Mitchell, W.: Reinventing the Automobile : Personal Urban Mobility for the 21st Century. Massachusetts Institute of Technology, Cambridge (2010)
11. Etherington, D., Kolodny, L.: Google's self-driving car unit becomes waymo (2016). https://techcrunch.com/2016/12/13/googles-self-driving-car-unit-spins-out-as-waymo/ Accessed 23 Jan 2017
12. Taylor, M.: Apple confirms it is working on self-driving cars (2016). https://www.theguardian.com/technology/2016/dec/04/apple-confirms-it-is-working-on-self-driving-cars Accessed 23 Jan 2017
13. Matthes, F., Neubert, C.: Wiki4eam - using hybrid wikis for enterprise architecture management. In: Proceedings of the International Symposium on Wikis and Open Collaboration (2011)
14. Rehm, S., Reschenhofer, T., Shumaiev, K.: Is design principles for empowering domain experts in innovation: Findings from three case studies. In: Proceedings of the International Conference on Information Systems (2014)
15. Reschenhofer, T., Bhat, M., Hernandez-Mendez, A., Matthes, F.: Lessons learned in aligning data and model evolution in collaborative information systems. In: Proceedings of the 38th International Conference on Software Engineering Companion, ICSE 2016, pp. 132–141. ACM, NY, USA (2016)
16. Iyer, B.R., Basole, R.C.: Visualization to understand ecosystems. Commun. ACM **59**(11), 27–30 (2016)
17. Satyanarayan, A., Russell, R., Hoffswell, J., Heer, J.: Reactive Vega: A streaming dataflow architecture for declarative interactive visualization. IEEE Trans. Vis. Comput. Graphics **22**(1), 659–668 (2016)
18. Roberto Lucchi, M.M., Elfers, C.: Resource oriented architecture and rest. Eur - scientific and technical research series, EUR 23397 EN - Joint Research Centre - Institute for Environment and Sustainability
19. Weiss, R.: Learning from strangers : the art and method of qualitative interview studies. Free Press, New York (1995)
20. Gläser, J., Laudel, G.: Experteninterviews und qualitative Inhaltsanalyse: als Instrumente rekonstruierender Untersuchungen. VS Verlag für Sozialwiss, Wiesbaden (2010)

The 2nd International Workshop on Semantic Web for Cultural Heritage (SW4CH 2017)

Introducing Narratives in Europeana: Preliminary Steps

Carlo Meghini, Valentina Bartalesi[(✉)], Daniele Metilli, and Filippo Benedetti

Istituto di Scienza e Tecnologie dell'Informazione "Alessandro Faedo" – CNR,
Via Giuseppe Moruzzi 1, 56124 Pisa, Italy
{carlo.meghini,valentina.bartalesi,daniele.metilli,
filippo.benedetti}@isti.cnr.it

Abstract. We present some preliminary steps towards the introduction of narratives as first-class citizens in digital libraries. The general idea is to enrich the digital libraries with events providing a rich contextualisation of the digital libraries' objects. More specific motivations are presented in the paper through a set of use cases by different actors who would benefit from using narratives for different purposes. We then consider a specific digital library, Europeana, the largest European digital library in the cultural domain. We discuss how the Europeana Data Model should be extended for representing narratives. We present a tool supporting the creation and the visualisation of narratives and we show how the tool has been employed to create a narrative of the life of the painter Gustav Klimt.

Keywords: Digital libraries · Narrative · Europeana · Ontologies

1 Introduction

Europeana[1] is the largest European digital library, containing descriptions of about 54 millions of cultural heritage objects — books, photographs, maps, movies, audio files — provided by more than 3500 of the most important cultural institutions across Europe. Europeana is a digital library for scholars, researchers, professionals and general users, providing a single access point to European cultural heritage. Indeed, the digital library collects digital objects representing cultural heritage items that themselves remain outside the Europeana data space [3]. The current access functionalities of digital libraries are largely influenced by Web search engines, whereby users express their information need in the form of a natural language query consisting of a few words, and in response the digital library returns a ranked list of relevant objects. No semantic relations between objects are reported, that is relations connecting objects in a way that reflects their meaning and their context. Europeana is no exception. Now, it is well known that events play a major role in the contextualization of objects, by providing an account of the happenings concerning the objects, in

---

[1] https://www.europeana.eu.

M. Kirikova et al. (Eds.): ADBIS 2017, CCIS 767, pp. 333–342, 2017.
DOI: 10.1007/978-3-319-67162-8_33

terms of the people, the places, the times and the objects that are involved in such events. In other words, events are the most natural candidates to act as the "semantic glue" filling the gaps between objects. Moreover, events are linked to other events to form "super-events" or stories. However, although events are included in the Europeana ontology, events do not show up in the Europeana database due to the fact that they do not show up in the catalogues of libraries, archives, museums and galleries where the descriptions collected by Europeana come from. Indeed, events are typically found in historical documents. But the automatic capturing of events and of their properties from such documents is an unsolved problem, and will probably remain so for a long time, given the difficulty that machines have to grasp the meaning of the languages used for inter-human communication (text, images, graphics, audio, audio-visual, and so on).

In our study, we aim at overcoming the limitations of the search functionality of current digital libraries by introducing *narrative* as a first-class concept in the data model of such digital libraries. The vision is that a user wishing to know what Europeana has about the Austrian painter Gustav Klimt would obtain in response not only the ranked list of objects (more or less) concerning Klimt that the digital library knows about, but also a narrative about Klimt, that is a semantic network linking such objects in a story that would work as a contextualisation of the objects themselves, and as such would provide the user with a larger and more significant amount of information. Narrative is a concept studied in several fields, from literary studies to cognitive science. Giving an account of the concept is beyond the scope of this paper. As a matter of fact, "narrative can be viewed under several profiles — as a cognitive structure or way of making sense of experience, as a type of text, and as a resource for communicative interaction" [7]. In our research, we intend a narrative as a semantic network, meaningful to the user, consisting of events related to one another, to the entities that compose the events (e.g. agents, places, time, physical objects) and to the digital objects through semantic relations.

As a first step towards the introduction of narratives in the search functionality of digital libraries, this paper presents an experiment we performed on Europeana. Using a semi-automatic tool to build narratives that we developed, we constructed a narrative about the life of Gustav Klimt[2]. This narrative, visualised in form of timeline, could be imported in Europeana and shown as a search result, thereby placing the objects related to Klimt in the more general context of the biography (or a part thereof) of their creator.

The paper is structured as follows: Sect. 2 reports some uses cases providing further evidence of the need of narratives as well as input to the future technical development. As a first step in this direction, Sect. 3 describes how the EDM needs to be extended in order to represent narratives. As a second step in the same direction, Sect. 4 presents a tool for building and visualising narratives that we developed. In Sect. 5 we describe how we created a narrative on the life of Gustav Klimt and how it could be imported in Europeana. Finally, in Sect. 6 we report our conclusions.

---

[2] https://dlnarratives.eu/timeline/klimt.html.

2   Use Cases

In this Section we consolidate our motivations by presenting some use cases of how narratives could improve the use of Europeana, but also of a digital library in general, by different kinds of users.

- *User: Scholar*

  Scholars, such as historians or biographers, can create and access narratives about the life and the works of the authors they study. They may provide their own texts from which the plot was extracted, and they may also be interested in expressing the primary sources supporting the plot. To this end, the construction of a narrative by a historian or biographer can be viewed as an inferential process, using evidence collected from sources to infer propositions that have been narrated in a text. Different scholars can create different narratives on the same topic. The data inserted by the scholar should be exported in CSV format in order to allow other scholars to make further analyses on the data.

- *User: High School Professor*

  The tool could be used by a professor as a learning tool. The professor may create a narrative on a topic of study and show it to the students through a timeline visualisation. For each created event, the professor should add her/his description of it using a specific field of the tool. She/he could also insert fragments from a text book, which are taken into account as secondary sources. The professor could also add some primary sources for the main events, in order to examine them in depth. Using the tool the professor can share her/his timeline with the students. The professor could also enrich the narratives with Wikimedia Commons[3] images or links to related digital objects included in digital libraries, in order to make the narrative more attractive for the students.

- *User: High School Student*

  The tool could be used to verify students' comprehension of history or literature taught by a professor, who could ask them to create a narrative on a particular topic using the tool. The students could enrich the narration by adding appropriate entities and providing textual descriptions of each event, using the interface of the tool. Alternatively, a quiz could be proposed to the students on the narrative created by the professor.

- *Exhibition or Museum Curator*

  A narration timeline could be used during a monographic exhibition in order to associate the works of an artist to her/his biography. The timeline could be used to help the visitors to better understand the life and works of a painter. A timeline could also be used in a museum context in order to show to the visitors the history of the museum and the acquisition of the main artworks. In the interface it could be useful to have a functionality that allows visualising only those artworks that are part of the museum's collections.

  Both for exhibitions or museums, the curator who builds a narrative should

---

[3] https://commons.wikimedia.org.

have the possibility to add her/his own new images of the objects kept in the museum. Furthermore, information about the location of the objects in the museum rooms could be added through a specific field of the tool.

The final user of a museum or exhibition is the visitor, thus, a Web interface should allow: (i) visualising a narrative on a timeline; (ii) visualising in the timeline the images of the museum objects; (ii) providing information about the location of the objects in the rooms of the museum.

– *User: Digital Curator*

A digital curator would be able to create narratives for the DL objects she/he would like to promote. For example, a narrative about the Versailles Conference may be defined by a digital librarian linking the information objects of the Versailles treaty (e.g. photographs of the people, the final declaration, etc.) to the event representing the Conference. Such event may further be divided into sub-events which will have to be properly placed on a temporal axis. This approach aims to build narratives by linking objects to one another using events. Curators may also be interested in representing the provenance of the digital objects. The final user of a digital curator's narrative may be a general user or an expert in the topic of the narration.

3   From the EDM to an Ontology for Narratives

The Europeana Data Model (EDM)[4] [5] is an ontology that allows data to be presented in different ways according to the practices of the various organisations that contribute cultural object descriptions to Europeana. The EDM provides two different approaches for descriptive metadata: "object-centric" and "event-centric". The object-centric approach focuses on the object described: information comes in the form of statements that directly express the features of the described object. Such features can be valued as literals or as resources denoting entities from the real world. A prominent example of an object-centric ontology is Dublin Core, which is largely used in the cultural heritage sector and beyond. Event-centric approaches consider that descriptions of objects should focus on characterizing the various events in which objects have been involved. The idea is that this approach would lead to establishing richer networks of entities — by representing the events that constitute an object's history — than with the object-centric approach, and in some cases, such as the description of archaeological findings, the only meaningful description. The event-centric approach underlies models such as the CIDOC CRM [4] and may suit the data of some (but of course not all) Europeana providers.

An event is represented in the EDM as an instance of the class edm: Event. The relations connecting events with the resources that characterize events are represented in the EDM using the three properties: (i) edm:happenedAt, which links an event to the place of its occurrence; (ii) edm:occurredAt, which links an event to the time span of its occurrence; and (iii) edm:wasPresentAt, holding between any resource and an event it is involved in. This is a very general

---

[4] http://pro.europeana.eu/page/edm-documentation#EDMmappingGuidelines.

property aimed at linking an event with the artifact(s) that were produced by the event, or with the people that participated (with possibly different roles) in the event, or with the concepts related to the event, and so on.

The concept of event is a core element of the narratives and more generally of the narratology theory [2]. Conventionally, an event is defined as an occurrence taking place at a certain time at a specific location. While the EDM provides the concept of event, it does not define any property between events. Thus, it is not possible to temporally, causally or mereologically order the events that compose a narrative. Furthermore, the EDM does not provide a classification of events by type, nor defines roles for the participants in an event. These are limitations that must be overcome to represent narratives in Europeana.

For the users who want to search for a specific topic, the Europeana portal provides a search engine on the textual values of the metadata of the digital objects included in the digital library. It is possible to refine the search through some facets provided by the portal: (i) collection (e.g. art, fashion, music); (ii) media type; (iii) copyright (free re-use, limited re-use or not re-usable); (iv) providing country; (v) language; (vi) aggregator (organisations performing a data aggregation role) and (vii) institution that provides the data. The portal does not currently allow a search based on events, nor does it supply a visualisation of the digital objects based on the events to which the objects are related.

We performed a SPARQL query through the Europeana SPARQL endpoint[5] in order to extract all the instances of the edm: Event class in the current Europeana database. The query returned zero results. This suggests, as already pointed out in the Introduction, that the data collected by Europeana is not organised according to the event-centric approach, nor is this information introduced by the data aggregators that work in the Europeana network.

In order to introduce narratives in Europeana, we developed an ontology for narratives. A complete description of the classes and relations of the ontology is available on-line[6]. To maximise its interoperability, our ontology was developed as an extension of the CIDOC CRM standard ontology. Since the CIDOC CRM underlies the EDM, our ontology automatically extends the EDM, re-using its definition of event and introducing new relations between events. In particular, the ontology for narratives allows:

- Representing the events that compose a narrative, linked to each other using three types of properties:
  - Temporal occurrence relation, associating each event with a time interval during which the event occurs; an event occurs before (or during, or after) another event just in case the period of occurrence of the former event is before (or during, or after) the period of occurrence of the latter event. We reused the intervals defined in Allen's temporal logic [1], which are adopted by the CRM.

---

[5] http://sparql.europeana.eu.

[6] A description of the classes and relations of our ontology is available at the following address: https://dlnarratives.eu/ontology.

- Causality relation, linking events that in normal discourse are predicated to have a cause-effect relation, e.g. the eruption of the Vesuvius caused the destruction of Pompeii. The causality concept is represented by introducing a new relation of causal dependency, named *causallyDependsOn*. The only causal property of the CRM, *P17 was motivated by*, cannot be used for narratives since it relates activities but not events. The causality relation is defined by the narrator according to his/her own interpretation of the events.
- Mereological relation, connecting events to other events that include them as parts, e.g., the birth of Klimt is part of the life of Klimt, represented using the CRM property *P9 consists of*.

– Linking an event with the related digital objects included in a digital library.
– Representing the inferential process of a narrator who reconstructs a narrative starting from the primary sources. In this way our ontology allows to represent the data provenance.

In order to populate the ontology, we developed a semi-automatic tool that produces as output an OWL graph [8]. The tool is described in Sect. 4.

4   The Narrative Building and Visualising Tool

In order to aid the user in constructing the narrative, we developed a narrative building and visualising tool (NBVT for short)[7]. NBVT is web-based and written in HTML5, CSS, and JavaScript with jQuery[8]. It interfaces with a CouchDB[9] database to store data and retrieve it on subsequent loadings. The communication with the database is handled by the PouchDB[10] library, which allows NBVT to store the inserted data locally and optionally synchronise it with a remote server. Figure 1 shows the main view of NBVT.

To simplify the insertion of data by the user and the subsequent population of the ontology, we decided to build NBVT on top of an existing knowledge base, in order to provide the user with a vast and detailed amount of resources with which to build the narrative. The knowledge base we opted for is Wikidata[11], a free collaborative knowledge base operated by the Wikimedia Foundation [9]. Wikidata was built by extracting structured knowledge from Wikipedia and other collaborative projects, e.g. Wikisource, Wikibooks, Wikiquote. It currently contains more than 25 million entities and allows exporting of the data through the Wikidata API and a SPARQL endpoint called Wikidata Query Service[12]. Wikidata encourages collaborative addition and editing of the data by its users through manual and automated means. NBVT takes as input data inserted

---

[7] https://dlnarratives.eu/tool.html.
[8] https://jquery.com.
[9] https://couchdb.apache.org.
[10] https://pouchdb.com.
[11] https://wikidata.org.
[12] https://query.wikidata.org.

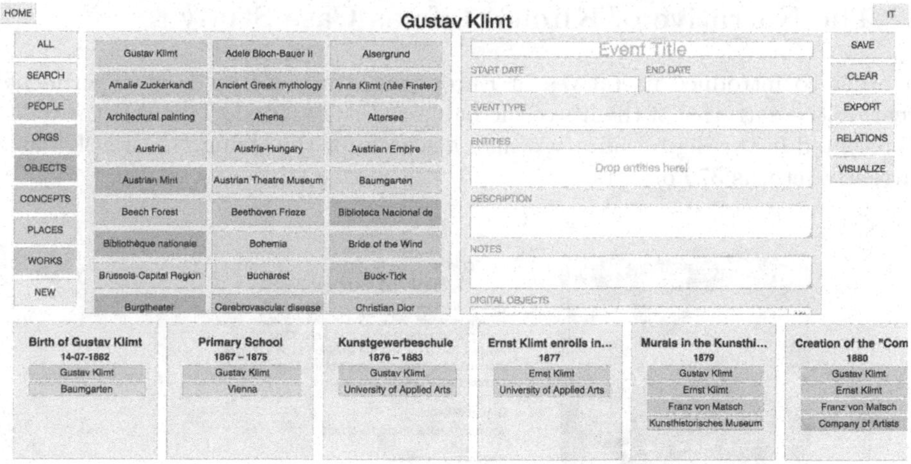

Fig. 1. The main view of NBVT: on the left side the entities automatically extracted from Wikipedia, on the right side the form for constructing an event, and at the bottom of the figure, the created events in chronological order.

manually by the user and imported automatically from Wikidata. It initially imports a few default events from Wikidata, such as births, deaths, company foundations, etc. The user then adds the remaining events of the narrative one by one, by inserting the following information:

- The title of the event
- The start and end dates of the event
- The event type
- A set of entities imported from Wikidata or defined by the user
- For each entity, one or more primary or secondary sources and, in the case of people, the role they played in the event
- A textual description of the event
- Optional textual notes
- One or more digital objects.

At any moment during the narrative construction, the user can switch to the relations view, which allows her/him to link the events through causal or mereological (part-of) relations through a drag-and-drop interaction.

The output of the web interface is an intermediate JSON representation of the narrative that is later converted to an OWL graph by a triplifier. The resulting representation is uploaded into a Virtuoso triple store [6], which can then be queried to produce visualisations such as tables, graphs, or timelines. In particular, we use the TimelineJS[13] library to build a rich-media timeline visualisation of the narrative. The collected data will also be accessible through a SPARQL endpoint.

---

[13] https://timeline.knightlab.com.

5   The Narrative of Klimt's Life as Case Study

In order to introduce narratives in Europeana, we performed an experiment creating the narrative of the life of the painter Gustav Klimt. This artist is well-represented in the digital library, where a search for the string "Gustav Klimt" currently returns 370 objects.

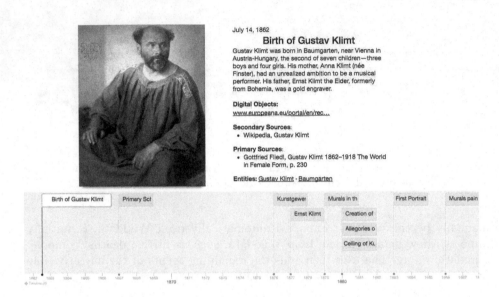

July 14, 1862

Birth of Gustav Klimt

Gustav Klimt was born in Baumgarten, near Vienna in Austria-Hungary, the second of seven children—three boys and four girls. His mother, Anna Klimt (née Finster), had an unrealized ambition to be a musical performer. His father, Ernst Klimt the Elder, formerly from Bohemia, was a gold engraver.

Digital Objects:
www.europeana.eu/portal/en/rec...

Secondary Sources:
• Wikipedia, Gustav Klimt

Primary Sources:
• Gottfried Fliedl, Gustav Klimt 1862–1918 The World in Female Form, p. 230

Entities: Gustav Klimt · Baumgarten

Fig. 2. An event in the timeline of Klimt's biography, showing the textual description of the event, the related digital object, the primary and secondary sources, the related entities and an image from Wikimedia Commons.

Since we are not art historians nor experts about Klimt, we decided to build the narrative based on the English Wikipedia page about the painter[14]. In the Wikipedia text we identified the main events of Klimt's life, and we reproduced them using NBVT. For each event, we defined the related entities (e.g. persons, physical objects, location, time). We also reported the Wikipedia fragments of text that describe the event and an image from Wikimedia Commons and related to the entities that compose the event. Furthermore, we reported the primary sources cited by Wikipedia, on the basis of which an event is placed in the narration. Figure 2 shows the first event of Klimt's life in our narration. When a user creates an event, NBVT suggests a list of digital objects extracted from Europeana. The list is built on the basis of the correspondence between the date of the created event and the value of the "Creation Date" field of the Europeana objects. Through this functionality, we were able to connect digital objects to the events of our narration, thereby creating a semantic network of events and

---

[14] https://en.wikipedia.org/wiki/Gustav_Klimt.

of the corresponding digital objects. This network is presented as a timeline in which the events are ordered chronologically. However, it is also possible to (i) visualise the events in a different order, based on their causal or mereological relations, (ii) highlight the events that contain one or more digital objects, as shown in Fig. 3, and (iii) filter the timeline to display only these events.

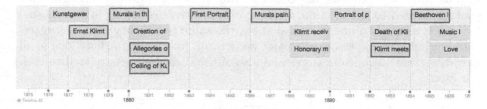

Fig. 3. Some highlighted events containing digital objects from Europeana.

The Klimt narrative is composed of a total of 54 events. 31 of them are connected with Europeana digital objects, and 18 are linked to more than one digital object. The total number of digital objects in the narrative is 127, that is 34% of all Klimt-related objects in Europeana. It should be noted that several Europeana objects are not related to Klimt's biography, e.g. posters, modern objects inspired by Klimt. We estimated that, using NBVT, the manual work for creating the narrative was about 3 person-days (7 h per day).

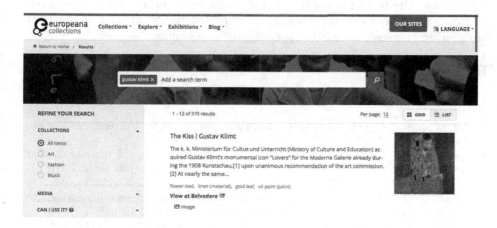

Fig. 4. The current results page of the search for "Gustav Klimt" in Europeana.

In order to integrate our narratives into Europeana, NBVT enriches them with metadata that describe the narrated topic. These metadata could be matched with those contained in Europeana to enhance its search functionality. When a user queries Europeana, she/he could obtain as response one or more narratives related to the topic of the search, along with the classical ranked list of digital objects. In the web interface, the timeline of each narrative could be visualised by adding a specific menu section, titled "Narratives", in the "Refine your search" menu on the left side of the page. Figure 4 shows the current results page of the search for "Gustav Klimt" in Europeana. Europeana also has an "Explore" section in its upper menu, providing a list of particular views on the Europeana data, e.g. views for people, for time periods, for sources. A new entry could be added to this section, showing all narratives collected in Europeana. If there is more than one narrative about the same topic, they could be aggregated by subject. Exploring this page, users could have an immediate idea of all narratives that Europeana stores and also of the digital objects related to each narrative.

6   Conclusions and Future Work

We have presented some use cases that would call for the introduction of narratives in digital libraries. We have then discussed how to extend the Europeana Data Model in order to realize such introduction in Europeana. A first experiment has been presented consisting in the creation of a narrative of the life of Gustav Klimt, the Austrian painter. Such creation has been performed by using a tool for building and visualising narratives. We plan to extend the experiment to other artists, thereby further enriching the Europeana database. We are also working on extending our narrative ontology in various ways: (1) representing and reasoning about the temporal relations between events; (2) representing in a richer way the text of narratives, called narrations, and the relation between narration fragments and their semantic counterparts, events and objects; (3) introducing narrative templates capturing recurring plots.

References

1. Allen, J.F.: Towards a general theory of action and time. Artif. Intell. **23**(2), 123–154 (1984)
2. Bal, M.: Narratology: Introduction to the Theory of Narrative. University of Toronto Press, Toronto (2009)
3. Concordia, C., Gradmann, S., Siebinga, S.: Not just another portal, not just another digital library: A portrait of Europeana as an application program interface. IFLA Journal **36**(1), 61–69 (2010)
4. Doerr, M.: The CIDOC conceptual reference module: an ontological approach to semantic interoperability of metadata. AI Magazine **24**(3), 75 (2003)
5. Doerr, M., Gradmann, S., Hennicke, S., Isaac, A., Meghini, C., van de Sompel, H.: The Europeana Data Model (EDM). In: World Library and Information Congress: 76th IFLA General Conference and Assembly, pp. 10–15 (2010)
6. Erling, O., Mikhailov, I.: RDF support in the Virtuoso DBMS. In: Networked Knowledge-Networked Media, pp. 7–24. Springer (2009)
7. Herman, D.: Basic elements of narrative. John Wiley & Sons (2011)
8. McGuinness, D.L., Van Harmelen, F., et al.: OWL web ontology language overview. W3C Recommendation **10**(10), 2004 (2004)
9. Vrandečić, D., Krötzsch, M.: Wikidata: a free collaborative knowledgebase. Commun. ACM **57**(10), 78–85 (2014)

Evaluation of Semantic Web Ontologies for Modelling Art Collections

Danfeng Liu, Antonis Bikakis[✉], and Andreas Vlachidis

Department of Information Studies, University College London, London, UK
{danfeng.liu.15,a.bikakis,a.vlachidis}@ucl.ac.uk

Abstract. The need for organising, sharing and digitally processing Cultural Heritage (CH) information has led to the development of formal knowledge representation models (ontologies) for the CH domain. Based on RDF and OWL, the standard data model and ontology language of the Semantic Web, ontologies such as CIDOC-CRM, the Europeana Data Model and VRA, offer enhanced representation capabilities, but also support for inference, querying and inter-linking through the Web. This paper presents the results of a small-scale evaluation of the three most commonly used CH ontologies, with respect to their capacity to fulfil the data modelling requirements of art collections.

1 Introduction

The voluminous, diverse and heterogeneous Cultural Heritage (CH) information that is available in museums, art galleries and other CH institutions, has recently led many of them to adapt advanced metadata models to better organise their data and enable the development of enhanced data services to the visitors of their physical collections and their digital counterparts. Amongst the several available metadata models developed for such purposes, Semantic Web ontologies are the most widely used, mainly because of their enhanced expressiveness, allowing to represent complex semantic relationships among CH entities, but also due to several other properties they enjoy such as extensibility, generality and inference support.

In the case of art collections, the data modelling challenges are even greater than in other fields of CH. Artworks are available in multiple formats (images, texts, etc.), they are multi-topical (art, science, etc.), multi-cultural and multi-targeted (different recipients [1]. It is therefore even more difficult to develop a data model that can efficiently capture all these different diversities and heterogeneities but in the same time remain simple in its use.

Motivated by the above observation, this paper attempts to address the following research questions:

Do the current CH ontologies meet the data modelling requirements of art collections and especially with respect to the following needs of art galleries:

(a) *cataloguing (collection & conservation management);*
(b) *display and publication of metadata (presentation of data);*
(c) *portals and system management.*

© Springer International Publishing AG 2017
M. Kirikova et al. (Eds.): ADBIS 2017, CCIS 767, pp. 343–352, 2017.
DOI: 10.1007/978-3-319-67162-8_34

Our study is focused on three CH ontologies, which are commonly used for modelling art collections metadata: the CIDOC Conceptual Reference Model (CIDOC-CRM), the Europeana Data Model (EDM) and VRA Core. Our methodology consists of selecting a characteristic sample of artworks and modelling its associated metadata using the three different ontologies, and then, based on the outcome of the first task, evaluating the three ontologies using a set of criteria associated to the data modelling needs and requirements outlined by the research questions.

The rest of the paper is structured as follows: Sect. 2 provides the necessary background information on the three ontologies. Section 3 describes our methodology in detail. Section 4 presents the results of the evaluation and Sect. 5 concludes.

2 Background

The CH domain was one of the first ones to adopt Semantic Web data models, methods and tools for modelling CH collections and publishing them online [2]. In this domain, SW technologies are mainly used for two purposes: the development of inner curation systems and the establishment of open collection databases. Ontologies and data models such as CIDOC-CRM, EDM and VRA Core are used to standardise the vocabularies for describing the relevant CH entities and their relationships, and in this way to enable interoperability among different CH institutions. Below, we provide some background information on these three data models, and specifically their most recent versions (CIDOC-CRM 6.2.1[1], EDM 5.2.7[2] and VRA Core 4.0[3]).

CIDOC Conceptual Reference Model (CIDOC-CRM). CIDOC-CRM is a formal structure developed by the International Committee for Documentation (CIDOC) of the International Council for Museums (ICOM) for describing the implicit and explicit concepts and relationships used in CH documentation [3]. Its event centric mechanism, which employs a vocabulary consisting of 82 classes and 262 properties and following the RDF semantics, enables the interrelation between people, things, places and time-spans through common events. In 2016, CIDOC-CRM became an ISO standard (ISO 21127:2006) for the interchange of cultural heritage information.

Europeana Data Model (EDM). EDM was developed in the context of the Europeana project as a Semantic Web-based framework for representing cross-domain collection metadata in museums, libraries and archives [4]. It is aligned to CIDOC-CRM in its definition of an event-centric model. To enhance interoperability, it reuses elements from other Semantic Web vocabularies, such as RDF, the Open Archives Initiative Object Reuse and Exchange (OAI-ORE) framework, the Simple Knowledge Organization System (SKOS) namespace, Dublin Core and the W3C Data Catalog Vocabulary (DCAT) [5]. It also introduces 11 new classes and 30 properties.

---

[1] http://www.cidoc-crm.org/Version/version-6.2.1.

[2] http://pro.europeana.eu/files/Europeana_Professional/Share_your_data/Technical_requirements/ EDM_Documentation/EDM_Definition_v5.2.7_042016.pdf.

[3] http://www.loc.gov/standards/vracore/schemas.html.

The Visual Resource Association Core Categories (VRA Core). VRA Core was developed for describing works of visual culture, collections, as well as images that document them [6]. It, therefore, represents three broad groups of entities, which are works, images, and collections. Having a much narrower scope than CIDOC-CRM and EDM, it consists of 19 elements including *Title, Record Type, Material, Creator, Measurements, Technique, Subject, Relation,* and *Rights*. It was originally developed as an XML Schema, but has recently become available as an RDF Schema.

3 Methodology

The sample we used for the evaluation consists of the four artworks presented in Table 1. The descriptions of the four artworks contain multiple kinds of information, meeting our aims for assessing the representation capabilities of the three ontologies. In the case of the paintings, the descriptions include their technical description, provenance, exhibition history, relevant bibliography, but also information about X-radiographs of the paintings. What is more interesting in the case of the two sculptures is their relationship: SA is the original work, and SB is the plaster cast of SA. SB is, however, considered, an artwork itself, created with certain art techniques.

Table 1. The sample of artworks used in our evaluation

Id	Artwork	Artist	Institution
PA	Self-Portrait (1659)[a]	Rembrandt	National Gallery of Art, Washington
PB	Queen Elizabeth I (1879)[b]	Unknown	National Portrait Gallery, London
SA	David (1501–1504)[c]	Michelangelo	Galleria dell' Accademia, Florence
SB	David (casted 1857)[d]	Unknown	V&A, London

[a]http://www.nga.gov/content/ngaweb/Collection/art-object-page.79.html
[b]http://www.npg.org.uk/collections/search/portrait/mw02082
[c]http://www.accademia.org/explore-museum/artworks/michelangelos-david/facts-about-david/
[c]http://collections.vam.ac.uk/item/O39861/david-plaster-cast-michelangelo/

The criteria and measures we used to evaluate the ontologies, presented in Table 2, were selected after studying the relevant literature on ontology evaluation [7–12]. The criteria were adopted from the application-based evaluation methodology proposed by Brank et al. [7]. According to this study, each of the criteria has a specific purpose: C1–C2 are associated to the cataloguing needs of a CH institution, C3 is related to portal & system management, while C4–C6 are related to the presentation and publication of metadata. The measures were adopted from the Ontology Quality Evaluation Framework proposed in [8] and the evaluation methodology of [12].

Table 2. Ontology evaluation criteria and measures

Criterion	Measure	Definition
Lexical, vocabulary & concept (C1)	**Accuracy**	Whether the ontology captures and represents correctly aspects of the real world
	Clarity	The effectiveness of the ontology in communicating the defined terms and their intended meaning
	Completeness/competency	Whether the ontology covers all essential and relevant concepts in the domain of interest
Hierarchy & taxonomy (C2)	**Conciseness**	Whether the ontology includes any irrelevant or redundant axioms
Computational-efficiency (C3)	**Interoperability**	Whether the ontology interacts or reuses axioms from other data models
User experience (C4)	**Ease of use**	Whether it is easy to operate the ontology and there is appropriate guidance
	Learnability	How easy it is to find the information needed to use, and whether there is any relevant documentation
Semantic relations (C5)	**Indexing and linking**	Whether the defined classes can act as indices to retrieve the requested information
	Inferencing	Whether the ontology can make implicit knowledge explicit through reasoning
Functional adequacy (C6)	**Consistent research and query**	Whether the ontology can achieve better querying and searching methods

4   Evaluation

4.1   Modelling the Four Artworks

The task of creating ontology-based descriptions of the four artworks using the three different ontologies consisted of two steps: The first step was to create the data model for each artwork using elements of each ontology. Examples of such data models are depicted in Figs. 1, 2 and 3. Figure 1 depicts the description of the creation of SA through three different production events, using terms from CIDOC-CRM. Figure 2 illustrates the use of EDM to describe the relationship between SA and its plaster copy, SB. Figure 3 demonstrates the use of aggregation in EDM for representing various web

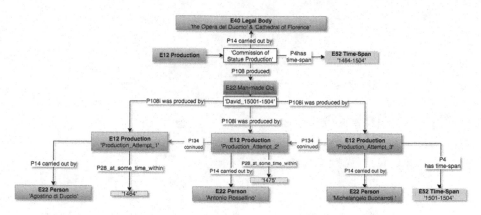

Fig. 1. The data model of PB using CIDOC-CRM

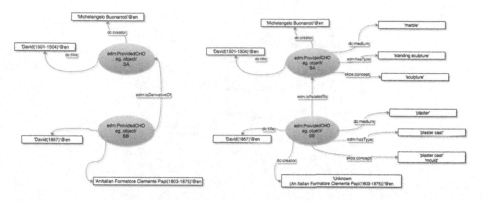

Fig. 2. Using EDM to model the relationship between SA and SB

resources related to SA. In the second step, we implemented the data models in Protégé[4] by adding appropriate individuals and property assertions with respect to the three ontologies. The second step aimed at verifying the data models, and also at examining the ontological inferences that could be made based on the semantics of each ontology. Figure 4 depicts the OWL/XML statements that describe different image resources related to PB, using terms from VRA Core.

4.2 Criteria-Based Evaluation

Based on the data modelling tasks described in Sect. 4.1, we assessed the three different ontologies using the measures we discuss in Sect. 3 and present in Table 1. The results of the evaluation are summarised in Table 3. Sections 4.2.1, 4.2.2 and 4.2.3 discuss in more detail the performance of each ontology.

---

[4] http://protege.stanford.edu/.

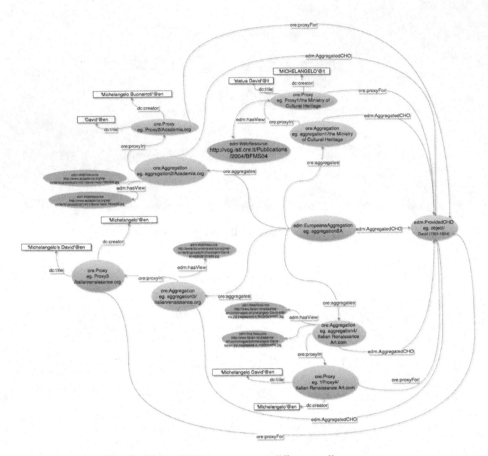

Fig. 3. Using EDM to aggregate different online resources

```
rdf:RDF
863       <!-- http://www.vraweb.org/vracore/vracore3.owl#PaintingA-Image1 -->
864 ▾     <owl:NamedIndividual rdf:about="http://www.vraweb.org/vracore/vracore3.owl#PaintingA-Image1">
865           <rdf:type rdf:resource="http://www.vraweb.org/vracore/vracore3.owl#Image"/>
866           <recordType rdf:resource="http://www.vraweb.org/vracore/vracore3.owl#image"/>
867           <type rdf:resource="http://www.vraweb.org/vracore/vracore3.owl#digital_image"/>
868           <description>the digital photo of Self_Portrait 1659 Rembrandt, coloured.</description>
869           <idNumber.currentRepository>Self-Portrait(1659)byRembrandt.jpeg</idNumber.currentRepository>
870           <rights>The National Gallery of Art, Washington, DC</rights>
871           <creator.corporateName>National Gallery of Art, Washington DC</creator.corporateName>
872           <creator.role>national gallery</creator.role>
873           <measurements.format>JPEG</measurements.format>
874           <measurements.resolution>3088 X 4000 px</measurements.resolution>
875           <technique>Scanning</technique>
876       </owl:NamedIndividual>
877
878       <!-- http://www.vraweb.org/vracore/vracore3.owl#PaintingA-Image2 -->
879 ▾     <owl:NamedIndividual rdf:about="http://www.vraweb.org/vracore/vracore3.owl#PaintingA-Image2">
880           <rdf:type rdf:resource="http://www.vraweb.org/vracore/vracore3.owl#Image"/>
881           <recordType rdf:resource="http://www.vraweb.org/vracore/vracore3.owl#image"/>
882           <type rdf:resource="http://www.vraweb.org/vracore/vracore3.owl#X-radiography"/>
883           <description xml:lang="en">the x-radiograph plate of the head of Self_Portrait 1659 Rembrandt, BW</description>
884           <idNumber.currentRepository>Self-Portrait(1659)byRembrandt2.jpeg</idNumber.currentRepository>
885           <rights>The National Gallery of Art, Washington, DC</rights>
886           <creator.corporateName>National Gallery of Art, Washington DC</creator.corporateName>
887           <creator.role>national gallery</creator.role>
888           <measurements.format>JPEG</measurements.format>
889           <technique>X-Radiography</technique>
890       </owl:NamedIndividual>
```

Fig. 4. Using VRA Core to model images of PB

Table 3. Summary of the evaluation of the three ontologies ('✗'denotes that the ontology doesn't exhibit good performance with respect to the corresponding criterion, '✓' denotes that the ontology performs well, and '★' that the ontology has excellent performance)

Measure	CRM	EDM	VRA
Accuracy	✓	★	★
Clarity	★	✗	★
Completeness/competency	✓	✗	✓
Conciseness	★	★	★
Interoperability	★	✓	✓
Ease of use	✓	✗	★
Learnability	★	✗	★
Indexing and linking	✓	★	✓
Inferencing	★	✗	✓
Consistent research and query	★	★	✓

4.2.1 CIDOC-CRM

CIDOC-CRM demonstrated its expressive power as a comprehensive ontology model for modelling cultural heritage information. It contains terms for capturing all aspects of the descriptions of the four artworks, structuring them in an event-centric way. For the specific scope of this experiment, there are, however, quite a lot of non-relevant to the task classes, which we did not use. Of course, the range of concepts it captures makes it applicable to a broad range of domains and applications and enhances its interoperability.

One problem we experienced with CIDOC-CRM was that it was not always easy to distinguish which class was more appropriate for modelling a specific entity, as some of its classes share very similar meanings. For example, both *E22 Man-made_Object* and *E19 Physical_Man_Made_Thing* seem to be appropriate for modelling artworks.

On the other hand, concepts in CIDOC-CRM are very systemically linked and designed. Once the user gets familiar with the logic and structure of the model, it is very smooth in its use, either for research or indexing. CIDOC-CRM is specified in a language-independent way, and has many expressions, amongst which one is in the RDF Schema vocabulary.

It is hard to judge the indexing ability of the ontology. On the one hand, it enabled us to represent nearly all data related to the four artworks. It allows, however, users to follow different data modelling approaches depending on the needs of the application, making the implementation of an indexing scheme for artwork difficult. For example, 'oil on canvas' (for PA) can be modelled in two different ways: as one individual, instance of *E55 Type*, linked to PA through the *P2 has type* property; or using two separate instances of *E55 Type,* 'oil painting' and 'canvas', linked to PA through the *P2 has type* property.

CIDOC-CRM is efficient in representing information related to custody and acquisition. Its expressivity sometimes, does not easily lend to a straightforward implementation of metadata for describing visual resources, e.g. information describing an image of the artwork, such as its dimension, time of creation, format, resolution and technique.

4.2.2 EDM

EDM is a metadata schema especially designed for Europeana.org so that its classes and properties are all appropriate for supporting the specific tasks of the project. However, it exhibits a poorly designed concept definition for some entities, which makes its use rather confusing. One example is the definitions of *edm:PhysicalThing* and *edm:ProvidedCHO*; edm:PhysicalThing refers to the persistent physical item that *edm:ProvidedCHO* represents. Therefore, the role of *edm:PhysicalThing* in data modelling is not clear and may confuse new users of EDM.

EDM is generic, including classes capturing high-level concepts related to artwork, but does not contain enough specialised classes for representing more specific metadata of CH artefacts. For example, we were only able to describe the location of the artworks at the granularity of country. It also lacks terms for modelling other relevant CH information, such as production process, acquisition, or technical reports. It is designed to focus on web resource aggregation and hence only collects the most basic information related to CH artefacts. Another important point is that EDM is only able to capture web resources. For example, for the case of the two sculptures (SA and SB), there was no way to describe their 'physical' characteristics in EDM.

The EDM vocabulary is rather difficult to understand and use, as its definition and guidelines to its users are unclear in some cases, and lack examples about the use of terms it imports from external vocabularies.

On the other hand, EDM is a simple and general model for the description of cultural heritage information, making it appropriate for supporting indexing and research queries. It can be used to extract exhibition records easily in a simple and clear representation. And its ability to model aggregated information (through *ore:Aggregation*, as shown in Fig. 3) is very useful, especially for the information management needs of a CH portal.

4.2.3 VRA Core

VRA Core was designed to enable object-centred descriptions of artwork with clear relationships between original work and its images. It smoothly models, for example, the numerous and complicated visual resources related to PB. It contains all essential terms to represent useful information about both the artwork and its images in an accurate, simple and concise way. For example, it uses *vra:stylePeriod.style* to describe the art style of the work, which is particularly useful for artwork. Moreover, it provides specific terms to describe provenance or acquisition data, such as *vra:location.formerRepository* and *vra: location.formerSite*.

With respect to the representation of dimensions, it is the most efficient among the three ontologies; it uses *vra:measurements.dimension, vra:measurements.format* and *vra:measurements.resolution*, which capture the different aspects of dimension for both physical objects and images.

Note that VRA Core does not provide an element to link a Work to its corresponding Image(s). This is described in [13] as a 'local implementation issue'. and may lead to problems when aggregating data from different resources. The solution is rather partly solved by introducing a generic property called *vra:relation*, which is used to describe relations between works and images with domain and range *vra:VisualResource* (a superclass of *vra:Work* and *vra:Image*). A subproperty of *vra:relation* is *vra:relation.depicts*, which links instances of *vra:Image* (domain) to instances of *vra:Work* (range). However, *vra:Work* and *vra:Image* are not disjoint classes, as some image might also be a work of art, such as photography. In our modelling task, it was very difficult to decide whether to define SB as an image, object or work in this context, while we also faced the same problem when attempting to represent any copied artwork of SB.

4.2.4   Evaluation Summary

In Table 4, we summarise the results of the evaluation with respect to the different research purposes that the ontologies can be used for.

Table 4. Summary of the evaluation of the three ontologies with respect to their research purposes

Research purpose	Criteria/metric	CRM	EDM	VRA
Institutional usage for cataloguing	Lexical, vocabulary, concept	✓	✗	★
	Hierarchy, taxonomy			
Portals & system management	Computational efficiency	★	✓	✓
Presentation of metadata	User experience	★	✗	★
	Semantic relations			

It is obvious that VRA Core and CIDOC-CRM better meet the cataloguing needs of CH institutions. We evaluate VRA Core as the best among the three ontologies for metadata presentation due to its efficient data modelling capabilities. EDM, on the other hand, appeared to be less strong for specialised cataloguing and data presentation. However, its support for modelling aggregation makes it very useful for the development and management of CH portals and information systems.

5   Conclusion

This paper evaluates the three popular Cultural Heritage ontologies with respect to their abilities to represent works of art. Our evaluation methodology consisted of selecting four characteristic examples of artwork, for which rich descriptions are available, creating descriptions of the four artworks using the three ontologies, and (based on the data modelling tasks) assessing the three ontologies using a set of evaluation criteria related to different uses of CH ontologies.

The main challenge of this evaluation was that, especially in the case of CIDOC-CRM and EDM, there were no detailed guidelines on how to use them to create the descriptions. Although, admittedly, the sample we used is rather small to generalize our conclusions, our experiment clearly identifies the main strengths and limitations of each ontology, and its results can be helpful for anyone who wants to semantically model similar CH information to support different kinds of applications. Our findings can be summarized as follows: CIDOC-CRM is a very general ontology, able to capture a great range of concepts related to the CH domain and in multiple different ways, according to the needs of the underlying application. EDM is more appropriate for creating and aggregating simpler semantic descriptions. VRA Core is the most appropriate among the three models for cataloguing purposes and for describing the relationships between different visual resources.

We plan to extend this work by considering more samples from a greater range of artwork types. Our endmost goal is to provide guidelines on how to best combine elements from the three (or other) ontologies in order to satisfy the set of criteria related to the design purposes of the CH ontologies in the best possible way.

Acknowledgements. This work was partially supported by CrossCult: "Empowering reuse of digital cultural heritage in context-aware crosscuts of European history", funded by the European Union's Horizon 2020 research and innovation program, Grant #693150.

References

1. Mantegari, G.: Cultural heritage on the semantic web: from representation to fruition. Ph.D. dissertation, Universita degli Studi di Milano Bicocca (2009). https://boa.unimib.it/handle/10281/9184
2. Hyvönen, E.: Publishing and Using Cultural Heritage Linked Data on the Semantic Web. Morgan & Claypool, Palo Alto (2012)
3. Doerr, M.: The CIDOC conceptual reference module: an ontological approach to semantic interoperability of metadata. AI Mag. **24**(3), 75–92 (2003)
4. Doerr, M., Meghini, C., Isaac, A., Hennicke, S., Gradmann, S.: The Europeana data model (EDM). In: World Library and Information Congress: 76th IFLA General Conference and Assembly, Gothenburg, Sweden, 10–15 August 2010
5. Europeana: EDM mapping guidelines V2.3 (2016). http://pro.europeana.eu/page/edm-documentation
6. The Library of Congress: VRA Core 4.0 schemas and documentation (2007). https://www.loc.gov/standards/vracore/schemas.html
7. Brank, J., Grobelnik, M., Mladenić, D.: A survey of ontology evaluation techniques. In: Proceedings of the Conference on Data Mining and Data Warehouses (SiKDD 2005) (2005)
8. Duque-Ramos, A., Fernández-Breis, J.T., Stevens, R., Aussenac-Gilles, N.: OQuaRE: a SQuaRE-based approach for evaluating the quality of ontologies. J. Res. Pract. Inf. Technol. **43**(2), 41–58 (2011)
9. Hlomani, H., Stacey, D.: Approaches, methods, metrics, measures, and subjectivity in ontology evaluation: a survey. Semant. Web J. **1**(5), 1–11 (2014)
10. Burton-Jones, A., Storey, V.C., Sugumaran, V., et al.: A semiotic metrics suite for assessing the quality of ontologies. Data Knowl. Eng. **55**(1), 84–102 (2005)
11. Vrandečić, D.: Ontology evaluation. Ph.D. thesis, Karlsruhe Institute of Technology, Karlsruhe, Germany (2010). http://www.aifb.kit.edu/images/b/b5/OntologyEvaluation.pdf
12. Vrandečić, D.: Ontology evaluation. In: Staab, S., Studer, R. (eds.) Handbook on ontologies, pp. 293–314. Springer, Heidelberg (2009). doi:10.1007/978-3-540-92673-3_13
13. Van Assem, M.: RDF/OWL representation of VRA (2005). https://www.w3.org/2001/sw/BestPractices/MM/vra-conversion.html

The CrossCult Knowledge Base:
A Co-inhabitant of Cultural Heritage Ontology and Vocabulary Classification

Andreas Vlachidis[1(✉)], Antonis Bikakis[1], Daphne Kyriaki-Manessi[2],
Ioannis Triantafyllou[2], and Angeliki Antoniou[3]

[1] Department of Information Studies,
University College London, London, England
{a.vlachidis,a.bikakis}@ucl.ac.uk
[2] Department of Library Science and Information Systems,
Technological Educational Institute of Athens, Athens, Greece
{dkmanessi,triantafi}@teiath.gr
[3] Department of Informatics and Telecommunications,
University of Peloponnese, Tripoli, Greece
angelant@uop.gr

Abstract. CrossCult is an EU-funded research project aiming to spur a change in the way European citizens appraise History, fostering the re-interpretation of what they may have learnt in the light of cross-border interconnections among pieces of cultural heritage, other citizens' viewpoints and physical venues. Exploiting the expressive power, reasoning and interoperability capabilities of semantic technologies, the CrossCult Knowledge Base models and semantically links desperate pieces of Cultural Heritage information, contributing significantly to the aims of the project. This paper presents the structure, design rationale and development of the CrossCult Knowledge Base, aiming to inform researchers in Digital Heritage about the challenges and opportunities of semantically modelling Cultural Heritage data.

Keywords: Cultural heritage · Ontology · Digital humanities · Semantic web · Vocabulary classification · CIDOC-CRM

1 Introduction

Without any doubt the era of digital distribution has introduced new exciting avenues for producing, accessing and consuming information. Within this realm, access to cultural heritage information has been significantly benefited by digital technologies, facilitating new ways of engaging with heritage and broadening public participation. Such advances not only enable an interactive engagement with heritage, but also reinstitute what we mean by heritage and how it can be accessed [1].

The CrossCult Project[1] realising the advances of digital technologies, particularly focused on the aspects of interactivity, recollection, and reflection, aims to spur a

---

[1] http://www.crosscult.eu.

© Springer International Publishing AG 2017
M. Kirikova et al. (Eds.): ADBIS 2017, CCIS 767, pp. 353–362, 2017.
DOI: 10.1007/978-3-319-67162-8_35

change in the way European citizens appraise History. By facilitating interconnections among pieces of cultural heritage information, public view points and physical venues, the project aims to foster the re-interpretation of history as we know it, which goes beyond the conventional siloed presentation of historical data, and focuses on aspects that are cross-cultural, cross-border, cross-religion, and cross-gender qualities.

A key contribution to this endeavour is the creation of a semantic knowledge base capable of interrelating a wide set of (existing and future) disparate digital cultural heritage resources. This paper discusses the scope of the CrossCult Knowledge Base, the design choices leading to the definition of its underlying Upper-level ontology, and the data-modelling outcome of a data sample. The Upper-level ontology delivers formalisms that describe the "world" of CrossCult, accommodating common conceptual arrangements, enabling augmentation, semantic-based reasoning and retrieval across disparate data resources.

Section 2 outlines relevant projects and the role of standard conceptual models for mediating semantic interoperability. Section 3 discusses the aims and design choices leading to the definition of the CrossCult Upper-level ontology. Section 4 presents the results of a data modelling exercise aimed at applying the conceptual arrangements and definitions of the CrossCult Upper-level ontology to a range of cultural heritage data resources. The discussion of a particular data modelling follows in Sect. 5, providing an insight the opportunities and limitations of the adopted modelling method. The last two sections highlight the most important lessons learned while defining and using the CrossCult Upper-level ontology and present the future steps towards finalising the semantic modelling endeavour.

2   Background

A fundamental problem area in dealing with Cultural Heritage data is to make the content mutually interoperable, so that it can be searched, linked, and presented in a harmonised way across the boundaries of the datasets and data silos [2]. In the sphere of contemporary information science, there is abundance of instruments for managing and modelling any kind of information including cultural heritage data. The Dublin Core (DC) Metadata Elements and DC Terms[2], the Simple Knowledge Organization System (SKOS[3]), the Functional Requirements for Bibliographic Record (FRBR[4]), the Europeana Data Model (EDM[5]), the CIDOC-CRM[6], the MIDAS Heritage standard[7], the Lightweight Information Describing Objects (LIDO[8]) and the VRA Core[9] to name

---

[2] http://dublincore.org/documents/dcmi-terms/.

[3] https://www.w3.org/2004/02/skos/.

[4] https://www.ifla.org/publications/functional-requirements-for-bibliographic-records.

[5] http://pro.europeana.eu/page/edm-documentation.

[6] http://www.cidoc-crm.org/.

[7] https://historicengland.org.uk/images-books/publications/midas-heritage/.

[8] www.lido-schema.org.

[9] https://www.loc.gov/standards/vracore/.

but a few, have been employed by numerous projects to harmonise access to content across disparate datasets [3]. Each model contains merits and limitations determined by its scope and origin. Some models are defined as nationally accepted standards, whereas others enjoy an international consent. Some models are domain independent and lightweight, others are more closely related to particular domain, some are described as harvesting metadata models and others present integrated manifestations.

In spite of the abundance of models and standards, the nature of cultural heritage data is such that does not simply lend to a straightforward cataloguing of information in the same way as warehouse data, administrational information or even library cata- logues [4]. Influenced by different scholarly disciplines and perspectives, the cultural heritage data contain an inherited variability that is reflected by a range of different types of historical objects with their different characteristics. Hence, it is crucially important semantic interpretation of cultural heritage data to be driven by real world concepts and events modelling data based on the relationships between empirically surfaced arrangements rather than artificial generalisations and fixed field schemas [5].

During the past decade, the CIDOC-CRM, a core ontology for cultural heritage data, has matured and gained a growing popularity among projects aimed at providing data aggregation and semantic harmonisation of cultural heritage information. Standing for Conceptual Reference Model (CRM) of the International Council of Museums (ICOM) – International Committee for Documentation (CIDOC), CIDOC-CRM is a well-established ISO standard (ISO 21127:2006) in the modelling of cultural heritage information [6]. It provides an extensible semantic framework that any cultural heritage information can be mapped to.

The applicability of the CIDOC-CRM in information systems of the broader cul- tural heritage domain is evident in the literature by numerous large-scale projects such as, the Oxford University CLAROS[10] project, the British Museum ResearchSpace[11] and the EU FP7 Ariadne Infrastructure[12]. The above projects integrate vast datasets of classical antiquity, museum exhibits and archaeological research respectively, pro- viding semantic interoperability and access to data based on the ontological and con- ceptual definitions of CIDOC-CRM. Specialisation of CRM instances to a terminological level is achieved by linking to external vocabulary sources, thesauri and classification schemes.

The CRM ontology provides a general mechanism for linking to terminological specialisations via the implementation of the E55 Type class, which enables connection to categorical knowledge commonly found in cultural documentation. A common implementation approach is to link CRM instances to thesauri concepts expressed as SKOS concepts. Simple Knowledge Organization System (SKOS) is a W3C recom- mendation designed for representation of thesauri, classification schemes, taxonomies, or any other type of structured controlled vocabulary [7]. It builds upon RDF and RDFS, and its main objective is to enable easy publication and use of such vocabularies

---

[10] http://www.clarosnet.org/.

[11] http://www.researchspace.org/.

[12] http://www.ariadne-infrastructure.eu.

as linked. SKOS structures can be linked to CIDOC-CRM instances to provide a specialised vocabulary.

3   Upper-Level Ontology – Definition and Requirements

The CrossCult Upper-level ontology is defined as a generic upper-level conceptual structure that captures common concepts and relationships across a diverse range of cultural heritage data. As such, the ontology delivers formalisms that describe the "world" of CrossCult; it accommodates common conceptual arrangements and enables augmentation, linking, semantic-based reasoning and retrieval across disparate data resources. In order to achieve its semantic interoperability aims the Upper-level ontology adopts a single and generic upper-level design, based on a robust ontological definition, enabling efficient semantic-based reasoning and retrieval, while being scalable to be extended formally to specialised conceptual needs when required.

Specified as a knowledge representation resource benefiting from maximum reuse of established semantic web resources and standards, the Upper-level ontology adopts the standard ontology for modelling cultural heritage data, CIDOC-CRM. The use of CIDOC-CRM guarantees integration under well-defined and interoperable semantics that support the generic aims of the upper-level structure whilst providing specialisations that can benefit the individual needs of pilots. On the other hand, CIDOC-CRM as a formal and generic structure of concepts and relationships is not tied to any particular vocabulary of types, terms and individuals. This level of abstraction, albeit useful for the semantics of the broader cultural heritage domain, does not cover the need for a finer definition of types, terms and appellations. The need for an additional level of vocabulary semantics is addressed by the use of thesauri and glossary supplementing the CIDOC-CRM with specialised terms.

3.1   Rationale and Design Choices

In the process of defining the ontological arrangements, the project reviewed the pilots' datasets and engaged in a series of meetings before concluding to a set of requirements and shared semantics across the four pilot's scenarios and data. The results led to the definition of the CC Upper-level ontology, which reuses terminology and maintains full compatibility with the widely-used standard in cultural heritage documentation CIDOC-CRM (ISO 21127:2006). The version of the upper-level ontology is a subset of CIDOC-CRM enhanced with additional semantics from the SKOS and FOAF [8] ontology.

The Upper-level ontology accommodates the range of shared semantics of the following commonly identified concepts across the four pilots; (a) Physical items, as is any museum artefact, painting, venue item or landmark, (b) Digital (audio-visual) content relating to one or more Physical Items, (c) Places of spatial focus, which could refer to the location of an object, a place of an event or a depicted place on a painting, (d) Time related definitions such as dates and periods, (e) Actor as a person or organisation related to a physical item by properties of ownership creation and illustration and (f) Reflective Topics carrying the semantics of subjects and topics of interest that drive the reflection and reinterpretation qualities of the application.

The Crosscult specific class *Reflective Topic*, acts as collection of primarily physical items (i.e. E22_Man-made Objects) which are aggregated under a common theme that enables interaction with the content, based on predefined reflection and reinterpretation threads. Instances of the class (threads) can be topics such as Immigration, Women in Society, Healing, Painting Style, etc. Each instance contains links to relevant subjects from the CCCS vocabulary enabling retrieval and cross-reference, narratives describing the topic, associations to reflection modules (e.g. quiz games and ratings), while it realises standard CIDOC-CRM relationships across individual physical items in terms of their location, material, date of production etc. For example, the individual CC2279 (Fig. 1), is a tombstone of the Middle Antonine period located at the Museum of Tripolis (Greece), and participating (cc.reflects) in the Reflective Topic *Woman Appearance*. Associations between individual physical items can be made through the use of a common reflection topic whereas other types relationships can be explored via the standard CIDOC-CRM properties.

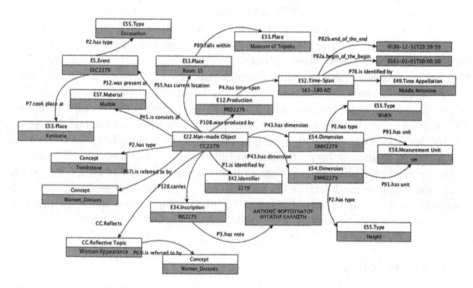

Fig. 1. Data model of museum exhibit 2279 (Archaeological Museum of Tripolis, Greece)

3.2 Vocabulary Requirements and Semantics

The upper-level ontology incorporates the SKOS semantics, specifically the SKOS Concept and Concept Scheme classes and their associated properties, to provide access to specialised vocabularies. In CrossCult this need is met by a custom built Classification scheme (CCCS) aiming at enhancing the concept representation of the reflective topics developed by the four pilots. This is supplemented by domain dependent vocabularies of geographical and chronological terms.

The CCCS supplements the CC ontology by providing an additional layer of semantics through a controlled vocabulary of concepts providing a concise representation of reflection themes and their interrelationships and guiding the reflective process through these interrelations. The vocabulary incorporated into CCCS accommodates the reflective topics and the relevant social and cultural terminology. In this sense the Classification Scheme can be used as a means for modelling vocabularies contributing to the cultural heritage domain.

The scheme aggregates terminology from standard thesauri resources such as, the Arts and Architecture Thesaurus of Getty (AAT), the EUROVOC, the UNESCO Thesaurus and the Library of Congress Subject Authorities (LC) vocabulary, whereas it incorporates a limited number of CrossCult specific terminology designed to accommodate specialized needs of the reflective process deriving from the pilots' scenarios and narratives. The vocabulary is organised and defined in a hierarchical order of broader-narrower term relationships, whilst CCCS terms can be employed both as "types" (instances of the E55 Type class) and as "propositional objects" (instances of E89) to describe the subjects related to individuals of the CC Upper-level ontology.

The reuse of standardised resources ensures the validity of the CCCS structure and the consistency in the use of its terms. To a lesser extent, project specific terminology has been incorporated into the CCCS and has been inter-weaved within its structure. To ensure the comprehensiveness of CCCS and to maintain the project specific focus of the terminology, the contributing terms are derived from the scenario descriptions of the four pilots and the descriptions of relevant cultural heritage objects, including their meaning, symbolism, materials, cultural context and creative techniques. The definition of the CCCS involved the following steps: (a) Identification of relevant vocabulary based on reviewing pilot scenarios and items involved for the building of the scenarios. This section relied heavily on cooperation with the historians, museum and venue curators and social scientists participating to the project as field experts; (b) Verification of vocabulary against authority thesauri and incorporation of authority terms as preferred terms when applicable; (c) Integration of mapped terms into the CCCS structure considering both original and CCCS hierarchies; (d) Further enhancement of CCCS vocabulary with related terms, suggested by the mappings with authority thesauri; (e) reviewing of CCCS structure and supplementing hierarchies as needed.

4   Data Modelling

Data modelling in the context of this paper refers to the specific process of applying the conceptual arrangements and definitions of the CrossCult Upper-level ontology to a range of disparate cultural heritage data resources. The origin of the data as well their coverage and granularity vary significantly.

Four distinct pilots contribute data to the CrossCult project covering a unique range of cultural heritage venues across Europe. From the large venue of National Gallery in London to the considerably smaller venue of the Archaeological Museum in Tripolis (Greece) and from the archaeological site of thermal springs in Montegrotto (Italy) to the historical points of interest in the cities of Luxembourg and Malta. Each pilot contributed data from about 25–30 unique items. The data sample describes museum exhibits, gallery items, archaeological sites and points of interest in terms of their unique identifier, associated descriptions, multimedia elements, and relevant keywords describing their content, use and/or symbolism.

The project ingests a wide range of diverse data associated to cultural heritage objects, events and subjects that span from antiquity to modern times and have a geographic span that runs across Europe. Hence, data is inherited to a wide array of formats, technologies, management and classification approaches relevant to each data provider or resource. The data modelling exercise relied on a rigorous set of Upper-level ontology definitions in order to express a diverse range of cultural heritage data on the same level of semantics and with the same degree of granularity.

Overall, the data modelling exercise delivered 80 uniquely identified items that are composed of 102 Physical Man Made Objects and 17 Physical Man Made Things. This translates to 3440 ontology (OWL) statements of named individual declaration and property assertion.

4.1 Method

The data modelling method addresses issues relating to the diversity of content types, data formats, and level of data detail. The process is abstracted into three main stages: (i) selecting and curating the source data for each pilot; (ii) data cleansing and normalisation, followed by data mapping to the Upper-level ontology; and (iii) automatic data assignment to CC ontology ensuring compliance with the model.

The **Manual Data Extraction** stage was dedicated to impose a data structure across a range of unstructured sample data available in text format. The volume of the data was not such to justify the development of a Natural Language Processing application for the automatic extraction of information from textual snippets. The task identified textual instances of relevant types (i.e. type of exhibit and related material), temporal and spatial information, dimensions, and other features of interest such as inscriptions or visual representations.

The **Semi-Automatic Database Construction** stage aimed at populating a set of relational database tables with structured data, from spreadsheets originating directly from the pilots or from the previous Manual Data Extraction stage. The relational database acted as a mediating layer between the semi-structured data files and the final OWL output feeding the routines of the Automatic Statements Generation stage with structured data. The database introduced a series of tables that stored the different types of CSV data, such as temporal, spatial, dimension, features, and other information associated to the cultural heritage data and conforming to CIDOC CRM structure of the ontology.

The **Final Automatic OWL Generation** stage, ingested the structured data of the relational database into the CC Upper-level ontology. The process employed a series of PHP routines driven by SQL queries for retrieving selected database records and declaring them as ontology individuals using OWL class and property assertions. The routines cater for the automatic generation of statements with respect to individual(s) declaration, class assertion, object property assertion, and data property assertion. String cleansing techniques were also applied for the generation of URI friendly values whereas in many cases complex SQL Join statements were used for retrieving record relationships across the database tables.

4.2 Data Modelling Example

The data modelling exercise delivered a representative example of pilot data with respect to the semantics of the Upper-level ontology. It managed to harmonise diverse data under a common semantic layer enriching their structure and enabling inference and retrieval. A leading modelling choice is the adoption of the specialised CIDOC CRM classes; E22.Physical Man Made Object and E24.Physical Man Made Thing, which provide a unified semantic view to a range of items of interest across the four pilots. This is augmented by the SKOS Concept and Concept Scheme classes drawing in the concepts incorporated in the CCCS. Hence, the range of artefacts, paintings, museum exhibits, monuments, and points of interest is modelled as instances of the aforementioned specialised classes.

The following example presents the modelling arrangements of museum exhibit 2279 from the Archaeological Museum of Tripolis (Greece). The museum contributes approximately 25 museums exhibits containing rich descriptions and information about their temporal, geometrical, spatial and contextual characteristics as seen below.

The example presents some specific requirements with respect to the modelling of the provenance of exhibits. The provenance information of the exhibit is accommodated by an E5.Event of type 'excavation' that took place in Kynouria (Greece). Figure 1 captures the semantics of the tombstone with respect to dimension, date of production, material, and location. The model accommodates relationships to conceptual characteristics that describe the artefact in terms of its reflective topic and subject keywords, these being the notion of death, funerary rites and funerary art through the ages, etc. It should be noted here that concepts through the structure of the CCCS can be enhanced at the direct terminology level, i.e. "tombstones" are part of "cemeteries" and are linked to "funerary sculpture".

2279: Marble pediment tombstone with a representation of a family (enface). The female figure bears a chiton and a cloak. The male figure and the boy bear a short chiton. On the architrave there is the inscription ANTIOXIC ΦΟΡΤΟΥΝΑΤΟΥ ΘΥΓΑΤΗΡ ΚΑΛΛΙΣΤΗ. Found in Herod Atticus villa in Loukou, Kynouria. Roman era work (middle Antonine era, 161 A.D - 180 A.D.). Dimensions: Height 1.60m, Width 0.82m. Location: Room 15, 1st floor

---

In addition, the inscription of the tombstone is modelled with precise semantics available from the upper-level ontology where the specialized property P128.carries, enables the relationship between the actual artefact and the carried inscription to be fully expressed. It is a different semantic relationship than the P62.depicts that is used for connecting an artefact with a depicted visual item. It is a fine distinction between depiction and carried inscription, demonstrating the flexibility and breadth of the ontology to deal with precise semantics when required.

The notion of women's dresses is given both as a reflective topic and a concept, as these coincide. The CCCS can lead the user of the app further to the "dress" as a "culture" element and what this expresses for "women's appearance".

5 Discussion

The design and development of the CrossCult Knowledge Base is not a straightforward data modelling exercise, but comes with some interesting research and practical challenges. The first challenge was the selection of the underlying ontology. Despite its growing popularity in the Cultural Heritage domain and its rich expressive capabilities, CIDOC-CRM was not an easy selection. Researchers with an Information Science background preferred solutions based on taxonomies or classification systems (e.g. Dublin Core), while software developers found CIDOC-CRM unnecessary complicated and verbose for the needs of the platform and mobile apps they develop. Considering the importance of modelling the relationships between the different cultural heritage resources used in the project, as well as the need for semantically linking such resources with external vocabularies and ontologies, we finally decided to adopt CIDOC-CRM.

Another critical challenge is related to the population of the ontology with appropriate individuals and statements describing the available cultural heritage resources. We presented the process of converting the available unstructured or semi-structured data into instances of the Upper-level ontology classes and statements using properties of the ontology. However, the mapping between the terms used by historians in the original descriptions of the resources and the elements of the ontology was not in many cases straightforward. Reaching a common understanding of the precise meaning of the original descriptions, and determining their mappings to the ontology required extensive communication between the ontology experts and the historians. By focusing on a representative sample from the four project pilots, we developed semi-automatic processes, which could then be re-used for all the pilot data.

The different backgrounds of the people who were involved in the development of the CrossCult Classification Scheme (information scientists, historians and museum experts) brought two more challenges to the project: how to determine the scope of the vocabulary, and how to come up with a commonly agreed structure. Two decisions that helped us address such challenges were: (i) to rely as much as possible to standard external vocabularies such as AAT; (ii) to setup and use an online environment for collaborative development and management of vocabularies, thesauri and taxonomies. Among others, the environment enables discussions on the terms and structure of the ontology, linking the vocabulary to external terms and creating RDF descriptions of the vocabulary.

6 Conclusion and Future Steps

The paper presented the main design decisions, tasks and challenges associated to the development of the CrossCult Knowledge Base. Apart from serving the specific aims of the project, the research we present in this paper, has three more general contributions to the Digital Heritage domain: (i) it demonstrates the use and deployment of standard cultural heritage ontologies, which have so far been used mainly for research purposes, in the context of user-oriented applications; (ii) it develops a vocabulary for historical reflection and integrates it into standard cultural heritage ontologies; (iii) it harmonizes datasets describing disparate cultural heritage resources, from museum exhibits and archaeological sites, to Points of Interest in urban settings.

We also presented a data modelling example, which demonstrated the semantic description of project pilots' data with respect to the semantics of the Upper-level ontology, which underlies the CrossCult Knowledge Base. The next stages will focus: (i) augmenting the data with media content and narratives that enhance their reflection and re-reinterpretation qualities; (ii) semantically enriching the resource descriptions with links to external standardised semantic web resources; (iii) further refining the scope and structure of Reflective Topics and their relation to keywords, narratives and other reflection proposals; (iv) extending the ontology to accommodate other project-related concepts, such as the pilots' venues and the users of the pilot apps.

Acknowledgements. This work has been funded by CrossCult: "Empowering reuse of digital cultural heritage in context-aware crosscuts of European history", funded by the European Union's Horizon 2020 research and innovation program, We would like to thank Louis Deladiennee (LIST), Kalliopi Kontiza (National Gallery), Yannick Naudet (LIST), Joseph Padfield (National Gallery) and Evgenia Vasilakaki (TEI-A) for their valuable comments and contributions during the course of this research.

References

1. Adair, B., Filene, B., Koloski, L. (eds.): Letting Go?: Sharing Historical Authority in a User-Generated World. Left Coast Press, Philadelphia (2011)
2. Hyvönen, E.: Publishing and using cultural heritage linked data on the semantic web. Synth. Lect. Semant. Web: Theor. Technol. **2**(1), 1–159 (2012)
3. Ronzino, P., Amico, N., Niccolucci, F.: Assessment and comparison of metadata schemas for architectural heritage. In: Proceeding of CIPA, 12 September 2012
4. Oldman, D., Doerr, M., de Jong, G., Norton, B.: Realizing lessons of the last 20 years: A manifesto for data provisioning & aggregation services for the digital humanities (a position paper). D-lib magazine, 20(7/8) (2014)
5. King, L., Stark, J.F., Cooke, P.: Experiencing the digital world: the cultural value of digital engagement with heritage. Herit. Soc. **9**(1), 76–101 (2016)
6. Doerr, M.: The CIDOC conceptual reference module: an ontological approach to semantic interoperability of metadata. AI Mag. **24**(3), 75–92 (2003)
7. Miles, A., Bechhofer, S.: SKOS simple knowledge organization system reference (2009). http://www.w3.org/TR/skos-reference/
8. Brickley, D., Miller, L.: FOAF vocabulary specification 0.99 (2014). http://xmlns.com/foaf/spec/

The Port History Ontology

Bruno Rohou[1,2(✉)], Sylvain Laube[1], and Serge Garlatti[2]

[1] Centre F. Viete (EA 1161), Université Bretagne Occidentale, Brest, France
bruno.rohou@univ-brest.fr
[2] IMT Atlantique, Lab-STICC, Univ. Bretagne Loire, 29238 Brest, France

Abstract. This paper presents a reference ontology, called PHO (Port History Ontology) as an outcome of a multidisciplinary research project in the field of Digital Humanities (History of Science and Technology or HST) and knowledge engineering. The PHO ontology is mainly based on the HST-PORT model representing the spatio-temporal evolution of ports (from digital Humanities) and the specialization of the CIDOC-CRM ontology. To ensure genericity and reusability, our ontology relies on case studies in comparative history of the ports of Brest (France), Mar del Plata and Rosario (Argentina).

Keywords: History of science and technology · Ontologies · CIDOC-CRM · DOLCE Lite

1 Introduction

Our research work is part of the research programs "History of port landscapes" and "Digital Humanities" developed within the framework of the research group PAM 3D Lab, where Center François Viéte (EA 1161) collaborates with Lab-STICC and CERV. One of those axes concerns the understanding of the scientific and technological evolution of ports (Brest in France, Mar del Plata and Rosario in Argentina) in the contemporary era with a methodological approach considering the port as a Large Technological System [6]. The aim is to build and validate new methods in digital humanities both for history and for development of the scientific, technological and industrial heritage applied to ports. From a computer perspective, our methods are based on knowledge engineering, ontologies and semantic web.

The hypothesis is to consider a port as a Large Technological System (or LTS) [4] whose spatio-temporal evolution as an artifact[1] [9] is part of studies in HST. This evolution can be considered as multi-scale (on both space and time). The harbor itself consists of a set of artifacts at various levels of granularity such as refit forms, jetties, wharves, cranes, moorings, or industrial production units (forges, rope factory, etc.). Periodizing the port in HST requires to highlight moments of breaks linked to the evolutions of some specific artifacts and thus to identify periods where the system is stable between two breaks. In addition,

---

[1] The word "artifact" will be used as equivalent to "human production".

© Springer International Publishing AG 2017
M. Kirikova et al. (Eds.): ADBIS 2017, CCIS 767, pp. 363–372, 2017.
DOI: 10.1007/978-3-319-67162-8_36

studying the life cycle of an artifact leads to analyze the nature of human activities. In the case of LTS, it is a matter of selecting relevant artifacts to periodize this system and then to characterize the entities in relation to them. Periodicizing a LTS and defining cycles of evolution of a port over a long period of time is thus tantamount to selecting relevant artifacts, i.e. artifacts that account for the studied periodization. Each of the artifacts forms tangible traces of port evolution at different scales of time and space.

The main contributions of our research work are: (i) A port periodization model, called HST-PORT, developed from the periodization model of ANY-PORT ports derived from the work of geographers [1]. It is based on five evolution phases and a model of human activities with relations between actors, knowledge and artifacts [7]; (ii) A reference ontology called Port History Ontology (PHO) whose classes, properties and structures are "derived" from the HST-PORT model. The latter is based on an approach proposed by G. Kassel [5] to design a formal ontology of artifacts, the CIDOC-CRM ontology, the OWL-Time ontology and the WGS84-pos ontology for geolocation.

The article begins by a brief state of the art on the CIDOC-CRM ontology and its extensions. We then present our HST-PORT periodization model exemplified with the Rosario harbor (Argentina). The design method for the PHO reference ontology is then described. We explain the alignment of PHO with DOLCE, OWL-TIME and WGS84-pos, but more particularly with CIDOC-CRM. The ports of Rosario and Brest are used to show how the evolution of the ports is modeled with the ontology PHO and how to query the knowledge base system to retrieve port evolutions. The article will end with a conclusion and perspectives.

2   CIDOC-CRM, Cultural Heritage and Ontologies

The modeling of knowledge in the field of cultural heritage has already been the subject of numerous studies leading to the creation of CIDOC-CRM for the management and enhancement of the museum heritage[2]. It is an ontology developed by the International Committee for Documentation (CIDOC) of the International Council of Museum (ICOM). The aim of the CIDOC-CRM ontology is to propose a model describing the concepts and the relations allowing to describe the cultural heritage of the museums. Its use concerns the tangible and intangible cultural heritage and more precisely "all types of material collected and exhibited by museums and related institutions", as well as "collections, sites and monuments related to fields such as history Social, ethnography, archeology, fine arts and applied arts, natural history, history of science and technology" [12]. Many communities have adapted CIDOC CRM to meet specific needs. In particular, the CRMgeo[3] Which combines the CIDOC CRM with GeoSPARQL

---

[2] "The CIDOC Conceptual Reference Model (CRM) provides definitions and a formal structure for describing the implicit and explicit concepts and relationships used in cultural heritage documentation", http://www.cidoc-crm.org/.

[3] CRMgeo: a Spatio-temporal model - http://www.ics.forth.gr/isl/index_main.php?l=e&c=661.

to provide an ontology providing a standard for the geolocation of cultural heritage or the CRMarchaeo[4] which is an ontology based also on CIDOC CRM interested in archaeological excavations. Another extension, called CRMba dedicated to archeology to encode metadata about the documentation of archaeological buildings [3,10,11]. The CRMdig [2] is an extension of the CIDOC-CRM ontology for capturing the modeling and the query requirements regarding the provenance of digital objects for e-science. Finally, a domain ontology dedicated to the conservation-restoration process of cultural objects was designed and focus on the development of the elements related to events affecting a cultural object [8]. In the medium term, our main goal is to "submit" an extension of CIDOC-CRM for history. Our current proposal is an CIDOC-CRM extension for port history based on our HST-PORT model.

3   The HST-PORT Model

The HST-PORT model uses the fundamental principles of the ANYPORT model. One of the contributions of this model is to show the evolution of the relationship between the port and the city. The generic model ANYPORT developed by Bird [1] allowed to periodize series of ports by studying the evolution in time of their port facilities. In ANYPORT, the study and description of the spatio-temporal evolution of a port is based on a set of major evolution phases and the breaks between these phases from the sources linked to the ports considered associated with the relevant port entities or facilities - as break traces. In the same way, our HST-PORT model is composed of:

- A set of major evolution phases: There are five main phases obtained from the artifacts' life cycle. The five phases are as follows: (i) Phase 0: emergence of a need/demand; (ii) Phase 1 or phase of project instruction: translation of this need into a technological problem, emergence of different solutions, choice of a solution; (iii) Phase 2 or phase of the production of the artifact chosen to solve the problem; (iv) Phase 3 or phase of use of the artifact accompanied by maintenance and repair phases; (v) Phase 4 or phase of obsolescence/disappearance of the artifact.
- A set of relevant entities: The considered entities are those that allow the modeling of human activities, that is to say Actors, Artifacts and Knowledge.

As the ANYPORT model, we assume that it is possible to describe the LTS evolution from artifacts and human activities. From the port sources, historians have to first find and analyze historical sources to identify the five phases described above. Secondly, they also have to find Artifacts and their relevant properties which form tangible traces of Port at different scales of time and space. For each of these artifacts, it involved Knowledge, actors and related

---

[4] CRMarchaeo: the Excavation Model - http://www.ics.forth.gr/isl/index_main.php?l=e&c=711.

activities. Now, we show the results of our analyzes of the primary and secondary sources (10,000 digital photos representing about 500 documents) from the ports of Brest, Mar del Palta and Rosario.

Example of Relevant Artifacts and Their Properties for Periodization. At this stage of research, jetties, wharves, cranes and grain elevators are relevant artifacts because their evolution over time reflects the evolution of ports and the different phases. However, these artifacts do not lead to the same periodization scale (jetties: a century, wharves: several decades, cranes from one year to several decades). Now, we take the example of the wharf among these artifacts to show how certain properties are revealing breaks. for this artifact, the revealing break properties are its characteristic dimensions (length and depth), the geolocation features, the used technologies and the other artifacts (like crane for example) present on the wharf. For the "National Wharves" in 1880 in Rosario, these properties are as follows:

(i) **Characteristic sizes**: The cumulative length of these platforms is 627 m and the maximum depth of 7 m. In general, wharves are made up of small docks (depth less than 3 m), platforms of medium mooring (3 m-5 m), and docks of great anchorage (beyond 6 m). The data on the length of the deep docks are as follows: 1880: 625 m; 1906: 1075 m; 1912: 3655 m; 1942: 3655 m. The evolution of length and depth produces a periodization of the history of this artifact, made of continuity and break.

(ii) **The geolocation references**: those of Rosario (Coordinates: -32.941397; -60.632886) and those of the geolocation points determining the contour of the wharves.

(iii) **Technologies**: This involves identifying the technology used to build it for each dock. It will depend directly on the know-how and knowledge available at the time of the construction activity of the artifact. From a practical point of view, we will distinguish the technology of the foundation (the submerged part of the wharf), the technology related to the elevation (the emerged part of the wharf). It can be a technology related to wood, stone, concrete.

(iv) **Other artifacts present**: two types of artifacts have been identified. The artifacts related to mobility (crane, voice transport, means of transport...) and those related to storage (warehouse, hangar, grain silos, etc.). For mobility-related artifacts, the relevant indicators for periodization are energy, maximum transport capacity and type of artifact (mobile crane, floating crane, fixed crane, etc.). For storage artifacts, the total capacity will be the used indicator.

These analysis, based on our HST-PORT model of port periodization, allowed us to design our PHO ontology using a design methodology that we will now explain. By analyzing and studying the different historical sources, the HST-PORT model enable us to acquire the different classes of artifacts and related activities, actors and knowledge to design the PHO ontology.

4   PHO Ontology Design

From a knowledge engineering perspective, our SHS model is consistent with the theoretical analysis of G. Kassel [5]. It associates with artefacts (of the same nature as ours): actions, skills and agents that correspond to our activities, knowledge and actors. Within this framework, we can consider human activities as Perdurants involving three classes of Endurants (artifacts, actors and knowledge). In addition, Kassel shows that it was more pertinent to specialize a formal ontology like DOLCE rather than others ontologies (Opencyc, SUMO, etc.) to define its formal ontology of artifacts. Thus, we specialized the DOLCE ontology in the same way. We will now specify the alignment of the main classes of the ontology PHO (Actors, Knowledge, Artifacts and Activities) with CIDOC CRM, DOLCE, OWL-TIME and WGS84-pos.

The PHO ontology design is based on a specialization of the DOLCE Lite ontology, the CIDOC-CRM ontology, the "OWL-Time" and "WGS84-pos" ontologies and the HST-PORT model. We will show in the first place how our PHO reference ontology is organized at the highest level of abstraction. Then, we will detail the contribution of CIDOC-CRM and its alignment with PHO. We will end this paragraph by describing the contribution of the HST-PORT model to the design of the PHO ontology.

4.1   Alignment with DOLCE

The left pane in Fig. 1 shows the alignment Of the classes "Actors", "Knowledge", "Artefacts", and "Activities" in relation to the two subcategories of "physical objects" and "Non physical objects" and to the Perdurants. Endurants are entities that persist in time. In the endurants, one distinguishes the "physical objects" and the "Non physical object". The former are spatially recognizable. "Artifacts" is a subclass of the "physical objects". The class of "Non physical object" are specialized into two subclasses: "Actors" and "Knowledge". First, the class of "Non physical objects" covers the domain of social entities and we can relate it to the "Actors" class of our ontology (see Fig. 1). Workers, engineers, decision-makers, etc. are considered as actors. Secondly, the class of "Non physical object" also covers knowledge, that is to say the procedures, the scientific knowledge implemented in a port. The term "Perdurant" is understood to mean entities that take place in time and "Endurant" lives by participating in a "Perdurant" during a Time Interval [5].

It can also be seen that the class "Artifacts" is a subclass of "WGS84-pos: SpatialThing" in order to spatially locate the actifacts. The main classes of PHO (Actors, Knowledge, Artefacts and Activities) are not only specializations of the DOLCE classes, but also CIDOC CRM classes specializations as we will now specify (multiple inheritance).

4.2   Alignment with CIDOC-CRM

The center and right pane in Fig. 1 shows the alignment of our classes with those of CIDOC CRM. The classes "Artifacts", "Actors" and "Knowledge" are

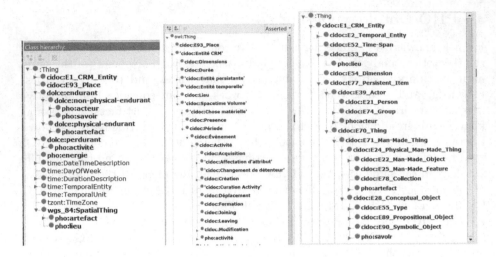

Fig. 1. PHO alignment with CIDOC-CRM, DOLCE and WGS84-pos

respectively subclasses of "E24 Physical Man Made Thing", "E39 Actor", "E28 Conceptual object". The Class "Activities" is a subclass of "E7 Activity" - central tree. More precisely:

Artifacts. Artifacts are human productions [9], constructed by the hand of human beings with a intention. A first list of relevant artifacts has been selected to periodize the history of the evolution of the ports: the wharf, the jetty, the lifting machine, the storage building (warehouse, silo...), communication channels.

This concept of artifact is found in a certain way in CIDOC-CRM in a class called "E70 thing". If the class "E70 Thing" regroups all kinds of objects, both objects made by man and objects created by nature; It makes no distinction between the origin of the objects. This class specializes in several subclasses including the "E71 Man-Made Thing" class. The class "E71" has many points in common with the "Artifact" class, the main one being to group man-made items. However, class "E71" does not differentiate between a truly constructed artifact and an artifact left in a project state, between an object that actually existed and an intangible object. Moreover, class "E71" makes no reference to the human activity responsible for the existence of the object. The CIDOC CRM therefore specified the "E71" class in a subclass called "E24 Physical Man-Made Thing". The class "E24" comes close to the definition of artifact given by Pomian. This is why we have placed the Artifact class as a subclass of the "E24" class of the CIDOC CRM and of the "Physical Objects" class of DOLCE.

Actors. In our model, the actors are human beings or a group of human beings. An actor will produce an activity through the use of an artifact. For the ports, the actors are very numerous: engineers, construction managers, workers, crane operators, machine operators, cargo handlers... They will all be responsible for an activity and will realize it thanks to their knowledge and their know-how. The notion of actor is also found in the CIDOC-CRM with the class "E39 Actor". To link the activities and actors, CIDOC-CRM foresees a specific relationship "P14 carried out by" between an activity and the actor who realizes it. The "Actor" class is therefore a subclass of the "E39-Actor" class of the CIDOC-CRM and of the "Non Physical Objects" class of DOLCE.

Knowledge. Knowledge can be theory, concept, but also procedure, regulation, technological know-how, etc. Knowledge is involved in the activity of the actor. The use of a particular technology in the artifact construction can be considered a good periodization indicator. If a wooden artifact is replaced by a reinforced concrete or a lifting device in the construction of a wharf, the replacement of one motive power by another means that there is a change in technology. The CIDOC-CRM also groups knowledge in the broad sense in the subclass "E28 Conceptual object". Thus, the class "Knowledge" is a subclass of the class "E28 Conceptual object" of the CIDOC-CRM and of the "Non Physical Objects" class of DOLCE.

Activities. Activities represent actions carried out by an actor, involving the artifact, following procedures or using knowledge or know-how. The life cycle of an artifact involves many human activities such as artifact design, construction, repair, use, and destruction. An activity takes place over a period of time; We can define its beginning and its end. It is the relationships between the artifact, the actor and the knowledge that it possesses that will enable the activity to be carried out. CIDOC CRM also uses the activity concept "E7 Activity" which specializes in the "Event" class of CIDOC CRM. The "Activities" class is therefore a specialization of the "E7-Activity" classes of the CIDOC-CRM and the "Perdurant" classes of DOLCE.

4.3   Events, Activities and Port Evolution

The CIDOC CRM ontology is an event model that accounts for changes of states, and therefore for port evolution. We will now show to examples: the Rosario port and the Brest port. The former enables us to show how our reference ontology PHO extension of CIDOC CRM and OWL-TIME, allows us to report on Rosario port evolution. The latter shows how to use SPARQL queries to observe Brest port evolutions.

The example is as follows: Argentina built in 1880 quays called "National Wharves" which were demolished and replaced in 1906 by modern quays called "Importation Wharves". Then, in 1912, the coastal trade platforms were opened for trade. In Fig. 2, the location of the port of Rosario can be seen, thanks to the "cidoc:P53 has formation or current location" property. The evolution of an artifact is modeled by activities that are limited in time - a beginning, an end. The property "cidoc:P110i was augmented by" allows us to define the addition of three new artifacts (three additions of quays: addition1, addition2, addition3) in 1880, 1906 and 1912.

Fig. 2. Port evolution of Rosario, "National Quays"

The other two additions modeled by the instances "pho:addition2 quai rosario" and "pho:addition3 quai rosario" are done in the same way as previously. On the other hand, the instance "pho:addition3 quai rosario" gives an account of the addition to the port of Rosario of a new quay line in 1912 represented by the instance "pho:quai rosario 1912". This platform line is composed of two platforms not having the same function. One of them will be to receive cabotage vessels and the other from ocean-going vessels. The "cidoc:P46 is composed of" property will therefore be used to account for the fact that an artifact may consist of "Artifacts (docks)". To represent the function of a wharf, the relation "cidoc:P19i was made for" is used in the "pho:quai rosario 1912" instance.

The Fig. 3 shows the results of two questions: (i) "On what dates were cranes added in the port of Brest? what are the number of cranes and the technology used?": that is to say the number of cranes, the technology (manual, steam and

?technology	?nb_cranes	?date_begin	?date_end
pho:technologie liée a energie manuelle	3	pho:en 1870	pho:en 1916
pho:technologie lié a la vapeur	1	pho:en 1890	pho:en 1916
pho:technologie lié a la vapeur	1	pho:en 1890	pho:en 1934
pho:technologie lié a la vapeur	2	pho:en 1890	
pho:technologie lié a la vapeur	2	pho:en 1890	
pho:technologie liee electricité	1	pho:en 1916	
pho:technologie liee electricité	1	pho:en 1916	
pho:technologie lié a la vapeur	5	pho:en 1916	
pho:technologie lié a la vapeur	4	pho:en 1916	pho:en 1922
pho:technologie liee electricité	6	pho:en 1920	
pho:technologie liee electricité	2	pho:en 1922	
pho:technologie liee electricité	6	pho:en 1928	
pho:technologie liee electricité	4	pho:en 1929	
pho:technologie liee electricité	12	pho:en 1929	

?port	?date	?entreprise	?technologie
pho:port de brest	1859^^xsd:qYear	pho:Entreprise GrandHomme	pho:technologie lie au bloc de béton
pho:port de brest	1865^^xsd:qYear	pho:Entreprise GrandHomme	pho:technologie lie au bloc de béton
pho:port de mar del plata	1909^^xsd:qYear	pho:SNTP	pho:technologie lie au bloc de béton
pho:port quequen	1921^^xsd:qYear	pho:Les Grands Travaux de Marseille	pho:technologie lie au bloc de béton
pho:port rosario	1902^^xsd:qYear	pho:entreprise Hersent	pho:techologie liée au caisson métallique
pho:port rosario	1906^^xsd:qYear	pho:entreprise Hersent	pho:techologie liée au caisson métallique
pho:port de mar del plata	1905^^xsd:qYear	pho:qardella cia	pho:tecnologie lie au bois
pho:port de mar del plata	1905^^xsd:qYear	pho:non connu	pho:tecnologie lie au bois
pho:port rosario	1906^^xsd:qYear	pho:entreprise Hersent	pho:tecnologie lie au bois
pho:port de brest	1921^^xsd:qYear	pho:non connu	pho:technologie inconnue

Fig. 3. Querying Brest port about cranes and several ports about quays

Fig. 4. SPARQL queries

electricity) in use and the dates (ii) "On what dates were docks built in the ports? What are the builders and the technologies used?": that is to say the different ports, the addition of cranes, the date and their technologies (concrete block, metallic crate, wood, unknown). The SPARQL queries corresponding to the previous questions are presented in the Fig. 4: the left one about cranes et the right one about quays.

5   Conclusion and Perspectives

The PHO ontology is the result of a multidisciplinary research on the modeling of knowledge at the SHS/STICC joint. In terms of scientific procedures, have been developed: (i) in SHS: the HST-PORT space-time evolution model; (ii) in knowledge engineering: PHO ontology based on the CIDOC-CRM reference ontology as a translation of the HST-PORT, PHO model which can be considered as a first element of a CIDOC-CRM extension in the domain Of the history of ports.

Beyond the theme of port history, this research project also aims to develop and validate reference methodologies in digital humanities. First of all, we see a very good correspondence between our concepts and those developed by Kassel in connection with DOLCE. On the other hand, if this work also shows an overlap with CIDOC-CRM (artifact/actor), several crucial points are locks to work in a thorough way: the question of the modeling of knowledge and activity on the one hand and, On the other hand, the fact that CIDOC-CRM certainly allows to describe events but presents an important gap for historians since it does not allow to describe states of an endurant. These points are now being worked on in a broader framework of a collaborative project to create an extension of CIDOC-CRM initiated by LARHRA (UMR 5190) in Lyon (by and for historians) and by creating a project Publication of digital corpus on the history of ports on the symogih.org platform[5].

---

[5] http://symogih.org/.

References

1. Claude, M.: C. J. Bird. The major seaports of the United Kingdom. Norois **46**(1), 243–245 (1965)
2. Doerr, M., Theodoridou, M.: CRMdig: a generic digital provenance model for scientific observation. In: TaPP (2011)
3. Guillem, A., Bruseker, G., Ronzino, P.: Process, concept or thing? Some initial considerations in the ontological modelling of architecture. Int. J. Digit. Libr. 1–11 (2016)
4. Hughes, T.P., et al.: The evolution of large technological systems. In: The Social Construction of Technological Systems: New Directions in the Sociology and History of Technology, pp. 51–82 (1987)
5. Kassel, G.: Vers une ontologie formelle des artfacts. In: 20me Journes Francophones en ingnierie des connaissances, 12 p. (2009)
6. Laube, S., Pourchasse, P., Querrec, S., Garlatti, R., Abiven, M.M.: Histoire compare des arsenaux de brest et venise du point de vue des sciences et des techniques: approche systmique et humanits numriques. In: Colloque International Les Arsenaux de Marine du XVIe sicle nos jours, Maison de la Recherche. Maison des Suds, Universit de Bordeaux-Montaigne (2016, to appear)
7. Laube, S., Rohou, B., Garlatti, S., de Marco, M.A.: Priodiser et comparer lvolution des ports: intrts croiss des humanits numriques et dune approche en histoire des sciences et des techniques applique aux ports de brest (france), mar del plata et rosario (argentine). In: Ports nouveaux, Ports pionniers, XIVe-XXIe sicles, Ive Colloque international du Rseau de recherche LA GOBERNANZA DE LOS PUERTOS ATLNTICOS, Lorient, Octobre 2016, to appear
8. Niang, C., Marinica, C., Leboucher, E., Bouiller, L., Capderou, C.: An ontological model for conservation-restoration of cultural objects. In: Digital Heritage, pp. 157–160. IEEE, Granada, Spain, September 2015
9. Pomian, K.: De l'exception humaine. Le Dbat (2014)
10. Ronzino, P., Amico, N., Niccolucci, F.: Assessment and comparison of metadata schemas for architectural heritage. In: Proceedings of CIPA (2011)
11. Ronzino, P., Niccolucci, F., Felicetti, A., Doerr, M.: Crmba a crm extension for the documentation of standing buildings. Int. J. Digit. Libr. **17**(1), 71–78 (2016)
12. Szabados, A.-V., Letricot, R.: L'ontologie cidoc crm applique aux objets du patrimoine antique. In 3e Journes d'Informatique et Archologie de Paris - JIAP 2012, Paris, France, June 2012. Version auteur de la communication prsente en juin 2012 lors des JIAP-2012, A paratre dans les Actes des 3e Journes d'Informatique et Archologie de Paris - JIAP 2012

A WordNet Ontology in Improving Searches of Digital Dialect Dictionary

Miljana Mladenović[1]([⊠]), Ranka Stanković[2], and Cvetana Krstev[3]

[1] EVox Solutions, Belgrade, Serbia
ml.miljana@gmail.com
[2] Faculty of Mining and Geology, University of Belgrade, Djusina 7, Belgrade, Serbia
ranka@rgf.bg.ac.rs
[3] Faculty of Philology, University of Belgrade, Studentski trg 3, Belgrade, Serbia
cvetana@matf.bg.ac.rs

Abstract. In this paper, we present a method for automatic generation of a digital resource, which connects all indirect synonyms of a dialect term to all indirect synonyms of a corresponding term in the standard language, aiming to improve the search of a digital dialect dictionary. The method uses SWRL rules defined in the Serbian WordNet ontology to identify sets of synonymous words. It also uses e-dictionaries to produce correct lemmas in standard language that users usually employ in searches. The method was applied and evaluated on verbs and a group of nouns derived from verbs (verbal nouns). We compared the results obtained by the system to those produced by humans and achieved the accuracy of 89.7%.

Keywords: Dialect dictionary · Ontology · WordNet · E-dictionary · South Serbian dialect

1 Introduction

The study of dialects has received a new impulse with the development of many software tools and digital resources for maintaining, enhancing, sharing, visualizing and analyzing digital dialect dictionaries. This includes: the development of digital dictionaries of dialects [1,4,13], the development of tools for the management of dialectal dictionaries [10], the software for data visualization and presentation of linguistic dialectal maps [11,12], the analysis of the geographical distribution of a language and geographical information relevant for linguistic research [9], the use of Semantic Web-based techniques for representing digital resources as knowledge based resources and as Linked Open Data (LOD) on the Web [6,15].

Digital dictionary of the South Serbian dialect[1], containing over twenty thousand terms, is the first comprehensive implementation [7] of a digital version of

---

[1] On-line at http://www.vranje.co.rs.

© Springer International Publishing AG 2017
M. Kirikova et al. (Eds.): ADBIS 2017, CCIS 767, pp. 373–383, 2017.
DOI: 10.1007/978-3-319-67162-8_37

Serbian dialect vocabulary, produced on the basis of traditional dialect dictionaries [16, 17]. This is the first digital resource for Serbian which, in addition to linguistic information, provides also: sound information (pronunciation) about terms and examples of the use of words or phrases as they are spoken in the dialect; graphic information about the geographical location using concepts of Google Maps; the etymological origin of the words, morphological information like part of speech, and additional semantic data. The content of the dictionary can be shared through social networks. Another important aspect of the dictionary is that it allows Web users to expand and complement it. Tools were developed to enable the search of terms in three ways: (a) search by a term, (b) search by logical queries created over metadata, (c) terms browsing by first letter. Search by a term implies that the user types a word or its part in the dialect. He/she can specify whether the search will be carried out for the terms in the dictionary which: start with the typed word, contain it or are equal to it. Search results offer information on the number of terms found in the dictionary that satisfy the given query. This kind of search is standard for on-line dictionary look-up, but it is based on the presumption that a user knows what she/he is looking for, which need not be the case for a user not familiar with a dialect. This problem often encountered by students of a foreign language can be solved by explaining terms not known in a foreign language by expressing the same concepts in a language they are familiar with. Two other ways of search (search by creating a logical query over metadata and browsing of terms by first letter) are more convenient for a novice, but they produce much more information than is expected or needed, which slows down the learning process.

In this paper we propose a method for connecting the standard language and the dialect, that would enable search over a digital dialect dictionary by using terms in the standard language. In Sect. 2, we discuss some previous approaches to searching digital dialect dictionaries. In Sect. 3, we represent resources used to improve searching performances of the digital dialect dictionary: Serbian morphological e-dictionaries used to produce all inflected forms of standard terms, and Serbian WordNet (SWN) ontology represented in OWL2 format, for which we define the rules expressed in Semantic Web Rule Language (SWRL) to be used to generate synonymous groups in the SWN ontology on the basis of the indirect synonymy relation. In Sect. 4, we propose a method for automatic generation of a digital resource which connects all indirect synonyms of a dialect term to all indirect synonyms of a corresponding term in the standard language. In Sect. 5, the method is evaluated on verbs and a group of nouns derived from verbs – verbal nouns. Finally, we offer some conclusions and suggest directions for future work in Sect. 6.

2   The Management of Digital Dialect Dictionaries

The challenging task of digitizing a dialect dictionary can be solved in different ways considering software platform, database storage, search framework and additional management tools. It can be a Web application that uses relational databases, such as Oracle used for storing southern Dutch dialects [13] and Joseph Wright's "English Dialect Dictionary" (EED) [8], MS SQLServer for storing South Serbian dialect dictionary data [7], or it can be one of the machine-readable and interoperable Semantic Web standards, such as RDF, SKOS and SKOS-XL which are used, for example, in the case of the dialect dictionary of the German language [15] and two Austrian dialect dictionaries [2]. When it comes to search techniques in a dialect dictionary, the metadata are usually used to search for specific dialect information. For example, the retrieval of information in EDD can be limited to the structural units in the entries (heads, definitions, citations, comments, variants, etc.), while logical filters, combining basic Boolean operators and metadata, can be used for advanced search options. Similar techniques are used for retrieving information from the South Serbian dialect dictionary data [7]. Some digital dialect dictionaries like Dutch [15] have records in their database that contain, besides *original headword*, additional fields – *dutchfield headword* and *search term in standard Dutch* – for advanced search options. In this paper we propose a method which enables a search, not only with one term in the standard language, but with a set of synonymous terms in order to improve the search.

3 Resources

3.1 Use of Morphological E-dictionaries

The first problem when searching for verbs in a dialect dictionary is the grammatical form of the headword of the lexical entry. Namely, the grammatical form of the headword of the verb lexical entry in the dialect dictionary is first person singular in the present tense, while the user's intent is to search for verbs using their infinitive forms, these being headwords for standard language dictionaries. To support that kind of search, it was necessary to add an infinitive form, that is, to lemmatize both a dialect verb and verbs in standard Serbian that were retrieved from its definition. For the lemmatization task we used Serbian morphological electronic dictionaries and grammars developed within the University of Belgrade Human Language Technology Group [14].

Morphological electronic dictionaries of Serbian for NLP have been under development for many years now. In the dictionary of lemmas (DELAS), each lemma is described in full detail so that the dictionary of forms containing all necessary grammatical information (DELAF), can be generated from it, and subsequently used in various NLP tasks. Serbian e-dictionaries of simple forms have reached a considerable size: they have more than 140,000 lemmas generating more than 5 million forms and 18,000 multi-word lemmas [5]. Dictionaries contain mostly standard language, but also some dialect lemmas, as well.

An e-dictionary of forms consists of a list of entries supplied with their lemmas, morphosyntactic, semantic and other information, so it is possible to attach lemma for all inflected forms in the dialect dictionary that match a form in the morphological e-dictionary. After separating all synonyms aligned with a dialect lexeme (from the standard language or dialect), infinitive forms were attached to the original form.

Among 4,152 filtered entries having a dialect form followed by a list of words or phrases in standard Serbian, 3,452 entries were verbs and the rest were verbal nouns (gerund). For 3,452 verb entries 7,353 synonyms were detected – related words or phrases in standard Serbian that describe dialect forms and that have the same meaning as corresponding dialect words. A few verb entry examples are:

- batalim | batalen, ostavim, napustim *to quit*
- batisujem | kvarim, upropašćujem *to ruin*
- bednim se | lepo se odevam, doterujem se *to dress up*
- begam | begaj, ja bega, ti bega, begajeći, bežim *to flee*

After lemmatization, we obtained the following result:[2]

- batalim_bataliti | batalen, ostavim_ostaviti, napustim_napustiti
- batisujem | kvarim_kvariti, upropašćujem_upropašćivati
- bednim se | lepo se odevam_odevati, doterujem_doterivati se
- begam_begati | begaj, ja bega_begati, ti bega_begati, begajeći, bežim_bežati

A lemma was assigned for 505 dialect forms out of 3,452 dialect forms given in the first person singular, present tense. Infinitive forms were assigned to 4,384 word forms in standard Serbian that were connected to dialect forms (out of 7,353). Word forms that were not lemmatized consisted of either another dialect form e.g. "ja znaja, ti znaja, znam_znati *to know*" not presented in e-dictioanries, or adjectives used to describe verbs e.g. "zgugurija se, zguguren, pogurim_poguriti se *to stoop, to be stooped*".

In the dialect dictionary the relation between verbal nouns and verbs was established in some entries: "šljakanje | od šljakati *to slap*", but it was not done systematically. Again, we used morphological e-dictionaries in which all verbal nouns are marked with a special marker which enabled us to establish missing connections between verbs and verbal nouns. In that way, 700 relations were established.

3.2 Calculating the Set of Near Synonyms by Using the WordNet Ontology

The development of the lexico-semantic resource Serbian WordNet (SWN), is based on the semantic network Princeton WordNet (PWN) [3]. Today, SWN is a set of more than 22,000 concepts called synsets where a concept is represented by the set of synonymous word forms that have the same or similar meaning in a given context. Synsets respect the syntactic categories noun, verb, adjective, and adverb and can be interconnected by semantic relations, while word forms can be connected by lexical relations. In SWN ontology there are currently 2,243 verb synsets defined as ontology individuals belonging to the VerbSynset class:

---

[2] Lemmatization was done using Unitex, the corpus processing system (http://unitexgramlab.org/).

```
<rdf:type  rdf:resource="&swn30;VerbSynset"/>
```

We wanted to define rules that can be used to generate synonymous pairs of verbs found in the SWN ontology that were not based only on the relation of direct synonymy. By doing that, we created a broader set of synonyms for each verb defined in SWN ontology. Relations that were used in finding the broader set of indirect synonyms are: *synonym, similar_to, also_see, verb_group, hyponym.* The total number of synset relations in SWN: *synonym* 22,162, *similar_to* 371, *also_see* 242, *verb_group* 191, *hyponym* 21,554. Figure 1 shows an example of synsets and relations between them in the SWN ontology which are used to define a set of indirect synonymous concepts of the verb "dopustiti – permit, allow, let, countenance" by using hyponymy-related verbs, synonyms of hyponymy-related verbs, verbs belonging to the same semantic group of verbs as the observed verb, and by verbs defined as semantically similar to the observed one.

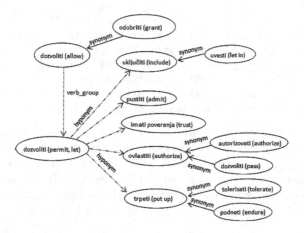

Fig. 1. Verb synsets in the SWN ontology which are mutually connected with relations that participate in finding a set of indirect synonyms of the verb "dopustiti - permit, allow, let, countenance".

Although the SWN ontology has been developing by using a free, open-source ontology editor, Protégé (http://protege.stanford.edu) and implementing SWRL rules, in this case we transformed SWN from OWL into Turtle format to be used in Java+Jena framework. Consequently, the automated reasoning rules for determining the existence of indirect synonymous pairs in the SWN ontology, in which the five previously mentioned relations are involved and which can be used for generating "indirectSynonymy" relation, are here presented in the form of the equivalent Jena rules:[3]

---

[3] We used the following software tools in this paper: Developing tool Eclipse Java EE IDE Luna and Apache Jena open source software development environment which allows for reasoning at the level of OWL 2 language by converting SWRL rules into the Jena rules format.

```
"[rule1:(?a eg.:label ?b)(?a eg.:synonym ?c)(?c eg.:label ?e) ->
(?b eg.:indirectSynonymy ?e)]"
"[rule2:(?a eg.:label ?b)(?a eg.:similar_to ?c)(?c eg.:label ?e) ->
(?b eg.:indirectSynonymy ?e)]"
"[rule3:(?a eg.:label ?b)(?a eg.:also_see ?c)(?c eg.:label ?e) ->
(?b eg.:indirectSynonymy ?e)]"
"[rule4:(?a eg.:label ?b)(?a eg.:verb_group ?c)(?c eg.:label ?e) ->
(?b eg.:indirectSynonymy ?e)]"
"[rule5:(?a eg.:label ?b)(?a eg.:hyponym ?c)(?c eg.:label ?e) ->
(?b eg.:indirectSynonymy ?e)]"
```

By looking at the rules from the set {rule2, ..., rule5}, it can be seen that each of them can be expanded with the synonymy relation, yielding the following expanded set of rules:

```
"[rule6:(?a eg.:similar_to ?c)(?a eg.:label ?b)(?c eg.:synonym ?d)
(?d eg.:label ?e) -> (?b eg.:indirectSynonymy ?e)]"
"[rule7:(?a eg.:also_see ?c)(?a eg.:label ?b)(?c eg.:synonym ?d)
(?d eg.:label ?e) -> (?b eg.:indirectSynonymy ?e)]"
"[rule8:(?a eg.:verb_group ?c)(?a eg.:label ?b)(?c eg.:synonym ?d)
(?d eg.:label ?e) -> (?b eg.:indirectSynonymy ?e)]"
"[rule9:(?a eg.:hyponym ?c)(?a eg.:label ?b)(?c eg.:synonym ?d)
(?d eg.:label ?e) -> (?b eg.:indirectSynonymy ?e)]";
```

We have carried out several experiments using different lengths of relation chains taken from the set of five given relations *synonym, similar_to, also_see, verb_group, hyponym* which enabled us to conclude that the sufficient length of a chain is 3. In that way we manually defined 24 rules (combinations of 4 relations taken 3 at a time) having the form illustrated by the following examples in which a chain is formed of relations *similar_to, also_see* and *synonym.*

```
"[rule10:(?a eg.:similar_to ?c)(?a eg.:label ?b)(?c eg.:also_see ?d)
(?d eg.:synonym ?e)(?e eg.:label ?f)->(?b eg.:indirectSynonymy ?f)]"
"[rule11:(?a eg.:verb_group ?c)(?a eg.:label ?b)(?c eg.:also_see ?d)
(?d eg.:synonym ?e)(?e eg.:label ?f)->(?b eg.:indirectSynonymy ?f)]"
"[rule12:(?a eg.:hyponym ?c)(?a eg.:label ?b)(?c eg.:also_see ?d)
(?d eg.:synonym ?e)(?e eg.:label ?f)->(?b eg.:indirectSynonymy ?f)]";
```

Restrictions that we introduced were the following: (1) a relation from the set of five given relations *synonym, similar_to, also_see, verb_group, hyponym* cannot be repeated more than once in a given rule; (2) the relation *synonym* has to be found only as the last one in a sequence of the given five relations. In this way we obtained 33 rules for reasoning about the existence of the *indirectSynonymy* relation and, after inferencing, 6,430 indirectSynonymy related pairs of verbs.

4   The Implementation of New Search Features

The proposed method for connecting standard language with the dialect dictionary relies on a table of sets of synonymous standard language words which are related to an equivalent set of dialect entries. This table is used as a part of the Web tool for an advanced search of the digital dictionary of the South Serbian dialect. Figure 2 shows resources and procedures included in the process of generating such a table, named *Expanded inverted index table*. The table was created in five steps.

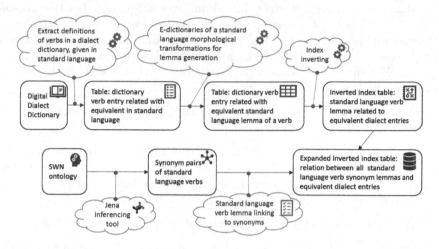

Fig. 2. Architecture of the system for building a resource that improves the dialect dictionary search tool.

(1) Automatic extraction of definitions of verbs and verbal nouns in the digital dictionary was performed by Transact-SQL stored procedures since dialect dictionary storing database is MS SqlServer. A total of 4,153 entries representing verbs, verbal nouns and their definitions were extracted, which represents about 20% of the size of the whole dictionary. Two examples of table rows (containing an entry and a definition) obtained in the first step are:

> isabim "(imp. isabi; aor. ja isabi, ti isabi; r.pr. isabija, -ila, -ilo) svr. iskvarim, upropastim."   *to ruin, to destroy*
> ačkam "(imp. ačkaj; impf. ačkašem) nesvr. kotrljam."   *to roll*

(2) E-dictionaries of the standard language were used for detecting verbs occurring in each definition and for transforming them from a form used in this dialect dictionary (the nominative case, singular, present tense, for example a verb *upropastim*) into the form used in contemporary dictionaries (the infinitive, for example *upropastiti*). Two corresponding examples obtained after the second step are (the infinitives generated by the transformation process are given in bold):

> isabim | isabi, ja isabi, ti isabi, isabija, iskvarim**_iskvariti**, upropastim**_upropastiti**
> ačkam | ačkaj, ačkašem, kotrljam**_kotrljati**

(3) An index table was created by inverting the table obtained in step (2) i.e. all dialect dictionary entries related to each infinitive representing a verb in standard language were found. In the case of verbal nouns, they were joined to verbs from which they were derived, and we subsequently treated them in the same manner as verbs. In the Subsect. 3.1, we noted that total of 4,152 headwords of the dialect were related to 4,384 infinitives of verbs in standard language, of which 4,252 infinitives are unique. For that reason, this is the length of the inverted table (total number of records). One of the records is shown below. It is related to the first example in step (2) and represents an infinitive of a verb *upropastiti* linked to the 8 entries found in the dialect dictionary whose definitions contained this verb in the form of the first person singular - *upropastim*.

> upropastiti | isabim, batišem, dokrajišem, istrovim, izabim, izakam, oznobim, profućkam

(4) In this step, the SWN ontology was used in order to implement inference rules defined in Subsect. 3.2. As a result, a table that in each row contains a verb lemma and its indirect synonyms (in the standard language), as shown in the next example, was obtained:

> upropastiti | unerediti, uništiti, uprskati, zabrljati, zakrmačiti, zasvinjiti

(5) In the final step, the tables obtained in steps (3) and (4) were joined aligning an infinitive of a verb in standard language, representing a set of synonyms in the dialect, with a verb in standard language representing a set of synonyms obtained from SWN ontology. The next example represents two joined sets of synonyms aligned with the verb *upropastiti*.

> upropastiti, unerediti, uništiti, uprskati, zabrljati, zakrmašiti, zasvinjiti | isabim, batišem, dokrajišem, istrovim, izabim, izakam, oznobim, profućkam

On the left side of the vertical line are synonymous words in the standard dictionary while on the right side are synonym words in the dialect. The use of this table by the dialect dictionary advanced search tool enables a user to type any of eight standard language words in order to obtain all equivalent synonymous words in the dialect.

5   Evaluation

The evaluation of the proposed method can be observed as the evaluation of a classification task, so we have performed an estimation of the accuracy of pairing digital dictionary entries with standard language entries, comparing the results obtained by the system with the results given by humans. Two language experts annotated the inverted table described in step 3 of Sect. 3. The table was divided into two equal parts and each annotator marked one of them. They used 3 marks for each record in the table: 1 – if an infinitive of the standard language has the same or similar meaning as verbs of the dialect in the same record;

2 – if it is not clear whether a standard language infinitive has the same or similar meaning as verbs of the dialect in the same record; 3 – if a standard language infinitive does not have the same or similar meaning as verbs of the dialect in the same record. At the same time, we created an automatic procedure to check if there were headwords of a dialect dictionary which were not related to any infinitive after inverting the table. This procedure classifies infinitives on those which take a part in relations (related) and those which do not (unrelated). When we compared human marks 1 with related, we obtained true positives. Human marks 2 and 3 compared to related gave false positives. Similarly, a comparison with the unrelated set produced false and true negatives. The confusion matrix is given in Table 1. Based on the confusion matrix, performance measures were calculated: precision $P = tp/(tp + fp) = 1.000$, recall $R = tp/(tp + fn)) = 0.874$, $F1 = 2PR/(P + R) = 0.933$, accuracy $= 0.897$. We can observe that the proposed method is completely precise, but the problem with the high value of false negatives lies in shortcomings that were found in the dialect dictionary. The most frequent are: writing errors, the lack of a verb in the definition, a verb is not written in a standard format (first case, singular, present tense), a verbal noun can not be linked to any verb. Also, some standard verbs from definitions were missing from e-dictionaries and some dialect verbs were misinterpreted by e-dictionaries.

Table 1. The confusion matrix of the process deciding whether dictionary entries are correctly aligned with standard language entries.

	System yes	System no
Expert yes	$tp = 3022$	$fn = 436$
Expert no	$fp = 0$	$tn = 784$

6   Conclusion

In this paper, we propose a method for improving searches of digital dialect dictionary by offering the possibility to search with terms in the standard language. The method uses SWRL rules defined in the ontology based on the semantic network Serbian WordNet to identify sets of synonymous words for each verb and verbal noun defined in the ontology. The method also uses e-dictionaries of the standard language to extract word forms defining verbs in the dialect dictionary and to transform them into lemmas (used for search). The method generates a table joining two sets of synonymous words – one originating from the dialect dictionary, another from e-dictionaries of standard language – for each verb extracted from a dialect dictionary. The evaluation of the method, treated as a classification, compared results obtained from the system to data provided by humans. The accuracy measure was acc $= 89.7\%$. In future work, we will experiment with other parts of speech and we will try to expand the set of ontological rules used in this system. Also we will use BabelNet and other resources in the Linguistic Linked Open Data cloud (LLOD), apart from WordNet.

References

1. Čavar, D., Geyken, A., Neumann, G.: Digital dictionary of the 20th century German language. In: Jezikoslovne Tehnologije za Slovenski Jezik: Proceedings of JS, pp. 110–114 (2000)
2. Declerck, T., Wandl-Vogt, E.: How to semantically relate dialectal Dictionaries in the Linked Data Framework. In: Proceedings of the 8th Workshop on Language Technology for Cultural Heritage, Social Sciences, and Humanities (LaTeCH 2014). pp. 9–12. ACL, Gothenburg, April 2014
3. Fellbaum, C.: WordNet: An Eletronic Lexical Database. The MIT Press, Cambridge (1998)
4. Karanikolas, N.N., Galiotou, E., Xydopoulos, G.J., Ralli, A., Athanasakos, K., Koronakis, G.: Structuring a multimedia tri-dialectal dictionary. In: Habernal, I., Matoušek, V. (eds.) TSD 2013. LNCS, vol. 8082, pp. 509–518. Springer, Heidelberg (2013). doi:10.1007/978-3-642-40585-3_64
5. Krstev, C., Vitas, D., Stanković, R.: A lexical approach to acronyms and their definitions. In: Mariani, Z.V.J. (ed.) Proceedings of 7th Language and Technology Conference, pp. 219–223. Fundacja Uniwersytetu im. A. Mickiewicza, Poznań, November 2015
6. McCrae, J., Aguado-de Cea, G., Buitelaar, P., Cimiano, P., Declerck, T., Gómez-Pérez, A., Gracia, J., Hollink, L., Montiel-Ponsoda, E., Spohr, D., Wunner, T.: Interchanging lexical resources on the Semantic Web. Lang. Resour. Eval. **46**(4), 701–719 (2012)
7. Mladenović, M.: Digital dictionary of the South Serbian Dialect. Infotheca **15**(1), 42–55 (2014)
8. Onysko, A., Markus, M., Heuberger, R.: Joseph Wright's 'English Dialect Dictionary' in electronic form: a critical discussion of selected lexicographic parameters and query options. Lang. Comput. **69**(1), 201–219 (2009)
9. O'Sullivan, D., Unwin, D.: Geographic Information Analysis. Wiley, Upper Saddle River (2010)
10. Pereira, S., Gillier, R.: TEDIPOR: thesaurus of dialectal Portuguese. In: Proceedings of the 15th EURALEX International Congress, Norway, pp. 267–281 (2012)
11. Petsas, S.: Visualising perceptual linguistic data. University of Edinburgh, Edinburgh (2009)
12. Sibler, P., Weibel, R., Glaser, E., Bart, G.: Cartographic visualization in support of dialectology in support of dialectology. In: Proceedings of the AutoCarto 2012: The International Symposium on Automated Cartography, Columbus, Ohio, USA (2012)
13. Van Keymeulen, J., De Tier, V.: The woordenbank van de Nederlandse dialecten (Wordbase of Dutch Dialects). In: Proceedings of the eLex 2013 Conference, Electronic lexicography in the 21st Century: Thinking Outside the Paper, Tallinn, Estonia, pp. 261–279 (2013)
14. Vitas, D., Popović, L., Krstev, C., Obradović, I., Pavlović-Lažetić, G., Stanojević, M.: The Serbian Language in the Digital Age. META-NET White Paper Series. Springer, Heidelberg (2012). doi:10.1007/978-3-642-30755-3. Rehm, G., Uszkoreit, H. (Series Editors)
15. Wandl-Vogt, E., Declerck, T.: Mapping a traditional dialectal dictionary with linked open data. In: Proceedings of the 3rd eLex Conference, Electronic Lexicography in the 21st Century: Thinking Outside the Paper, Tallinn, Estonia, pp. 460–471 (2013)

16. Zlatanović, M.: Rečnik govora juga Srbije: (provincijalizmi, dijalektizmi, varvarizmi i dr.). Učiteljski fakultet, Vranje (1998)
17. Zlatanović, M.: Rečnik govora juga Srbije: (provincijalizmi, dijalektizmi, varvarizmi i dr.). Aurora, Vranje (2011)

When It Comes to Querying Semantic Cultural Heritage Data

Béatrice Markhoff[1]([⊠]), Thanh Binh Nguyen[2], and Cheikh Niang[3]

[1] Université Francois Rabelais de Tours, Laboratoire d'Informatique, Tours, France
beatrice.markhoff@univ-tours.fr
[2] Université d'Orléans, INSA CVL, LIFO, Orléans, France
thanh-binh.nguyen@univ-orleans.fr
[3] Agence Universitaire de la Francophonie, Paris, France
cheikh.niang@auf.fr

Abstract. As more and more cultural institutions publish their data using the web-of-data semantic level, there is a need for novel applications for exploring, analyzing, mining and visualizing such data. A first step for these applications is to be able to query the linked open data. In this paper we survey the different existing systems for this purpose, providing examples from our experiences in the Cultural Heritage domain, and discussing some possible mutual enrichment between some of them.

Keywords: Semantic web · Querying · SPARQL endpoint · Federated-query system · Ontology-Based Data Integration · Cultural Heritage

1 Introduction

When considering research papers in Digital Humanities and Cultural Heritage (CH) conferences that deal with semantic web topics, most of them report either on solutions for building ontologies and metadata for CH, or for data curation, mapping for integration, and enrichment using semantic web standards and technologies. Indeed, since the end of 2000's, many projects rely on semantic web technologies to expose CH data in the Linked Open Data (LOD [2]): a survey is given in [15], where it is shown that the commonly used process is to convert museum catalogs into open RDF triplestores, through an extract-transform-load process. Moreover, proposals described for instance in [7] rely on schema-level alignments, using existing LOD resources (DBPedia, Geonames) and reference ontologies, including the *CIDOC-CRM*. Originally designed for the semantic integration of information from museums, libraries, and archives, this later has become an extensible semantic framework that a wide range of CH resources can be mapped to [8,19]. Based on such resources, nowadays many cultural institutions provide big Knowledge Bases (KBs) on the web, that can be used

T.B. Nguyen—Supported by a PhD grant Orléans-Tours.

M. Kirikova et al. (Eds.): ADBIS 2017, CCIS 767, pp. 384–394, 2017.
DOI: 10.1007/978-3-319-67162-8_38

by semantic web applications: the British Museum[1], EUROPEANA[2], the Smithsonian American Art Museum[3], the BnF[4] and so on.

There is a need now for *semantic web applications*, to help humans exploring this huge knowledge network, for performing data analysis tasks, and even for data mining, to extract new knowledge which may complete or correct the existing KBs. To this end, semantic web application designers first have to consider solutions for accessing semantic web data source(s). When analyzing research papers on querying the semantic web, lots of them were first only dealing with querying one RDF/RDFS/OWL resource, until the W3C SPARQL specifications were published (formal studies are still published recently, that deal with more expressive solutions). From the end of the 2000's, two main approaches for querying semantic web data sources can be distinguished: the first one addresses the linked-data's specificity, it is represented by the LOD Query systems, while the second one, in general motivated by users' needs, corresponds to the Ontology-Based Data Integration (OBDI) systems. LOD Query systems range from a single SPARQL Endpoint to Full-Web Query systems based on links traversal [12], and include the Federated Query systems, surveyed in [20,23]. These latter propose a single interface to perform a query on a fixed set of SPARQL Endpoints. This is also the case for OBDI systems, which fulfill the needs of a community to access a set of sources [6,17].

Linked Open Data sources are published on the web according to the principles introduced in [2], and now officially formalized by the W3C[5]. The LOD-initiators' vision of how applications shall use web data, called *follow-your-nose*, may be summarized as follows: data providers provide data, while data consumers discover, select and tailor data to their needs. We give a short state of the art of LOD querying approaches in Sect. 2. It allows us to notice that there is a need for an upper application level upon such systems, for taking into account specific user needs, and in particular the need to safely rely on a tailored view of the sources. Offering a defined view and a single access point to several sources is the purpose of data integration systems [14], and this is why we recall the principles of existing solutions based on semantic web data integration in Sect. 3. Building such solutions is now a well defined process when the sources' owners contribute to the integration task. But in the LOD open space, sources are not supposed to contribute to any specific integrated system, a priori. Nevertheless, we believe it is feasible to enhance semi-automatic building of mediator systems for accessing several CH data providers on the LOD, by adapting techniques used for querying the LOD, and by integrating them at the mediator level.

---

[1] http://collection.britishmuseum.org/.
[2] http://labs.europeana.eu/api/linked-open-data-sparql-endpoint.
[3] http://americanart.si.edu/collections/search/lod/about/.
[4] http://data.bnf.fr/sparql/.
[5] https://www.w3.org/TR/dwbp/.

2 LOD Querying Systems

LOD querying approaches can be classified according to the scope of the queried data: one single source, a finite set of query-federated sources and the full web. We will present their main features following this classification.

2.1 Single SPARQL Endpoint

The cultural Knowledge Bases evoked in Introduction are central repository infrastructures, in the same way as the huge, general purpose, and multilingual KBs such as Yago or BabelNet, that are automatically constructed by harvesting public knowledge-sharing platforms [26]. It means that they collect data from different sources and integrate it into a single repository before query processing. Such central repository infrastructures are supported by Triple Store Management Systems[6], and provide in general an ontology as the schema to design queries. Notice that, in the Cultural Heritage field, in general an ontology also plays the role of global schema for integrating the collected data (see Sect. 3, and this is the very purpose of the CIDOC-CRM's design. The data managed by triple stores is queried in SPARQL via a SPARQL endpoint, a web service that implements the SPARQL protocol defining the communication processes as well as the accepted and output formats (e.g. RDF/XML, JSON, CSV, etc.). The major advantage of central repository infrastructures is the direct availability of locally stored data, which enables optimized query evaluation techniques. However, when harvested from independent sources, the queried data is not always up to date, which may be a serious drawback in the dynamic context of the web of data. Very few initiatives exist that tackle the challenge of allowing humans (besides applications) to explore cultural heritage Knowledge Bases, even for a single provider. To our knowledge, the best example of such solution is the ResearchSpace project[7], that uses Metaphactory, the Metaphacts end-to-end plateform[8]. The ResearchSpace project is led by the British Museum [19].

2.2 Full-Web Querying

Full-Web query systems refer to approaches where the scope of queries is the complete set of Linked Data on the Web [20]. Instead of extracting, transforming, and loading all data from a fixed set of sources before querying it, here all relevant data for a query is discovered during runtime execution. The query evaluation is initialized from a single triple pattern as starting point and, in an iterative process, relevant data is downloaded by dereferencing URIs which are used to identify Linked Data documents on the web. Parts of the query are iteratively evaluated based on downloaded data, and additional URIs are added, which are dereferenced in the next iteration step. Indeed, as an RDF

---

[6] https://www.w3.org/wiki/LargeTripleStores.

[7] http://www.researchspace.org/.

[8] http://www.metaphacts.com/application-areas/cultural-heritage.

resource may be referred by multiple URIs from multiple independent sources which may use different ontologies to model their RDF knowledge bases, for instance URIs can be co-referenced via the *owl:sameAs* property, data from different sources are connected together, and as a consequence applications can potentially traverse the whole Web of Data by starting from one point. The evaluating process terminates when there are no more URI with potential results to follow. Fully relying on the Linked Data principles, the only requirement is that the needed data should correctly comply with those principles. This method potentially reaches all data on the web, and the freshest data. But, as the Web of Data is an unbounded and dynamic space, the querying evaluation may also not terminate, so practical experiments [13] actually restrict the range of queries to a finite part of the Web of Data.

2.3   Federated-Query Systems

A federated-query system refers to a unique interface for querying data from multiple independent given data sources[9], based on a federation query engine that decomposes the incoming user query into sub-queries, distributes them to data sources, and constructs the final result by combining answers from each source. The query processing in a federation framework comprises four phases performed in the following order: query parsing, data source selection, query optimization and query execution. Query parsing transforms the initial query into a set of triple patterns. Data source selection is the most studied phase, as it highly determines the overall performances of Federated Query systems. Even if no existing solution is directly based on the Full-Web query principles, *owl:sameAs* links are taken into account when determining the relevant sources containing relevant results for each triple pattern of the query, in order to avoid sending all of them to all participating sources, which is essential to avoid network overloading. Many techniques are proposed to deal with this challenge, they are categorized as index-free, index-assisted and hybrid [1,22,24,25].

FedX [25] is an index-free federated engine. It sends SPARQL ASK queries to data sources at runtime to discover potential sources. Thank to the simplicity of ASK queries which return boolean values, relevant data sources who can answer parts of the query triple patterns are quickly identified, but this can become expensive when the number of triple patterns and the number of data sources grow, so a cache mechanism is used to save the relevance of each triple regarding each data sources. In contrast to the index-free fashion, index-assisted approaches as DARQ [22] rely on statistics to create indexes for all predicates and types used in the queries, concerning their presence in data sources. As using only indexed summaries and possibly out-of-date indexes does not guarantee a result set completeness, those systems must deal with indexes maintenance. Hybrid

---

[9] Notice that this is different from the W3C recommendation of a Federated Query extension for SPARQL 1.1, for executing queries distributed over different SPARQL endpoints by specifying the distant endpoint using the SERVICE keyword, which supposes that the query author has to manage all this low-level knowledge.

systems as ANAPSID [1] combine indexed data and ASK queries. Proposed enhancements generally consist in integrating, or dynamically querying, more knowledge about data sources. For instance, some sources are automatically discarded by making use of the URI authorities in HiBISCuS [24].

The query optimization phase aims to eliminate unnecessary data transfers between the federated-query system and the sources by (i) using caching, (ii) choosing the appropriate join method, (iii) ordering and grouping the triple patterns. The resulting execution plan is processed in the last phase, the query execution. In the Federated Query approach, queries are answered based on the up-to-date data at original sources. In the web context, this is a major advantage compared to centralized materialized approaches. This is also the case for Ontology-Based Data Integration (OBDI) systems [6,21]. In the next section, we recall their principles and report some experiences of building and using OBDI systems in Cultural Heritage projects.

We do not know any existing web information system for Cultural Heritage based on Full-Web or Federated-Query solutions. But it is very interesting to notice that, the more Federated Query systems store information in cache or index, about their sources on the one hand, and the user queries on the other hand, the more they resemble to traditional integration systems. In particular, indexes storing the relationships between query predicates and source predicates play the same role as GAV or LAV mappings. Compared to the expressive power of OBDI systems, Federated Query systems lack the ability to compile more knowledge into the user query in order to get more complete and more correct results with respect to the user's needs. Nevertheless, the capabilities they developed for automatically harvesting knowledge, about sources and about queries, may be reused to facilitate and enhance OBDI systems building.

3   Ontology-Based Data Integration Systems

The knowledge that OBDI systems add to the user query is stored in their ontology, that is their global schema. Remember that a data integration system $\mathcal{J}$ is a triple $\mathcal{J} = \langle \mathcal{G}, \mathcal{M}, \mathcal{S} \rangle$, where [14]:

- $\mathcal{G}$ is the expected global schema.
- $\mathcal{S}$ is the source schemas, i.e. schemas of the sources where data are stored.
- $\mathcal{M}$ is the mappings between $\mathcal{G}$ and $\mathcal{S}$, i.e. a set of assertions establishing the connection between the elements of the global schema and those of the source schema.

We first briefly recall the fundamentals for defining and using a global schema, which takes here the form of a web ontology. Next, we analyze requirements for sources to become parts of a web based semantic mediation system, through the mappings, which represent a crucial part of a data integration system. Finally, we recall the principles of query architectures based on the OBDI paradigm.

3.1   Web Ontology as Global Schema

Remember that two data integration architectures exist, the *data warehouse* and the *mediation* systems. We already mentioned examples of the first type of architectures in the semantic web context: Yago and BabelNet. They rely on an ontology as the global schema. The main advantage of this architecture is the efficiency of query evaluation. Its main drawbacks are its costs in term of storage, and in terms of refreshment process, as updates performed on the original data sources must be propagated to the warehouse. On the contrary, in the mediation approach, data are kept in sources and information is retrieved dynamically from original databases, at query time. The integration is virtual in the sense that data stay in sources, but the user who interacts with the mediator, via the global schema, feels like interacting with a single database. The challenges of this solution are related to its query-answering process: when a query q_g is posed in terms of the global schema $\mathcal{G}$, the system must reformulate it in terms of a suitable set of queries q_s posed to the sources, send each computed sub-query q_{si} to the involved source $\mathcal{S}_i$, and compose the received results into a final global answer for the user. Mediation solutions fit the open, volatile and distributed web context.

Both of the data warehouse and mediation approaches require the design of a shared global schema: web ontologies play this role in the semantic web context. OBDI systems combine semantic web and data integration principles for overcoming semantic heterogeneity in order to share and efficiently reuse data among autonomous interconnected stakeholders. The following three main OBDI architectures are traditionally used [6]:

- *The single-ontology approach*, where source data is directly mapped to a global ontology. For sources having different views of the shared domain, finding a consensus in a minimal ontology commitment is known to be a difficult task.
- *The multiple-ontologies approach*, where source data is described with its own local ontology, and local ontologies are organized as a peer-to-peer system. This approach requires the construction of mappings between local ontologies and the lack of a common vocabulary between them can make this task difficult.
- *The hybrid approach*, which combines the two previous ones, using local ontologies that are mapped to a common top-level vocabulary, alleviating thereby the definition of inter-ontology mappings.

The global ontology building is a difficult task that usually requires human experts, but there already exist many resources and techniques that can facilitate it. The ResearchSpace [19] and the ARIADNE project for COINS presented in [10] are good examples of *single-ontology* OBDI systems that chose the data warehouse approach: data is mapped and transformed from the source schemata to RDF triples, compliant with the CIDOC-CRM schema. In [18], a *single-ontology* OBDI mediator is presented, where the global ontology was devised by experts, based on the CIDOC-CRM with extensions and thesaurus of the conservation-restoration domain. A solution for automating the global ontology building in *an hybrid OBDI system* was used in the query architecture presented in [3], that was designed for a project of prosopography in the Renaissance. The European project EP-Net also well demonstrated the usefulness of Ontology-Based Data Access and OBDI for easing the access of scholars to historical and cultural data, distributed across different data sources, including public semantic resources on the web [5].

3.2   Semantic Web Source

In single-ontology and in hybrid OBDI systems, in order to provide the unified global query-interface, the mediator relies on mappings $\mathcal{M}$, between the global schema $\mathcal{G}$ and the local schemas $\mathcal{S}$. Mappings are used for rewriting the global query into a union of queries that match the local schemas. These mappings can be directed either from entities in the global schema to entities in the local sources - Global As View (GAV) mappings, or from entities in the local sources to the ones in the global schema - Local As View (LAV) mappings [14]. LAV mappings require more sophisticated inferences to resolve a query on the global schema than GAV mappings, but they make it easier to add or retrieve data sources to the mediation system. Some hybrid solutions exist, for instance in [3] the overall architecture uses both LAV mappings between the sources and the domain reference ontology and GAV mappings between the global ontology and the sources.

To participate in a web-ontology-based mediation system, a source must anchor the data that it wants to share into the global system. To this end, when the source is not yet a semantic web resource, a required step is to build an ontological representation of its data, at least for the part that should contribute to the integration process. There are several ways to expose data at the semantic level, either by using suitable tools such as Ontop [4], X3ML [16], or Karma [11], or by implementing ad hoc wrappers. For instance in [18], a solution based on Ontop is implemented for the first source (a relational DB), while an ad hoc wrapper is used to export in RDF the second source (a set of MS-Word files). Once the source can be queried in SPARQL, mappings $\mathcal{M}$ can be defined and used to implement the distributed query system.

3.3   Mediator Querying System

We focus here on a concrete implemented example, but keeping the presentation sufficiently formal to be reusable. In [3] each submitted query is reformulated based on the knowledge contained in the global ontology, and on the integration knowledge (i.e. knowledge about sources, including the mappings). In a nutshell, given the global ontology $\mathcal{O}_g$ and a set of source repositories $\mathcal{S}$, this rewriting processes a global query q_g, expressed over $\mathcal{O}_g$, by reformulating it into a union of sources queries $\mathcal{Q}_s$, after having compiled the suitable knowledge in $\mathcal{O}_g$ into q_g. More precisely, the query q_g, intended to extract a set of elements from the distributed semantic databases, is resolved following these steps:

1. Apply the *Consistent* algorithm [21] with q_g as canonical instance, for verifying its consistency w.r.t constraints expressed in $\mathcal{O}_g$.
2. Apply the *PerfectRef* algorithm [21] with q and $\mathcal{O}_g$ as input, for integrating the ontological constraints into the initial query.
3. Let $\mathcal{Q}_s$ be the union of queries resulting from Step 2, apply an adapted *MiniCon* processing in order to distribute $\mathcal{Q}_s$ to the involved sources. Each source evaluates the received queries over its semantic repository.
4. Let $ans(\mathcal{Q}_s)$ be the set of answers received from sources, build the global answer $ans(\mathcal{Q}_g)$.

The first step avoids evaluating queries that could only lead to an empty result. The second step is a reasoning task performed over the global schema $\mathcal{O}_g$. In [3], as positive assertions are used for expressing the GAV mappings, each obtained sub-query q_s in $\mathcal{Q}_s$ is expressed using terms of a specific source, allowing the *MiniCon* to distribute them and compute the global answer afterwards. It is interesting to notice that, compared to Federated Query systems presented in Sect. 2.3, the difficult challenge they face for data source selection does not exist with OBDI, because the necessary knowledge is stored in the mediator.

4   Conclusion

When it comes to querying semantic web Cultural Heritage data, this is useful to have an overview of semantic web querying systems in order to build tailored services for users. We surveyed the existing solutions for querying the semantic web, those for LOD querying on the one hand, and those based on OBDI on the other hand. It is important to notice that, in the Cultural Heritage field, existing semantic data management systems are OBDI systems, in general based on the CIDOC-CRM or some extensions, and that most of them are centralized materialized RDF data triple stores, that are queried via a SPARQL endpoint. Nevertheless, in order to go closer to The Dream of a Global Knowledge Network for Cultural Heritage [9], it is necessary to be able to also rely on the *decentralized* Linked Data that is *distributed* over the WWW.

To this end, we recalled the alternative to centralized materialized data warehouses, that traditionally exists among data integration systems: the mediators, which allow the user for querying several legacy data sources without extracting and loading data. Before that, we sketched the principles of existing Full-Web and Federated-Query systems, noticing that they have the general tendency to propose some automatic generation of knowledge, about sources and about queries, this knowledge being stored in cache, or index, or summary, or ranking table as in [13]. According to the so-called *follow-your-nose* query principle [2], these systems are not devised to offer an integrated view of several resources, on contrary they explicitly leave the data integrating effort to their users, but some of their methods for harvesting knowledge, and for optimizing query execution plans, may be re-used to improve the automation of web OBDI mediators building, and also to improve their query performances. Big KBs that exist on the semantic web, such as Yago or BabelNet, are harvested from public web knowledge-sharing platforms, which involves information extraction techniques to produce RDF data, sometimes from natural language texts. Their creators also devised methods and techniques [26] that could be re-used for projects dedicated to Cultural Heritage.

Now that many Cultural Heritage institutions have opened their data to the semantic web level, we are convinced that building OBDI systems is the best way to develop applications for connecting Cultural Heritage data, for many different user needs. This is already done in many projects [5,10,18,19]. Moreover, even though we know that centralized materialized data warehouses are the most efficient solution for operating complex computations on big data, we also believe that semantic mediators are the best solutions in order to deal with the decentralized and highly evolutive features of the web. Our survey highlights some elements for making their construction simpler and their operation more efficient, even for querying existing public semantic web resources.

References

1. Acosta, M., Vidal, M.-E., Lampo, T., Castillo, J., Ruckhaus, E.: ANAPSID: an adaptive query processing engine for SPARQL endpoints. In: Aroyo, L., Welty, C., Alani, H., Taylor, J., Bernstein, A., Kagal, L., Noy, N., Blomqvist, E. (eds.) ISWC 2011. LNCS, vol. 7031, pp. 18–34. Springer, Heidelberg (2011). doi:10.1007/978-3-642-25073-6_2
2. Bizer, C., Heath, T., Berners-Lee, T.: Linked data - the story so far. Int. J. Semant. Web Inf. Syst. **5**(3), 1–22 (2009)
3. Bouchou, B., Niang, C.: Semantic mediator querying. In: International Database Engineering and Applications Symposium (IDEAS), pp. 29–38. ACM (2014)
4. Calvanese, D., Cogrel, B., Komla-Ebri, S., Kontchakov, R., Lanti, D., Rezk, M., Rodriguez-Muro, M., Xiao, G.: Ontop: answering SPARQL queries over relational databases. Semant. Web **8**, 471–487 (2017)
5. Calvanese, D., Liuzzo, P., Mosca, A., Remesal, J., Rezk, M., Rull, G.: Ontology-based data integration in EPNet: production and distribution of food during the Roman Empire. Eng. Appl. Artif. Intell. **51**, 212–229 (2016). Mining the Humanities: Technologies and Applications
6. Cruz, I.F., Xiao, H.: Ontology driven data integration in heterogeneous networks. In: Tolk, A., Jain, L.C. (eds.) Complex Systems in Knowledge-Based Environments: Theory, Models and Applications. SCI, vol. 168, pp. 75–98. Springer, Heidelberg (2009). doi:10.1007/978-3-540-88075-2_4
7. Damova, M., Dannells, D.: Reason-able view of linked data for cultural heritage. In: Dicheva, D., Markov, Z., Stefanova, E. (eds.) Third International Conference on Software, Services and Semantic Technologies S3T 2011, vol. 101, pp. 17–24. Springer, Heidelberg (2011). doi:10.1007/978-3-642-23163-6_3
8. Doerr, M.: Ontologies for cultural heritage. In: Staab, S., Studer, R. (eds.) Handbook on Ontologies. International Handbooks on Information Systems, pp. 463–486. Springer, Heidelberg (2009). doi:10.1007/978-3-540-92673-3_21
9. Doerr, M., Iorizzo, D.: The dream of a global knowledge network – a new approach. J. Comput. Cult. Herit. **1**(1), 5:1–5:23 (2008)

10. Felicetti, A., Gerth, P., Meghini, C., Theodoridou, M.: Integrating heterogeneous coin datasets in the context of archaeological research. In: Proceedings of the Workshop on Extending, Mapping and Focusing the CRM Co-Located with 19th International Conference on Theory and Practice of Digital Libraries, Poznań, Poland, pp. 13–27, 17 September 2015

11. Gupta, S., Szekely, P., Knoblock, C.A., Goel, A., Taheriyan, M., Muslea, M.: Karma: a system for mapping structured sources into the semantic web. In: Simperl, E., Norton, B., Mladenic, D., Della Valle, E., Fundulaki, I., Passant, A., Troncy, R. (eds.) ESWC 2012. LNCS, vol. 7540, pp. 430–434. Springer, Heidelberg (2015). doi:10.1007/978-3-662-46641-4_40

12. Hartig, O.: Querying a web of linked data: foundations and query execution. In: SIGWEB Newsletter, vol. 2014, no. Autumn, pp. 3:1–3:2 (2014)

13. Hartig, O., Özsu, M.T.: Walking without a map: ranking-based traversal for querying linked data. In: Groth, P., Simperl, E., Gray, A., Sabou, M., Krötzsch, M., Lecue, F., Flöck, F., Gil, Y. (eds.) ISWC 2016. LNCS, vol. 9981, pp. 305–324. Springer, Cham (2016). doi:10.1007/978-3-319-46523-4_19

14. Lenzerini, M.: Data integration: a theoretical perspective. In: PODS, pp. 233–246 (2002)

15. Marden, J., Li-Madeo, C., Whysel, N., Edelstein, J.: Linked open data for cultural heritage: evolution of an information technology. In: Proceedings of the 31st ACM International Conference on Design of Communication, SIGDOC 2013, pp. 107–112. ACM (2013)

16. Minadakis, N., Marketakis, Y., Kondylakis, H., Flouris, G., Theodoridou, M., de Jong, G., Doerr, M.: X3ML framework: an effective suite for supporting data mappings. In: Proceedings of the Workshop on Extending, Mapping and Focusing the CRM Co-Located with 19th International Conference on Theory and Practice of Digital Libraries, Poznań, Poland, pp. 1–12, 17 September 2015

17. Niang, C., Marinica, C., Leboucher, E., Bouiller, L., Capderou, C., Bouchou, B.: Ontology-based data integration system for conservation-restoration data (OBDIS-CR). In: 20th International Database Engineering & Applications Symposium, IDEAS 2016, Montreal, Canada, pp. 218–223, 11–13 July 2016

18. Niang, X., Marinica, C., Markhoff, B.B., Leboucher, E., Laissus, F., Malavergne, O., Bouiller, L., Darrieumerlou, C., Capderou, C.: Supporting semantic interoperability in conservation-restoration domain the parcours project. ACM Journal on Computing and Cultural Heritage (JOCCH) - Special Issue on Digital Infrastructure for Cultural Heritage 10, 16 (2017)

19. Oldman, D., Doerr, M., de Jong, G., Norton, B., Wikman, T.: Realizing lessons of the last 20 years: a manifesto for data provisioning & aggregation services for the digital humanities (a position paper). D-Lib Mag. 20(7/8) (2014)

20. Özsu, M.T.: A survey of RDF data management systems. Front. Comput. Sci. 10(3), 418–432 (2016)

21. Poggi, A., Lembo, D., Calvanese, D., Giacomo, G., Lenzerini, M., Rosati, R.: Linking data to ontologies. In: Spaccapietra, S. (ed.) Journal on Data Semantics X. LNCS, vol. 4900, pp. 133–173. Springer, Heidelberg (2008). doi:10.1007/978-3-540-77688-8_5

22. Quilitz, B., Leser, U.: Querying distributed RDF data sources with SPARQL. In: Bechhofer, S., Hauswirth, M., Hoffmann, J., Koubarakis, M. (eds.) ESWC 2008. LNCS, vol. 5021, pp. 524–538. Springer, Heidelberg (2008). doi:10.1007/978-3-540-68234-9_39

23. Saleem, M., Khan, Y., Hasnain, A., Ermilov, I., Ngonga Ngomo, A.-C.: A fine-grained evaluation of SPARQL endpoint federation systems. Semant. Web **7**, 493–518 (2016)

24. Saleem, M., Ngonga Ngomo, A.-C.: HiBISCuS: hypergraph-based source selection for SPARQL endpoint federation. In: Presutti, V., d'Amato, C., Gandon, F., d'Aquin, M., Staab, S., Tordai, A. (eds.) ESWC 2014. LNCS, vol. 8465, pp. 176–191. Springer, Cham (2014). doi:10.1007/978-3-319-07443-6_13

25. Schwarte, A., Haase, P., Hose, K., Schenkel, R., Schmidt, M.: FedX: optimization techniques for federated query processing on linked data. In: Aroyo, L., Welty, C., Alani, H., Taylor, J., Bernstein, A., Kagal, L., Noy, N., Blomqvist, E. (eds.) ISWC 2011. LNCS, vol. 7031, pp. 601–616. Springer, Heidelberg (2011). doi:10.1007/978-3-642-25073-6_38

26. Weikum, G., Hoffart, J., Suchanek, F.M.: Ten years of knowledge harvesting: lessons and challenges. IEEE Data Eng. Bull. **39**(3), 41–50 (2016)

ADBIS Doctoral Consortium

Preference-Based Stream Analysis for Efficient Decision-Support Systems

Lena Rudenko[(⊠)]

University of Augsburg, 86135 Augsburg, Germany
lena.rudenko@informatik.uni-augsburg.de

Abstract. Stream query processing is an important development trend as more time-oriented data is produced nowadays. It is not easy to find relevant and interesting content in large amount of data. Furthermore users want to have personalized results of stream data processing which correspond to their preferences. In this paper I present first research results achieved during my work on my doctoral thesis. I also discuss open issues and challenges on the way to my goal - the development of a preference-based stream analyzer for efficient decision-support.

1 Introduction

Today, data processed by humans as well as computers is very large, rapidly increasing and often in form of data streams. Users want to analyze this data to extract personalized and customized information in order to learn from this ever-growing amount of data, e.g., [1–3]. This data comes as a flow or stream. Many modern applications such as network monitoring, financial analysis, infrastructure manufacturing, sensor networks, meteorological observations, or social networks require query processing over data streams, e.g., [4–6]. Therefore stream data processing is a highly relevant topic today.

A stream is a continuous unbounded flow of data objects made available over time. These objects are very different and can be both simple numbers and complex data, depending on the kind of stream provider. Due to the continuous and potentially unlimited character of stream data, it needs to be processed sequentially and incrementally. However, queries on streams run continuously over a period of time and return different results as new data arrive. That means, looking on stream data twice is not possible. Hence, analyzing streams can be considered as a difficult and complex task which is in the focus of current research.

For my doctoral thesis I want to combine the stream and database analysis to create efficient decision support systems. Let us imagine the following situation:

Example 1. A user wants to visit a football match in an unfamiliar city. He wants to park at the arena, but on the way he reads in Twitter that the parking is full. The user is not familiar to this city and needs help to find alternative parking.

Supervisor: Markus Endres, University of Augsburg, 86135 Augsburg, Germany.

M. Kirikova et al. (Eds.): ADBIS 2017, CCIS 767, pp. 397–409, 2017.
DOI: 10.1007/978-3-319-67162-8_39

In this situation the user has to manage different kinds of information which should help him to make his decision (park his car) - the streamed data from Twitter (message about full parking) and data stored in the database (information about parking near the arena). The user gets the list from decision support system containing free parkings near the stadium and associated travelling routes. He chooses one offered variant and parks his car there.

On the way to create such systems many issues and tasks have to be solved - evaluation of stream data with user defined preferences, developing of efficient algorithms to compute preference-based queries on stream data, combination of database and stream data analysis, etc. In this paper I present the results I've obtained so far:

(P1) An approach to *analyze data streams with the support of user preferences*. Queries in the context of preferences are soft constraints that should be fulfilled as closely as possible. If exact matches are not available, optimal alternatives are better than nothing. This feature of soft constraints is very important for our stream analysis approach. Despite the really huge amount of stream records, it often happens that the exact matches cannot be found. The hard condition approach provides in this case an empty result and therefore can not satisfy the user.

(P2) The *Preference Continuous Query Language* (PCQL) for connecting and querying streams in Preference SQL [3], a SQL extension which supports preference queries. I outline also my open-source stream processing application which allows users to construct queries in a quick and easy manner and evaluate them on various selectable streams of data.

I also present open challenges which I will address in the future:

(F1) An algorithm for *efficient preference stream processing*, which exploits the lattice structure constructed by a preference query. Existing stream evaluation approaches have a quadratic worst-case runtime because of tuple-to-tuple comparisons to find the best objects. The proposed algorithm is based on a lattice structure representing the Better-Than relationships that must be built only once for effective Pareto preference computation on data streams.

(F2) An idea to *combine database and stream analysis* to get the best and most relevant results. The data stored in database can be changed while the user waits for the query answer. Thus, possibly he has not the best results. But the database changes can be handled as event stream and the user can get notification every time, if the stored data has been changed.

The remainder of this paper is organized as follows: Sect. 2 highlights related work. Section 3 recapitulates essential concepts of the used preference model and Preference SQL query language. In Sect. 4 I describe the developed results for preference based stream analysis and in Sect. 5 future challenges and work.

2   Related Work

Many scientists all over the world try to process and to analyze streams to extract important information from such continuous data flows. Babu and Widom [7], e.g., focus primarily on the problem, how to define and evaluate continuous queries over data streams. Faria et al. [8] describe various applications of novelty detection in data streams, and Krempl et al. [9] discuss challenges for data stream mining such as: protecting data privacy, handling incomplete and delayed information, or analysis of complex data. In [10] the authors examine the characteristics of important preference queries (Skyline, top-k and top-k dominating) and review algorithms proposed for the evaluation of continuous preference queries under the sliding window streaming model. However, they do not present any framework for preference-based stream evaluation. An important research direction associated with stream processing is data stream clustering. This issue is discussed, e.g., in [11–14]. In [15] the authors investigate queries over continuous unbounded streams for applications like network monitoring, financial analysis, and sensor networks. For this they present the Stanford Data Stream Management System (STREAM).

All previous work rely on analyzing the content of a stream or describe stream processing systems. In this paper we filter data by a preference-based approach, which never leads to an empty result and only extracts the most important information w.r.t. the user's preference.

3   Preference Background

Preference modeling have been in focus for some time, leading to diverse approaches, e.g., [16]. A preference $P = (A, <_P)$ is a strict partial order (SPO) on the domain of A. Thus $<_P$ is irreflexive and transitive. Some values are considered to be better than some others. Two values not ordered by the strict partial order $<_P$ are regarded as *indifferent*, i.e. $\neg(x <_P y) \land \neg(y <_P x)$. As strict partial orders are transitive, better-than relations in this preference model are, too.

The *maximal objects* of a preference $P = (A, <_P)$ on an input database relation R are all tuples that are not dominated by any other tuple w.r.t. the preference. These objects are computed by the preference selection operator $\sigma[P](R)$. It finds all best matching tuples t w.r.t. the preference P, where $t.A$ is the projection to the attribute set A.

$$\sigma[P](R) := \{t \in R \mid \neg \exists t' \in R : t.A <_P t'.A\} \tag{1}$$

The evaluation follows a Best-Matches-Only (BMO) query model that retrieves exact matches if such objects exist and best alternatives else.

3.1   Preference Constructors

To express simple preferences targeting only one attribute diverse base preference constructors are defined. There are base preference constructors for *numerical*, *categorical*, *temporal*, and *spatial* domains. Subsequently, I present some selected constructors which are often applied by users.

For example, the *categorical (positive) preference* $POS(A, \text{POS-set})$ expresses that a user has a set of preferred values, the POS-set. The *negative preference* $NEG(A, \text{NEG-set})$ is the counterpart to the POS preference. It is possible to combine these preferences to POS/POS or POS/NEG. The $AROUND(A, value)$ preference expresses a preferred numerical value. From two values, the one with less deviation from the specified value is preferred. Note that the dominance criterion of the base preferences is based on a SCORE function $f : dom(A) \rightarrow \mathbb{R}_0^+$, where $f(v)$ describes how far the domain value v is away from the optimal value. Dominated tuples have higher function values, i.e., $\mathbf{x} <_\mathbf{P} \mathbf{y} \iff f(x) > f(y)$. For more details we refer to [3].

Complex preferences determine the relative importance of preferences and combine base or again complex preference constructors. In the *Pareto preference* $P := P_1 \otimes P_2 = (A_1 \times A_2, <_P)$ all preferences are of equal importance, it is defined as:

$$(x_1, x_2) <_p (y_1, y_2) \Leftrightarrow (x_1 <_{P_1} y_1) \wedge (x_2 <_{P_2} y_2 \vee x_2 = y_2))$$
$$\vee (x_2 <_{P_2} y_2) \wedge (x_1 <_{P_1} y_1 \vee x_1 = y_1))$$

While in a *Prioritization preference* $P := P_1 \ \& \ P_2$ the preference $P_1 = (A_1, <_{P_1})$ is more important than $P_2 = (A_2, <_{P_2})$. For a more formal definition and more detailed information we refer to [3,16].

Example 2. Let us consider the presidential election in France in spring 2017. A lot of authors posted messages on Twitter to this theme. The preference for the tweets in German or in English can be expressed by a POS constructor: $P_1 := POS(\text{tweet\_language}, \{de, en\})$. Equally important for me is the number of followers of the author - it must be around 20000. This wish can be expressed by $P_2 := AROUND(\text{followers}, 20000)$. Both base preferences can be combined to Pareto preference $P := P_1 \otimes P_2$ to find the most relevant tweets.

3.2 Preference SQL

The Preference SQL query language is a declarative extension of SQL by strict partial order preferences, behaving like soft constraints under the BMO query model. Syntactically, Preference SQL extends the `SELECT` statement of SQL by an optional `PREFERRING` clause, cp. Fig. 4. The keywords `SELECT`, `FROM` and `WHERE` are treated as the standard SQL keywords. The `PREFERRING` clause specifies a preference which is evaluated after the `WHERE` condition (Fig. 1).

Example 3. The Pareto preference in Example 2 can be expressed in Preference SQL as follows, where `Election` is the database relation which contains information about the presidential election in France.

```
SELECT * FROM Election
PREFERRING tweet_language IN ('de', 'en')
PARETO followers AROUND 20000;
```

```
SELECT        ... <projection, aggregation>
FROM          ... <table_reference>
WHERE         ... <hard_conditions>
  PREFERRING ... <soft_conditions>
```

Fig. 1. Simplified preference SQL query block.

In the above example IN expresses a POS preference. The keyword PARETO states a Pareto preference, and PRIOR TO would lead to a Prioritization (not shown).

4   Preference-Based Stream Analysis

In this section I describe first results of preference-based stream processing.

I want to analyze the data streams employing the user's defined preferences. My approach is based on University prototype of the Preference SQL system [3]. This system was developed to run queries against bounded data sets that are stored persistently in a relational database, but I would like to analyze the unbounded stream data. Hence, I need a framework that transforms the data from a stream into a Preference SQL processable format. In this section I first describe my processing framework, then I give an overview of the *Preference Continuous Query Language* (PCQL) I have developed, and finally I present my application to query the streams using preferences.

4.1   Stream Processing Framework

To transform the data into a format which is compatible with Preference SQL, I use *Apache Flink*[1]. For the implemented framework and application prototype I decided to use Twitter as an example data source, because it has an open API and a lot of free accessible data (tweets). In addition, Apache Flink provides a Twitter Streaming API connector for direct connection to this social network.

When processing data streams with Preference SQL, two fundamental tasks must be carried out:

1. The data must have a Preference SQL processable format, because Preference SQL works on attribute based data, but streams are often encoded in various formats, e.g., as JSON[2]-objects as in Twitter.
2. The result computation must be adapted to stream properties, since the dataflow is continuous. There is no "final" result after some data of the stream is processed. Hence, the result must be calculated and adjusted as soon as new data arrive.

The lower part of Fig. 2 provides an overview of the system architecture. The incoming data stream is processed by Apache Flink, where the StreamProcessor

---

[1] https://flink.apache.org/.

[2] http://www.json.org/.

transforms the data into a Preference SQL readable format. The `DataAccumulator` builds chunks of objects, which can be processed by Preference SQL. It is composed of the client application to submit user's queries, parser and optimizer to construct query execution plan. Preference SQL evaluates the user preference on these data chunks and presents the result to the user.

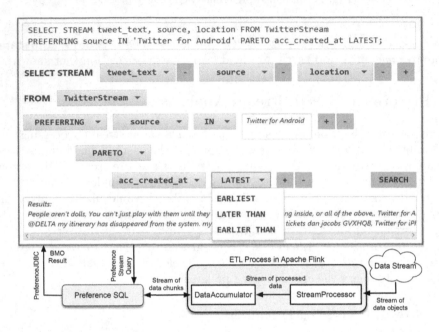

Fig. 2. Application for stream analysis with Preference SQL and ETL process in Apache Flink.

4.2   Object Representation in a Preference SQL Processible Format

The objects delivered by a data stream are encoded and not compatible with Preference SQL. That means the data must be structured and needs an attribute based format like `attributeName` = *attributeValue*. The transformation of the stream objects to a list of single attribute values occurs in the `StreamProcessor`, cp. Fig. 2. For this one has to implement the mapping of the object in the stream to a table structured format. The data types of the attributes can be extracted from the stream objects.

After the object encoding and preparation the endless stream must be split into finite parts to be processed in Preference SQL. This is done by grouping objects into chunks. This concept is similar to moving window, described in [17]. The grouping occurs in the `DataAccumulator`, see Fig. 2. It takes a stream of processed objects as input and provides a stream of chunks as output. Each chunk itself is finite and can be processed with the Preference SQL system.

In the last step the data chunks are analyzed within Preference SQL. The input is a stream of chunks, the output contains the most preferred objects w.r.t. the preference specified by a user, i.e., the BMO-set.

4.3   Finding the BMO-set of a Stream

To find the best results in each chunk the Block-Nested-Loop Algorithm (BNL) [18] is used. This algorithm is based on tuple-to-tuple comparison approach and has a worst-case complexity of $O(n^2)$, where n is the number of input objects.

Example 4. A user wants to find the tweets in *German* or alternatively in *English* with hashtags *#Macron* or *#LePen*. About the other candidates he needs no information. **Language** and **hashtags** are equally important for him.

The evaluation of this preference with BNL on one chunk consisting of four tweets is schematically represented in Fig. 3.

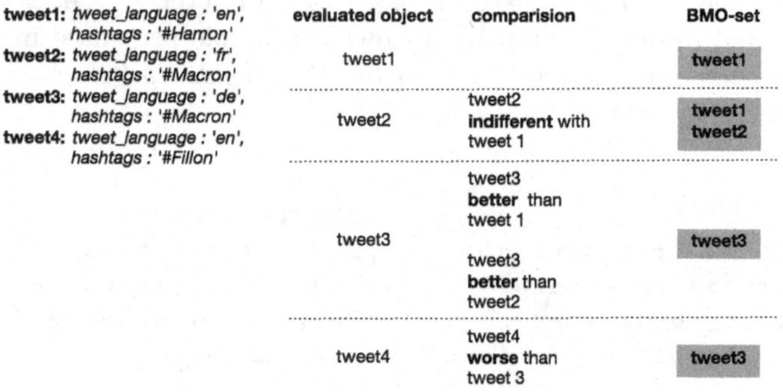

Fig. 3. Example of data evaluation with BNL.

In the first step **tweet1** is added to the current BMO-set. In the second step **tweet2** is compared to **tweet1**. They are *indifferent* and both are in the new BMO-set. The next candidate **tweet3** must be compared with all objects from the current BMO-set. It is *better* than **tweet1** and *better* than **tweet2**. The *worse* candidates are removed from BMO-set, the *better* one is added to it. Finally, **tweet4** is compared to **tweet3**. It is *worse* than **tweet3** and will not be added to the BMO-set. The final result consists of **tweet3**.

But the BMO-set of one chunk is not the final result, because when new data arrives it may occur that the next chunk contains better objects w.r.t. the user preference. Hence, the current temporary BMO-set must be compared to the objects from the next chunks in the ETL process, and so on.

More detailed description of this concept: Let $c_1, c_2, \ldots$ be the chunks provided by the ETL processor. The Preference SQL system evaluates the user preference P on the first chunk, i.e., $\sigma[P](c_1)$. Since c_2 could contain better objects we also have to compare the new objects from c_2 with the current BMO-set, i.e., it will be computed $\sigma[P](\sigma[P](c_1) \cup c_2)$, and so on. However, this leads to a computational overhead if c_2 is large. Therefore I apply a pre-filter to c_2, i.e., I first compute $\sigma[P](c_2)$ and afterwards apply the preference selection to the union of $\sigma[P](c_1)$ and $\sigma[P](c_2)$, since the following holds [19]:

$$\sigma[P](c_1 \cup c_2) = \sigma[P](\sigma[P](c_1) \cup \sigma[P](c_2)) \qquad (2)$$

This leads to a correct result and is more efficient than the first approach.

However, it is still very inefficient. A large number of comparison leads to a quadratic worst-case to run time. To make the response time better, I want to develop and to implement a lattice based stream algorithm to evaluate Pareto queries in linear run time. I refer to the idea described in [20–22] and want to adapt this algorithm to stream data. For more details I refer to Sect. 5.1.

4.4 Preference Continuous Query Language

Now I want to present a preference stream query language for stream data.

The Preference SQL syntax was described in Sect. 3. Preference based stream queries have some differences. Syntax of Stream Preference SQL is illustrated in Fig. 4a, in Fig. 4b one can find the following example presented in this query language: *the user wants to read tweets about the presidential election in France. The text of these tweets must not be French.*

```
SELECT  STREAM <projection>
FROM    <stream_reference>
WHERE   <hard_conditions>
 PREFERRING <soft_conditions>
```

(a) Stream Preference SQL syntax.

```
SELECT STREAM *
FROM TwitterStream
WHERE tweet_language <> 'fr'
 PREFERRING hashtag IN
 ('#franceelection');
```

(b) Stream Preference SQL example.

Fig. 4. Stream preference SQL query block.

The WHERE- and PREFERRING-blocks have no changes compared with Preference SQL (see Sect. 3). In the FROM-clause we refer to the stream using the keyword STREAM, instead to a database relation. The stream name is defined in the stream connector implementation. For more details see [23].

During my work a demo application was developed in the context of preference based stream processing [24]. Figure 2 shows its main view. My application illustrates the query building process over streams and allows users to construct queries and to evaluate them easily.

5 Challanges and Future Work

I want to concentrate on two general directions in the near future:

(1) Development, implementation and test of a cost efficient lattice-based algorithm for stream evaluation w.r.t users preferences.
(2) Combined analysis of stream data and data stored in databases to get the most relevant result.

5.1   Efficient Preference Stream Processing

Processing of preference based queries requires high efficient evaluation algorithms, especially on time-oriented data streams. In my approach I use the BNL Algorithm (see Sect. 4.3) which was brought in a database context with the *Skyline operator* in [18]. This algorithm is not very efficient, because it is based on tuple-to-tuple comparison. The new stream objects received later can match the user preferences better than the objects already recognized in previously computed BMO-sets. New objects have to be compared with all from the current BMO-set. The continuous compare process is the most expensive operation of preference-based stream evaluation. Therefore I want to design and implement an algorithm which avoids tuple-to-tuple comparison and the associated quadratic worst-case runtime of BNL. It will be also interesting to study memory requirements of this algorithm.

Standard Lattice Skyline. I refer to the algorithms *Hexagon* [21] and *Lattice Skyline* [20] as well as [22] which exploit the *lattice* induced by a Pareto query over *discrete* domains to compute the best objects. Visualization of such lattices is often done using *Better-Than-Graphs* (BTG) (similar to Hasse diagrams), graphs in which edges state dominance. Each node in the BTG contains objects which are mapped to the same feature vector by a scoring function f. All values in the same node are *indifferent*. The algorithm in general consists of three phases:

(1) *Construction phase* for initialization of the BTG.
(2) *Adding phase*: Tuples are read and mapped to the nodes.
(3) *Removal phase*: Remove dominated and empty nodes.

The elements of the dataset D that compose the BMO-set is build up by those nodes in the BTG that have no path leading to them from another non-empty node. All other nodes have direct or transitive edges from the BMO-set nodes, and therefore are dominated. The worst case complexity of such lattice algorithms is linear w.r.t. the number of input tuples and the size of the BTG [20].

For the implementation of such algorithms the lattice is usually represented by an array, where each position stands for one node in the lattice. The array stores the *empty*, *non-empty*, and *dominated* node states. The nodes are visited in a breadth-first traversal order (**BFT**). Non-empty nodes cause a depth-first traversal (**DFT**), where the dominance flags are set. Finally those nodes represent the BMO-set which are both non-empty and non-dominated.

In Fig. 5 one can see the BTG constructed for the query from Example 4 and assignment of some tweets to the graph's nodes. The best objects are on the top of BTG, the worst ones on the bottom.

Stream-based Lattice Skyline (SLS). Stream Lattice Skyline has some changes in comparison to the standard version:

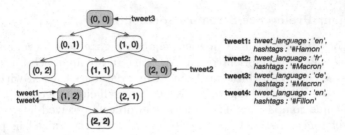

Fig. 5. BTG for the query from Example 4.

(1) *Construction phase* does not change.
(2) *Adding phase*: Since the stream data is infinite, the adding phase, in which the objects are mapped to their nodes, runs after the start continuously, thereby in parallel to the BFT and DFT.
(3) *Removal phase*: The dominated and empty nodes are not removed from the lattice, only the objects associated with these nodes are deleted. It is unknown which object will be received from stream later, therefore the whole array is required. Moreover the BMO-set found by SLS is the result set only for the current data chunks. The next chunk can contain better objects and the BMO-set will change. The objects corresponding to the non-empty, non-dominated nodes will not be removed from the BTG. Therefore the evaluation of objects, received in later chunks, requires "only" the BTG-throughput, not the expensive candidates comparison with all the objects in BMO-set.

Unfortunately, the SLS algorithm presented above is restricted to Pareto preferences. To evaluate basic preferences using this approach, our algorithm must be adjusted as described in [25].

Challenges of SLS. My approach has some open challenges. For BTG building (*construction phase*) the domain maximum of each base preference composing the Pareto preference is needed. Streams have no meta-data about their objects, so the domain knowledge is missing completely or for some dimensions. To handle the unrestricted domains the SLS algorithm has to be adapted. The BTG will be constructed for the attributes with known domains and for the unknown one its *locally optimum value* (lov) will be stored in the non-empty nodes during the *adding phase*. During the *removal phase* the objects belonging to dominated nodes will not be discarded immediately, but they have to be compared with regard to the lovs of the attribute with unrestricted domain. For the details I refer to [20, 22]. I want to expand the idea with locally optimum value for more than one attribute for SLS. Last but not least future work includes numerous experiments on the performance of SLS.

5.2   Combination of Stream and Database Analysis

Despite increasing importance of stream data analysis, queries over data saved in databases still play an important role, such as looking for some product on Amazon, searching for suitable flight to holiday place or choice of nearest parking. Changes in the data occur often, also during the query process. The user is not notified about these changes and the obtained query result is not consistent with the data from database.

Example 5. Let us consider the following situation: The user wants to buy a car and he is looking for a model on the web-page of some car dealer. Actually, he wants a red model, but the black one is cheaper. He takes the black car and thinks, he has a good deal. What the user doesn't know, is that the dealer has a special offer for the desired red car now. The offer is added to the database during the user's offering process but after the query processing and result searching. The provided result set doesn't include the new offer. The user has also no notification about the changes and he buys a black car instead of the desired red one.

Notifications provided in form of stream data might help us to avoid the described situation. Each data change will be stored in the database but also a related event will be added to the notification stream. In the simplest case the stream event has time stamp and reference to changed relation. The user will be informed, if some changes occur in the queried relation, regardless of whether the data was added, removed or changed. It is better to save in the stream event not only the affected relation, but also kind of changes (adding, removing, updating etc.) and the related attribute. In this case the user will be informed only about changes relevant for his query.

Let us consider again Example 1. The analysis of stream data helps the user to get the information about full arena parking. City's existing parkings with position and number of places are stored in the database. Query answer is additionally influenced by the free-places-information received as data stream. The notifications about changes at the parkings can help the user to make a decision and to choose one parking from list offered by a decision support system.

Such combination of stream and database data analysis can be very productive for effective decision making. This is, what I want to accomplish at the end of my doctoral thesis.

6   Conclusion

In this paper I present the contributions I've made so far: An approach, framework and application to analyze and query data streams with the support of user preferences as well as a Preference Continuous Query Language (PCQL) for connecting and querying streams in Preference SQL. I describe open challenges I am going to work on in the next time to accomplish my goal - development of a preference-based stream analyzer for efficient decision-support systems.

References

1. Stefanidis, K., Koutrika, G., Pitoura, E.: A survey on representation, composition and application of preferences in database systems. ACM Trans. Database Syst. **36**(3), 19:1–19:45 (2011)
2. Golfarelli, M., Rizzi, S.: Expressing OLAP preferences. In: Winslett, M. (ed.) SSDBM 2009. LNCS, vol. 5566, pp. 83–91. Springer, Heidelberg (2009). doi:10. 1007/978-3-642-02279-1_7
3. Kießling, W., Endres, M., Wenzel, F.: The preference SQL system - an overview. Bull. Tech. Comm. Data Eng. IEEE CS **34**(2), 11–18 (2011)
4. Chen, J., DeWitt, D.J., Tian, F., Wang, Y.: NiagaraCQ: a scalable continuous query system for internet databases. In: Proceedings of SIGMOD 2000, pp. 379–390. ACM, New York (2000)
5. Bonnet, P., Gehrke, J., Seshadri, P.: Towards sensor database systems. In: Tan, K.-L., Franklin, M.J., Lui, J.C.-S. (eds.) MDM 2001. LNCS, vol. 1987, pp. 3–14. Springer, Heidelberg (2001). doi:10.1007/3-540-44498-X_1
6. Sankaranarayanan, J., Samet, H., Teitler, B.E., Lieberman, M.D., Sperling, J.: Twitterstand: news in Tweets. In: Proceedings of ACM 2009, pp. 42–51 (2009)
7. Babu, S., Widom, J.: Continuous queries over data streams. SIGMOD Rec. **30**(3), 109–120 (2001)
8. Faria, E.R., Gonçalves, I.J.C.R., de Carvalho, A.C.P.L.F., Gama, J.: Novelty detection in data streams. Artif. Intell. Rev. **45**(2), 235–269 (2016)
9. Krempl, G., Žliobaite, I., Brzeziński, D., Hüllermeier, E., Last, M., Lemaire, V., Noack, T., Shaker, A., Sievi, S., Spiliopoulou, M., Stefanowski, J.: Open challenges for data stream mining research. In: SIGKDD 2014 Explorations Newsletter, vol. 16, no. 1 (2014)
10. Kontaki, M., Papadopoulos, A.N., Manolopoulos, Y.: Continuous processing of preference queries in data streams. In: Leeuwen, J., Muscholl, A., Peleg, D., Pokorný, J., Rumpe, B. (eds.) SOFSEM 2010. LNCS, vol. 5901, pp. 47–60. Springer, Heidelberg (2010). doi:10.1007/978-3-642-11266-9_4
11. Silva, J.A., Faria, E.R., Barros, R.C., Hruschka, E.R., de Carvalho, A., Gama, J., Clustering, D.S.: Data stream clustering: a survey. ACM Comput. Surv. **46**(1), 13 (2013)
12. Baruah, R.D., Angelov, P., Baruah, D.: Dynamically evolving clustering for data streams. In: IEEE Conference on EAIS 2014, pp. 1–6 (2014)
13. Gu, X., Angelov, P.P.: Autonomous data-driven clustering for live data stream. In: SMC 2016, pp. 1128–1135. IEEE (2016)
14. Kastner, J., Endres, M., Kießling, W.: A pareto-dominant clustering approach for pareto-frontiers. In: Proceedings of the Workshops of the EDBT/ICDT 2017 Joint Conference, Venice, Italy (2017)
15. Arasu, A., Babcock, B., Babu, S., Datar, M., Ito, K., Nishizawa, I., Rosenstein, J., Widom, J.: STREAM: the stanford stream data manager. In: Proceedings of SIGMOD 2003, pp. 665–665. ACM, New York (2003)
16. Kießling, W.: Foundations of preferences in database systems. In: Proceedings of VLDB 2002, Hong Kong SAR, China, pp. 311–322. VLDB Endowment (2002)
17. Babcock, B., Datar, M., Motwani, R.: Sampling from a moving window over streaming data. In: Proceedings of SODA 2002, Philadelphia, USA, pp. 633–634 (2002)
18. Börzsönyi, S., Kossmann, D., Stocker, K.: The skyline operator. In: Proceedings of ICDE 2001, pp. 421–430. IEEE Computer Society, Washington (2001)

19. Hafenrichter, B., Kießling, W.: Optimization of relational preference queries. In: Proceedings of ADC 2005, Darlinghurst, Australia, pp. 175–184 (2005)
20. Morse, M., Patel, J.M., Jagadish, H.V.: Efficient skyline computation over low-cardinality domains. In: Proceedings of VLDB 2007, pp. 267–278 (2007)
21. Preisinger, T., Kießling, W.: The hexagon algorithm for pareto preference queries. In: Proceedings of the 3rd Multidisciplinary Workshop on Advances in Preference Handling in Conjunction with VLDB 2007, Vienna, Austria (2007)
22. Endres, M., Kießling, W.: High parallel skyline computation over low-cardinality domains. In: Manolopoulos, Y., Trajcevski, G., Kon-Popovska, M. (eds.) ADBIS 2014. LNCS, vol. 8716, pp. 97–111. Springer, Cham (2014). doi:10.1007/978-3-319-10933-6_8
23. Rudenko, L., Endres, M., Roocks, P., Kießling, W.: A preference-based stream analyzer. In: Workshop STREAMEVOLV 2016, Riva del Garda, Italy (2016)
24. Rudenko, L., Endres, M.: Personalized stream analysis with PreferenceSQL. In: BTW 2017, Stuttgart, Germanym, pp. 181–184 (2017)
25. Endres, M., Preisinger, T.: Beyond skylines: explicit preferences. In: Candan, S., Chen, L., Pedersen, T.B., Chang, L., Hua, W. (eds.) DASFAA 2017. LNCS, vol. 10177, pp. 327–342. Springer, Cham (2017). doi:10.1007/978-3-319-55753-3_21

Formalization of Database Reverse Engineering

Nonyelum Ndefo[✉]

Free University of Bozen-Bolzano, Piazza Domenicani 3, 39100 Bolzano, Italy
`ndefo@inf.unibz.it`

Abstract. During the life cycle of an Information System, the original design of the database may be difficult to acquire. The database schema may have been continuously modified and drifted away semantically from the intended design, or perhaps no conceptual modelling method was employed at all. A conceptual schema offers a much richer description of a domain than the database schema, this makes it important for, among other reasons, the maintenance of semantic consistency of the database at runtime. Database reverse engineering involves the retrieval of domain semantics from an existing set of database schemas into a conceptual schema. Though several research works exist on creating database reverse engineering methodologies, the process in itself has never been properly and fully formalised with an emphasis on its correctness. This paper introduces the ongoing research surrounding database reverse engineering and the goal of rendering a formal reverse engineering framework. The expected formalism will be valuable to database and domain experts as a sound foundation for implementing better reverse engineering tools.

Keywords: Reverse engineering · Relational database · Conceptual modelling · Mapping · Schema transformation

1   Introduction

The concept behind reverse engineering has been used for different causes across computer science disciplines. In the development and maintenance of software systems, reverse engineering tries to reconstruct design information from an implemented system [17]. In data management practices, reverse engineering has been studied in close connection with databases, more specifically with relational databases.

Database Reverse Engineering (DBRE) is a means to recover domain semantics encoded in a specified database and represent them as a high-level conceptual schema that corresponds to the possible design specifications of the database schema [5].

1.1   DBRE Motivation

In general, the reasons to reverse engineer a database may include data migration from legacy systems, data integration, data exchange at conceptual level,

© Springer International Publishing AG 2017
M. Kirikova et al. (Eds.): ADBIS 2017, CCIS 767, pp. 410–421, 2017.
DOI: 10.1007/978-3-319-67162-8_40

semantic acquisition for system maintenance, for domain (re)documentation, and for query answering using the conceptual schema [7].

To fully comprehend the DBRE cause, an understanding of database design procedure is also necessary. Database design, or *forward engineering*, is a process which maps elements in a conceptual schema to elements in a database via a series of tasks such as requirement analysis, entity and relationship identification through dependencies (constraints), key selection, normalisation, etc.

During the database design process, while the specification transitions from the conceptual schema to the logical and physical schemas, it is often found that domain knowledge has been left implicit or lost within the database. This can be attributed to the fact that the relational schema is semantically poorer than most conceptual schemas, and it cannot explicitly express certain domain information, e.g. generalisation and specialisation hierarchies, unions and disjointness of entity sets [12].

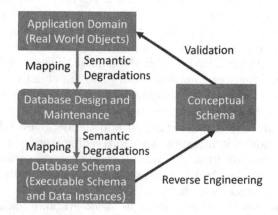

Fig. 1. Semantic loss during Database Design [5]

Furthermore during the life cycle of the database system, the structure of the database is prone to changes over the years of its usage and is subject to semantic degradation as described in Fig. 1. Whether this degradation is caused by the poor semantic nature of the relational database, tricky original design or the use of certain optimisation practices e.g. de-normalization of some relations to increase performance, the database users remain at risk of losing the understanding of its contents.

With no concrete knowledge of the initial design strategy used to forward engineer the database system, no detailed documentation through which continuous modifications to the database could have been tracked, and the unavailability of the original database developers, the problem becomes not only non-trivial but also tedious [8]. As one of the main benefits of a conceptual schema is to provide users which an abstract and eminently clear representation of the domain, the DBRE process is required to carefully implement the backward transitioning of the forward engineering process, filling up the perceived semantic gaps in the database schema to construct a conceptual schema.

The rest of this paper is organised as follows: Sect. 2 gives an overview of the DBRE approaches, also highlighting the need for a completely formal framework. In Sect. 3, several related works are reviewed including the state of the art research on DBRE and some studies on schema transformation in database systems. The plan for formalising the entire DBRE process is detailed in Sect. 4, including the choice of conceptual schema and formal language to illustrate the formalism, and important concepts of interest concerning schema transformation. Finally, Sect. 5 provides a summary and additional information about the research.

2   Overview of DBRE Approaches

The DBRE approaches usually perform what is referred to as *semantic acquisition* as part of the first step of the process. This is done by analysing the database schema along with other sources such as the schema's DDL, data instances, and the SQL queries derived from application programs that use the database, in the hope of obtaining a richer description of the domain. These approaches use the mentioned sources (singly or combinatorially) and are able to attain additional semantic information from them, this includes all information about elements, which are basically relations[1] and constraints.

After these sources have been explored and the database schema has been enriched with new-found information, then succeeding steps can occur: the classification and conceptualisation of elements. In the conceptualisation step, each element within the relational schema is transformed into a corresponding element in the conceptual schema by a set mappings.

Here, the term *conceptual schema* is used loosely and considers high-level data models which describe semantics of data in similar ways, for instance, in the Entity Relationship (ER) model with its use of *entities*, which are viewed as *classes* in the Unified Modelling Language (UML) class diagrams, *concepts* in ontology languages, and *objects* in Object Role Modelling (ORM). Even though conceptual schemas in general can have various degrees of expressiveness, it can be proven that at least most of the common elements can be accurately translated among these similar data models. A survey of these approaches shows that a preferred target schema for re-engineering database schemas is the ER model or its upgraded version Enhanced Entity Relationship (EER) [1,6,11,14,15,18,21]. Other choices have been Object Modelling Technique (OMT) [19], ERC+(an extension of the ER model with multivalued objects and multi-instantiation) [3], and First Order Logic (FOL) ontologies [4,13,14]. Table 1 shows a nicely structured overview of certain similarities and contrasts among of the existing approaches.

---

[1] Relations consist of typed attributes, with each attribute belonging to its own finite attribute domain.

Table 1. Comparison of DBRE approaches

Approach	Source schema	Other sources	Target schema	Formal Verification	Extracted constraints	Automation
Astrova [4]	Relational schema in 3NF	Data instances	Ontology for the Semantic Web	None	Key attributes, Inclusion, Disjointness, Intersection constraints	Semi-automated
Alhajj [1]	Schema in 3NF	Data instances	EER	None	Key attributes, Cardinality, Inclusion constraints	Automated
Andersson [3]	Schema (no assumptions on normal forms)	SQL queries, DDL, Data instances	ERC	None	Key attributes, Inclusion constraints	Non-automated
Chiang et al. [6]	Schema in 3NF	Data instances, DDL	EER	None	Keys, Inclusion constraints	Semi-automated
Petit et al. [18]	Schema in 3NF	Data instances, SQL equi-join queries	EER	None	Key attributes, Inclusion constraints, Disjointness constraints	Semi-automated
Lammari et al. [12]	Schema	DDL, SQL queries, Data instances (if necessary)	EER	None	Key attributes, Inclusion and Intersection constraints	Automated (or Semi-automated, if necessary)
Lin et al. [13]	Schema in 3NF	Data Instances	OWL DL ontology	None	Key attributes, Classes, Data-type Property, Object Property, Cardinality, Inclusion constraints, Disjointness, Equivalent Classes, Covering	Automated
Signore et al. [21]	Schema assumed to be in 3NF	Application program code with embedded SQL	ER Model	None	Key attributes, IS-A hierarchies	Non-automated
Markowitz & Makowsky [15]	Schema in BCNF	None	EER	Proof sketches for mappings	Key attributes, Inclusion constraints	Non-automated
Lubyte & Tessaris [14]	Schema in 3NF	None	DLR-Lite (Description Logic) Ontology	Proof for equivalence preserving mapping	Key attributes, Inclusion, Disjointness, Covering, Mandatory and Optional Participation, Functionality, Role Typing	Automated
Blaha & Premerlani [19]	Schema (no assumptions on normal forms)	Data instances	OMT	None	Key attributes, inclusion constraints, Cardinality	Non-automated
Johannesson [11]	Schema in 3NF		Conceptual schema defined as a pair of language L and a set IC of typing, mapping and generalization constraints	Proof sketch for schema dominance	Key attributes, Inclusion constraints	Non-automated

2.1 Problems with Existing Approaches

More compactly, a DBRE setting should consist of a source schema, a target schema and a well-defined transformation process between the schemas. Closely inspecting these approaches, it is discovered that they are mostly non-automated and involve human interaction not just for a final user validation but often times mid-process. Therefore the mappings induced by the transformations most likely rely on heuristic techniques and informal rules dependent on the expertise and view point of the users. Consequently, each approach, if used on the same database schema may channel towards a relatively diverse target schema, with probable semantic inconsistencies.

The above claim is attributed moreover to the little effort demonstrated in presenting a guaranteed measure for correctness post-transformation, thus rendering the validity of the reverse engineering process questionable.

Take for instance the case of classifying an association between two relations, say *SSN* and *PERSON*. Relation *SSN* holds records of taxation details for each person while relation *PERSON* holds basic information for the each person.

$$PERSON[\textbf{ssn}, name, address...]$$
$$SSN[\textbf{ssn}, tax\_category, issuing\_state, ...]$$

Both relations have the same key *ssn*, used to uniquely identify each tuple in both *PERSON* and *SSN*. If no information about referential integrity is known, based on the set of heuristic rules in [6], this association is justified as a subtype inclusion constraint because the relations share the same key attribute, and is therefore mapped accordingly. Whereas Lin et al. [13] argue that it may not be so that "a *PERSON* is a *SSN*" or vice versa just for this reason and consider a "more natural interpretation" which portrays this association as a one-to-one or one-to-many binary relationship.

Another usually overlooked issue concerns the handling of nullable attributes (null values) within DBRE. Since the relational and conceptual schemas are of different signatures, the interpretation of nullable attributes in the conceptual schema needs to be properly addressed with respect to preserving information. This aspect is yet to be thoroughly treated whether formally or informally.

We believe that providing documented formal descriptions for the process would help in resolving these issues and other related ones. Then we can establish the most logically plausible mappings which rightly preserve source schema semantics in the target schema, and verify that a DBRE transformation can be indeed correct.

However, before delving into our plans to address these issues, we consider it necessary to review relevant literature in the reverse engineering context through which substantial contributions have been produced. In the next section some notes on these notable contributions are presented.

3   Related Work

Since the 1990s, studies concerning DBRE have been carried out, of which the most referenced have been listed in Sect. 2.

In [11], the database schema in 3NF is translated into a conceptual schema represented by a pair consisting of a language and a set of typing, mapping, and generalisation constraints. The methodology investigates object structures and how they can be identified in the database schema based on the correlation between keys and inclusion constraints. The work concludes by defining the notion of *schema dominance* as a measure of correctness for the reverse engineering method. As future work, the paper suggests a potential line of research: *investigating the influence of views on conceptual modelling*. We plan to improve on this work by investigating this schema dominance on different reverse engineering scenarios, after defining mappings (views) of elements in one schema based on the signature of the other schema.

The work of [15] identifies object structures in a database, taking as input the relational schema in Boyce-Codd normal form (BCNF) consisting of key constraints and key-based inclusion constraints. The output of the approach is an EER model. This method argues against the informal mappings presented in other works and introduces some formal mapping descriptions for a procedure which determines convertibility of a BCNF normalised relational schema into the EER model, although BCNF is considered as one of the stricter normal forms in the context of realistic database schemas.

The most recent research works steer towards reverse engineering a relational database to logic-based ontologies. In [14], an ER model is extracted from an existing relational database, with mappings defined between the extracted schema and the relational database by associating views over the latter as a means to access the underlying data. The approach subtly exploits the Global-as-View (GAV) mappings from data integration. It highlights important points such as obtaining an equivalence preserving schema transformation and uses description logic to express semantics of the schema. Our approach contributes to this by considering in addition reverse mappings similar to the Local-as-View (LAV) to verify correctness DBRE in terms of losslessness.

Astrova [4] focuses on reverse engineering a relational database into an OWL ontology for use on the Semantic Web. The approach confirms semantics by analysing not only the correlation between primary and foreign keys, but also the correlation among data values and attributes. The reverse engineering process is simple and shows some mapping for relations, attributes, relationships, and constraints. However it is not formally presented and there is no verification for information consistency. The work claims to show an extraction of more semantics from the relational database compared to other approaches.

The work in [13] describes a DBRE-based automatic algorithm to extract an OWL-DL ontology with richer semantics from a given relational database. Compared to the others mentioned above, this approach goes a step further by illustrating the correspondences among semantics of the ER model, database schema, and OWL ontology.

In general, the reverse engineering approaches could rely heavily on human intervention, be semi-automated or fully automated. However it is to be noted that in any system, before complete automation can occur for any process there should be a set of formal definitions properly set in place. Though these works have contributed significantly to this field, they will be used as building blocks for rendering a more complete framework.

Since schema transformation is a fundamental part of DBRE, it is useful to study transformation practices involving relational databases.

Hainaut et al. in [9] offer a generic technique for schema transformation for a database, describing schema transformation as consisting of two types of mappings: *structural mapping* and *instance mapping*. The first ensures that elements in the source schema can be replaced by elements in the target schema, and the latter ensures correspondences between instances in both schemas. It also introduces transformation concepts such as *schema reversibility* and *semantics preservation*.

McBrien and Poulovassilis in [16] present a schema transformation approach based on combining both GAV and LAV mapping approaches. The coined name, BAV, meaning *Both-as-View* leverages the benefits of both GAV and LAV views. The concept of reversibility is used to maintain here as a desired aspect of the transformation process.

The work presented in [2] aims to discover semantic matches between two databases along with their already existing conceptual schemas. The approach described here finds the mappings based on the correspondences (or links) between attributes and relations.

Qian, in [20], emphasizes that the correctness of a schema transformation is achieved when the instances and constraints in the source schema are preserved. Three transformation properties are highlighted: instance, constraint, and information preservation. The last is a consequence of the first two. The work defines an information preserving transformation as in terms of containment between the source and target schemas denoted by s and t respectively i.e. to determine if $s \subseteq t$ and $t \subseteq s$. And if these containments hold then s is said to be equivalent to t, denoted by $s \equiv t$.

This section has presented concise descriptions of other research works related to the one proposed. As stated earlier, schema transformation is a pivotal aspect in DBRE, and any plan to produce a useful method should focus primarily on how semantic equivalences are identified between the source and target schemas and how eventual maps are created. This therefore justifies the following section which discusses the details of the proposed formalism.

4   Formalisation Plan of DBRE

By definition, one understands that DBRE consists of three main aspects i.e. the source schema, schema transformation, and the target schema. The approach to formalise DBRE rests on grasping the semantics of each of the individual aspects and how they relate. This section gives brief details on the source and target schemas, and the formalisation plan.

4.1   Source Schema

We will provide various realistic scenarios of database schema as examples. We assume that semantic acquisition has already been carried out on the source schema by analysing the input sources mentioned in Sect. 2, and the schema is now complete with all intended semantics from the domain i.e. all possible information about relations and constraints are known. At this point, the schema is also presumed to have been decomposed into a sufficient normal form, ideally 3NF, so that the conceptual schema elements can be seamlessly identified and classified prior to conceptualisation.

4.2   Target Schema

For the target schema, ORM [10] will be used to describe the semantics extracted from the source schema. Modelling in ORM is fact-based i.e. data is described as elementary facts, which are predicates asserting that an object type participates in a role (relationship). The model is attribute-free, representing the relationship between an object and its attributes also as roles. The example in Fig. 2 illustrates two fact types: one binary and the other ternary. The binary role *seeks* connects the object types *Student* and *Degree*. In the ternary fact, the role *...was awarded...on...* connects the same object types but now with another object type *Date*.

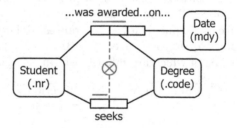

Fig. 2. Example ORM diagram (http://www.orm.net)

Integrity constraints and other constraints can be expressed clearly. Given the same example in Fig. 2, the bars over the roles indicate uniqueness constraints over each entity participating in those roles. The other constraint seen in the diagram is the exclusion constraint (circled 'x') between the roles *seeks* and *...was awarded...on...* indicating disjointness between the instances in these roles.

An advantage of using ORM over the other similar conceptual schemas, besides the fact that it has a standard notation and its ability to model relatively more features, is the possibility to automatically derive natural language verbalisation for its conceptual schemas using the Natural ORM Architect (NORMA) tool within Microsoft Visual Studio development environment. The verbalisation texts are expressed with regards to FORML, the controlled language for ORM and as far we know, an automatic verbalisation feature with such convenient readability is yet to be embedded in any of the other common modelling tools. The following verbalisation texts explain Fig. 2:

- **Student** was awarded **Degree** on **Date**.
- *For each* **Student** and **Degree**, that **Student** was awarded that **Degree** on *at most one* **Date**.
- **Student** seeks **Degree**.
- Each **Student** seeks *at most one* **Degree**.
- It is possible that *more than one* **Student** seeks *the same* **Degree**.
- For each **Student** and **Degree**, *at most one* of the following holds: that **Student** seeks that **Degree**; that **Student** was awarded that **Degree** on *some* **Date**.

This feature is very useful for verification with non-technical users who may have trouble understanding the non-trivial technical terminology in the diagram.

4.3   Schema Transformation

Before discussing the schema transformation aspect, we briefly describe the language intended for expressing the semantics. For the purpose of this research and the nature of the schemas in question, a language which is able to clearly express relational elements is most suitable.

Relational Algebra. Relational algebra is a procedural query language that manipulates instances of relations in a database schema by applying operators in order to also produce as output relational instances. Relational algebra is the backbone of the database query language SQL. Below are the main operators used by the algebra, some of which are inherited from set theory:

$\cap$	intersection	π	projection
$\cup$	union	σ	selection
$-$	minus	$\bowtie$	join
$\times$	cartesian product	$\leftarrow$	rename

We use the formula below to illustrate an example for the syntactic structure of a relational algebra query expression:

$$stellar\_students \leftarrow \pi_{nr,\ age}$$
$$\sigma_{Degree.code=Student.degree\_code\_FK} \left(Student \times Degree \right) \qquad (1)$$
$$and\ Degree.final\_score>98$$

Apart from query expressions, the procedural system of the algebra can also be employed to describe constraints over relations. Key constraints, inclusion constraints, and functional dependencies are examples of important constraints which can be captured by relational algebra. Constraints can be seen as boolean expressions that are valid i.e. always true. To describe constraints, we depend on the full power of relational algebra, plus assertions between an expression and the empty set $\emptyset$. Take for example the model in Fig. 2. We interpret *seeks* and *...was awarded...on...* simply as binary and ternary relations respectively. To show the exclusion constraint between the two roles, we use the following expressions:

$$Temp \leftarrow \pi_{nr\_FK,\ degree\_code\_FK}(wasAwardedOn)$$
$$Temp \cap Seeks = \emptyset$$

Expressions in relational algebra can also be translated into *relational calculus* which is a variant of first-order logic.

Transformation. The transformation aspect of DBRE represents the core of the entire process, and therefore requires the profoundest consideration. Here, we summarise the plan to formalise this aspect.

A semantically enriched source schema s is a pair (Σ_s, λ_s), where Σ_s is the signature of the schema consisting of a set of relations, and λ_s is a set of constraints in s. The target schema t is a pair (Σ_t, λ_t) where Σ_t is the signature of the conceptual schema consisting of objects connected by roles, and λ_t is a set of constraints in t. Conventionally, a DBRE schema transformation yields a set of mappings μ for each relation r_s such that $r_s \in s$ and is transformed into a corresponding object o_t such that $o \in t$, without violating λ_s.

After conceptualisation of each relational term into a corresponding conceptual term, a mechanism is required to check the correctness of the transformation. Our research work particularly contributes to this by expanding the transformation process. We propose including inverse mappings as a means to check correctness, therefore defining mappings not only in the direction $s \rightarrow t$ but also from $t \rightarrow s$, for every element in both schemas.

More formally, we describe the semantics of the mapping below:

The mapping μ is a function symbol such that $\mu : r_s \rightarrow v_t$, where v_t is an associated view in t i.e. a relational algebraic expression for the corresponding o_t which associates with r_s. Then we say that μ^{-1} is the inverse of μ such that $\mu^{-1} : o_t \rightarrow v_s$, where v_s is an associated view in s i.e. a relational algebraic expression for the corresponding r_s which associates with o_t. Therefore we may also state that μ is a bijection. In μ, each r_s is mapped to a v_t expression in a LAV-like manner. In μ^{-1}, the reverse holds, where each object o_t is mapped to a v_s expression, as would GAV mappings.

Though s and t describe the same domain, they are projected differently. Defining views over them will require decomposing and joining of these projections. For this reason we need to ensure that both μ and μ^{-1} can be carried out in such a way that the semantics of the data and their constraints are not lost between them. This leads to investigating the schema transformation property known as *losslessness*.

We say that μ and μ^{-1} together are lossless if and only if $\lambda_s \models \lambda_t[o_t^i/v_s(o_t^i)]$ and $\lambda_t \models \lambda_s[r_s^i/v_s(r_s^i)]$. More verbosely, this means that λ_s entails λ_t following the substitution of each o_t according to v_s, the defined view for o_t over the language of s, and λ_t entails λ_s following the substitution of each r_s according to v_t, the defined view for r_s over the language of t.

If this losslessness property is always true, then we can assert that $\mu^{-1}(\mu(r_s)) = r_s$, for each r_s. This means that $\mu^{-1} \circ \mu$ which maps from $r_s \rightarrow o_t \rightarrow r_s$ constitutes the identity function id_{r_s} for all r_s, and $\mu \circ \mu^{-1}$ which maps from $o_t \rightarrow r_s \rightarrow o_t$ constitutes the identity function id_{o_t} for all o_t.

Intuitively, with the evidence of a lossless transformation we can then affirm that the DBRE process is correct and complete, then a possibility opens up for building queries over the conceptual schema itself.

5   Summary

The main contribution of this research will be the formalisation of the entire DBRE process, with focus on ensuring correctness on the conventional methods.

The submission of this paper marks the conclusion of the first 6 months of the ongoing PhD research work. The past months have been dedicated to literature study of which we have been able to produce substantially. Having covered the significant grounds for this research work, the immediate next step is to demonstrate DBRE in action, covering various natural scenarios, with the affiliated proofs to support its correctness.

References

1. Alhajj, R.: Extracting the extended entity-relationship model from a legacy relational database. Inform. Syst. **28**(6), 597–618 (2003)
2. An, Y., Borgida, A., Miller, R.J., Mylopoulos, J.: A semantic approach to discovering schema mapping expressions. In: 2007 IEEE 23rd International Conference on Data Engineering, pp. 206–215, April 2007
3. Andersson, M.: Extracting an entity relationship schema from a relational database through reverse engineering. In: Loucopoulos, P. (ed.) ER 1994. LNCS, vol. 881, pp. 403–419. Springer, Heidelberg (1994). doi:10.1007/3-540-58786-1_93
4. Astrova, I.: Reverse engineering of relational databases to ontologies. In: Bussler, C.J., Davies, J., Fensel, D., Studer, R. (eds.) ESWS 2004. LNCS, vol. 3053, pp. 327–341. Springer, Heidelberg (2004). doi:10.1007/978-3-540-25956-5_23
5. Chiang, R.H.L., Barron, T.M.: Quality issues in database reverse engineering: an overview. In: Proceedings for Operating Research and the Management Sciences, pp. 185–189, June 1995
6. Chiang, R.H.L., Barron, T.M., Storey, V.C.: Reverse engineering of relational databases: extraction of an eer model from a relational database. Data Knowl. Eng. **12**(2), 107–142 (1994)
7. Hainaut, J.-L.: Introduction to database reverse engineering. LIBD Lecture Notes (2002)
8. Hainaut, J.-L.: Research in database engineering at the university of namur. ACM SIGMOD Rec. **32**(4), 124–128 (2003)
9. Hainaut, J.-L., Tonneau, C., Joris, M., Chandelon, M.: Schema transformation techniques for database reverse engineering. In: Proceedings of the 12th International Conference on ER Approach, pp. 353–372. Springer, Arlington-Dallas (1993)

10. Halpin, T.: Object-role modeling: principles and benefits. Int. J. Inform. Syst. Model. Des. **1**(1), 33–57 (2010)
11. Johannesson, P.: A method for transforming relational schemas into conceptual schemas. In: Proceedings of the 10th International Conference Data Engineering, 1994, pp. 190–201. IEEE (1994)
12. Lammari, N., Comyn-Wattiau, I., Akoka, J.: Extracting generalization hierarchies from relational databases: a reverse engineering approach. Data Knowl. Eng. **63**(2), 568–589 (2007)
13. Lin, L., Zhuoming, X., Ding, Y.: Owl ontology extraction from relational databases via database reverse engineering. JSW **8**(11), 2749–2760 (2013)
14. Lubyte, L., Tessaris, S.: Automatic extraction of ontologies wrapping relational data sources. In: Bhowmick, S.S., Küng, J., Wagner, R. (eds.) DEXA 2009. LNCS, vol. 5690, pp. 128–142. Springer, Heidelberg (2009). doi:10.1007/978-3-642-03573-9_10
15. Markowitz, V.M., Makowsky, J.A.: Identifying extended entity-relationship object structures in relational schemas. IEEE Trans. Softw. Eng. **16**(8), 777–790 (1990)
16. McBrien, P., Poulovassilis, A.: Data integration by bi-directional schema transformation rules. In: Proceedings of the 19th International Conference on Data Engineering, 2003, pp. 227–238. IEEE (2003)
17. A Müller, H., Jahnke, J.H., Smith, D.B., Storey, M.-A., Tilley, S.R., Wong, K.: Reverse engineering: a roadmap. In: Proceedings of the Conference on the Future of Software Engineering, pp. 47–60. ACM (2000)
18. Petit, J.-M., Toumani, F., Kouloumdjian, J.: Relational database reverse engineering: a method based on query analysis. Int. J. Coop. Inform. Syst. **4**(02n03), 287–316 (1995)
19. Premerlani, W.J., Blaha, M.R.: An approach for reverse engineering of relational databases. In: Proceedings Working Conference on Reverse Engineering, pp. 151–160, May 1993
20. Qian, X.: Correct schema transformations. In: Apers, P., Bouzeghoub, M., Gardarin, G. (eds.) EDBT 1996. LNCS, vol. 1057, pp. 114–128. Springer, Heidelberg (1996). doi:10.1007/BFb0014146
21. Signore, O., Loffredo, M., Gregori, M., Cima, M.: Reconstruction of ER schema from database applications: a cognitive approach. In: Loucopoulos, P. (ed.) ER 1994. LNCS, vol. 881, pp. 387–402. Springer, Heidelberg (1994). doi:10.1007/3-540-58786-1_92

Supporting Conceptual Modelling in ORM by Reasoning

Francesco Sportelli[✉]

Free University of Bozen-Bolzano, Bolzano, Italy
fsportelli@unibz.it

Abstract. Object-Role Modelling (ORM) is a framework for modelling and querying information at the conceptual level. It comes to support the design of large-scale industrial applications allowing the users to easily model the domain. The expressiveness of the ORM constraints may lead to implicit consequences that can go undetected by the designer in complex diagrams during the software development life cycle. To avoid these issues we perform the reasoning on ORM diagrams in order to detect relevant formal properties, such as inconsistencies or redundancies, that cause a software quality degradation leading to an increment of development times and costs.

In this paper we present an extension of ORM formalisation by Derivation Rules, which are additional ORM constructs that capture some relevant information of the domain that cannot be expressed in standard ORM.

Moreover, we provide a tool (UCM Framework) which enables reasoning on conceptual modelling software along with an implemented case of study (ORMiE).

Keywords: ORM · Conceptual modelling · Reasoning · Rules

1 Overview

Conceptual modelling is a critical step during the development of a database system. It is the detailed description of the universe of discourse in a language that is understandable by users of the business domain. Object-Role modelling (ORM) is a conceptual language for modelling, which includes a graphical and textual language for specifying models, a textual language for formulating queries, as well as procedures for constructing ORM models, and mapping to other kinds of models like UML and ER. ORM is fact-oriented, i.e., it models the information in a way that it can be verbalized using sentences that are easily understandable by domain experts and even for those who are not familiar with IT in general. Unlike ER and UML, fact-oriented models are attribute-free, treating all facts (sentences) as relationships (unary, binary, ternary etc.) and this makes it more stable and adaptable to changing business requirements. For example, instead of using the attributes Person.isSmoker and Person.hiredate, fact-oriented models use the fact types Person smokes and Person was hired on Date [1].

© Springer International Publishing AG 2017
M. Kirikova et al. (Eds.): ADBIS 2017, CCIS 767, pp. 422–431, 2017.
DOI: 10.1007/978-3-319-67162-8_41

The ORM constraints expressiveness may lead to implicit consequences that can go undetected by the designer in complex diagrams; this may also lead to various forms of inconsistencies or redundancies in the diagram itself that give rise to the degradation of the quality of the design and/or increased development times and costs. The approach used to solve this issue involves the automated reasoning in order to detect inconsistencies and redundancies. Moreover, ORM diagrams can be equipped with Derivation Rules which are additional ORM constructs which are able to express knowledge that is not expressible with standard ORM. The usage of those rules brings to a further complexity so the goal of this paper is to detect a decidable fragment in order to perform the reasoning task even on those ORM diagrams equipped with those rules.

We also introduce the state of the art starting from the first ORM formalisation until the most recent one. Then we introduce an overview of the ORM language, focusing on its main features like fact-oriented, the verbalisation and the graphical notation, providing also the running example that will be shown in the rest of the paper. The section concerning the research has three subsections:

1. UCM Framework, a tool designed to activate automated reasoning on conceptual modelling software;
2. ORMiE, a tool which performs reasoning on ORM diagrams using UCM Framework;
3. Derivation Rules, where we show the results achieved so far concerning the formalisation.

We conclude the paper presenting the list of what is still to be done in the frame of the PhD.

2 Motivation

The ORM formalisation has been treated in years of research and still today it is a topic of interest. Formalising such language allows to activate reasoning procedures, carried out by Description Logics reasoners. Since the reasoning is able to detect relevant formal properties, such as inconsistencies or redundancies, the purpose of the reasoning is to support the modeller during the modelling which is a delicate process during the development of a software or a database. Without this support such issues could lead to a degradation of the quality of the design and an increase of development times and costs. Especially in large diagrams those issues are hard to spot by naked eye, so the need of this approach is crucial to prevent mistakes during the software or database development.

Although several papers presented their own ORM formalisation, no one has taken into account the formalization of derivation rules so far. Derivation rules are new ORM constraints which are able to express knowledge that is beyond normal ORM capabilities, but this feature leads to an increase of expressiveness of the diagrams. For this reason, the challenge is to identify a decidable fragment in order to extend the reasoning even on those ORM diagrams equipped with those rules.

Another challenge has more a methodological flavour, which involves the development of a system that is able to extend any conceptual modelling applications with reasoning services. The direct impact of this research involves the modellers, the developers and those who need a support tool which easily checks the consistency of conceptual schema in order to save time during the development life cycle.

3   Related Work

The ORM formalisation started with Terry Halpin's PhD Thesis [2]. In the context of design conceptual and relational schemas, Halpin formalized the NIAM language that is the ancestor of ORM. In his thesis there is the first attempt to formalize a modelling language in order to perform the reasoning task, so the main objective is to provide formal basis for reasoning about conceptual schemas and for making decision choices. After the spreading of ORM and its implementation in NORMA [3,4], ORM became more popular so the logicians' community took into account the possibility to formalize this very expressive language.

In 2007, Jarrar formalizes ORM using $\mathcal{DLR}_{ifd}$ [5], an extension of Description Logics introduced in [6]. The paper shows that a formalisation in OWL $\mathcal{SHOIN}$ would be less efficient than $\mathcal{DLR}_{ifd}$ because some ORM constraints cannot be translated (predicate uniqueness, external uniqueness, set-comparison constraints between single roles and between not contiguous roles, objectification n-ary relationships). In [7], Jarrar encodes ORM into OWL $\mathcal{SHOIN}$. Another formalisation of ORM in $\mathcal{DLR}_{ifd}$ was done by Keet in [8].

In 2009 OWL2 was recommended by W3C Consortium as a standard of ontology representation on the Web bringing some benefits: it is the recommended ontology web language; it is used to publish and share ontologies on the Web semantically; it is used to construct a structure to share information standards for both human and machine consumption; automatic reasoning can be done against ontologies represented in OWL2 to check consistency and coherency of these ontologies.

An ORM formalisation based on OWL2 is proposed by Franconi in [9], where he introduces a new linear syntax and FOL semantics for a generalization of ORM2, called ORM2plus, allowing the specification of join paths over an arbitrary number of relations. The paper also identifies a "core" fragment of ORM2, called ORM2zero, that can be translated in a sound and complete way into the ExpTime-complete Description Logic $\mathcal{ALCQI}$. In [10] is provided a provably correct encoding of a fragment of ORM2zero into a decidable fragment of OWL2 and it is discussed how to extend ORM2zero in a maximal way by retaining at the same time the nice computational properties of ORM2zero.

The most recent paper related to ORM formalisation is [11] where Artale introduces a new extension of $\mathcal{DLR}$, namely $\mathcal{DLR}^+$. This paper is strictly connected with this work because the logic $\mathcal{DLR}^+$ it is meant to represent n-ary relationships which are suitable for languages like ORM. The ORM implementation we use is an ongoing work based on $\mathcal{DLR}^+$. In particular, the decidable fragment we use is $\mathcal{DLR}^\pm$, obtained by imposing a simple syntactic condition on the appearance of projections and functional dependencies in $\mathcal{DLR}^+$. In the paper is also provided an OWL encoding and it is proved that $\mathcal{DLR}^\pm$ captures a significant fragment of ORM2.

Since this work is also focused on the formalisation of derivation rules, we need to mention OCL. OCL stands for Object Constraint Language, it is the declarative language for describing rules that apply to UML diagrams for defining constraints in order to support the conceptual modelling, like Derivation Rules for ORM. In [12] has been provided a formalisation of a fragment of this language and has been also proved the equivalence between relational algebra and the fragment with only FOL features, namely OCL_{FO}.

We conclude this section stating that, to best of our knowledge, we use $\mathcal{DLR}^\pm$ in order to build a decidable ORM mapping into OWL. In particular, we focus on Derivation Rules formalisation.

4   ORM

ORM stands for Object-Role Modelling. It is a language that allows users to model and query information at the conceptual level where the world is described in terms of *objects* (things) playing *roles* (parts in relationships) [13]. The idea behind ORM and its approach is that an object-role model avoids the need to write long documents in ambiguous natural language prose. It's easy for non-technical sponsors to validate an object-role model because ORM tools can generate easy-to-understand sentences. After an object-role model has been validated by non-technical domain experts, the model can be used to generate a class model or a fully normalised database schema. ORM main features are:

- *fact-oriented*, all facts and rules are modelled in terms of controlled natural language (FORML) sentences easy to understand even for non-technical users;
- *attribute-free*, unlike ER and UML, makes it more stable and adaptable to changing business requirements;
- *graphical*, it has a graphical notation implemented by the software NORMA;
- *formalised*, it has a clear syntax and semantics, so reasoning on an ORM diagram is enabled.

Unlike ER or UML, ORM makes no use of attributes in its base models; although this often leads to larger diagrams, an attribute-free approach has advantages for conceptual analysis, including simplicity, stability, and ease of validation. Attribute-free models with a controlled natural language facilitate model validation by verbalisation and population. Model validation should be a collaborative process between the modeller and the business domain expert who best understands the business domain. All facts, fact types, constraints and derivation rules may be verbalised naturally in unambiguous language that is

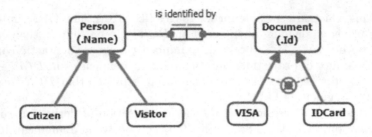

Fig. 1. ORM diagram example

easily understood by domain experts who might not be experts in the software systems ultimately used for the physical implementation.

The meaning of the diagram in Fig. 1 is the following: a person can be a citizen or a visitor; each person is identified by one and only one document which can only be either a visa or an id card. The entities are depicted by smooth rectangles and the relationships by a sequence of tiny boxes according to the cardinality relationship. The purple dot represents the mandatory constraint, the dash on the tiny rectangle box is the uniqueness constraint, the equivalent of the relational keys. The arrows among entities represents the ISA relationship. Finally, the circle with the cross inside means disjointness; the one with another circle inside means covering; the combination of this two is the circle we see between Visa and IDCard. The notation (.Name) and (.Id) inside Person and Document it is a graphical shortcut provided by NORMA for top level entities. Intuitively, it means that each person has a name and each document has an id. The corresponding FORML verbalization is the following:

```
Person is an entity type.
Citizen is an entity type.
Visitor is an entity type.
Document is an entity type.
VISA is an entity type.
IDCard is an entity type.
Person is identified by Document.
Each Citizen is an instance of Person.
Each Visitor is an instance of Person.
Each VISA is an instance of Document.
Each IDCard is an instance of Document.
Each VISA is an instance of Document.
Each IDCard is an instance of Document.
Each Person has exactly one Document.
For each Document, at most one Person has that Document.
For each Document, exactly one of the following holds: that Document is
some VISA; that Document is some IDCard.
```

This feature turns out to be helpful during the modelling phase especially when the non-IT stakeholders interact with the software engineers in order to reach a mutual comprehension about the meaning of the diagram. For example, if the non-IT stakeholder detects unexpected sentences which do not reflect the software specifications, it is easy for the modeller to modify the interested part.

5   Contributions

The main goal of the research concerns to enrich the conceptual modelling by reasoning in order to detect constraints which can lead to unexpected software behaviours, or to infer new knowledge. This research is characterized by the synergy of the methodological and the theoretical aspects. UCM Framework (Universal Conceptual Modelling Framework) and ORMiE (ORM Inference Engine) are tools developed to implement the ORM language in order to perform the reasoning over ORM schemas. Moreover, the theoretical part is focused on the formalisation of ORM Derivation Rules which has been also implemented in the aforementioned tools.

5.1   UCM Framework

Usually conceptual modelling tools do not take into account the problem of checking whether the semantics of the conceptual schema is consistent or not. To tackle this situation we developed UCM Framework which activates reasoning on conceptual modelling applications. UCM Framework has several features: it provides API for developers, reasoning services, the import/export of ontologies and diagrams in different languages like ORM, UML and ER. In Fig. 2 is shown the architecture of the framework. Each conceptual modelling application communicates with the core system using a specific driver, both for input and for the output where the inferences are encoded. The input schema is first encoded in a data structure (UCM Model) by API services, then the reasoning is performed by Fact++ reasoner [14]. After that, the inferences are stored into another data structure (UCM Inferred Model) by API and inferences are delivered to the destination application by drivers. Using this approach one can easily integrate this framework in order to enrich its conceptual modelling application by reasoning. Currently, two applications use this framework: Menthor [15] and ORMiE.

5.2   ORMiE

ORMiE (ORM Inference Engine) is an extension of NORMA, which is the official ORM-based Microsoft Visual Studio conceptual modelling tool [3]. ORMiE uses UCM Framework, so it is just one example how UCM Framework works on a target ORM-based software. ORMiE activates automated reasoning over ORM diagrams providing an interface where mistakes, redundancies or more in general new inferred knowledge are shown. It takes advantage of all nice features from Visual Studio Framework being such a powerful tool for those who need to model a domain following the ORM methodology. Moreover, ORMiE is able to perform the reasoning on those ORM diagrams equipped with Derivation Rules.

5.3   Derivation Rules

Derivation Rules are special ORM constructs which express knowledge that would otherwise not be expressible by standard ORM, so their expressibility is far reaching than

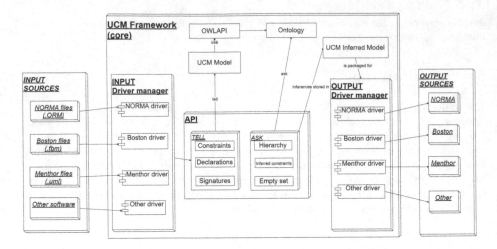

Fig. 2. UCM framework architecture

standard ORM. Their purpose is to derive new information from other information, like triggers, stored procedures and views in SQL. The goal is to enable the reasoning on those rules in order to extend the reasoning even on those ORM diagrams equipped with Derivation Rules. There are two kind of Derivation Rules: the Subtype Derivation Rules and the Fact Type Derivation Rules. A Subtype Derivation Rule defines all the instances which belongs to a subentity by a set of constraints defined in the rule definition. The reason because those rules are applied on subentities is because in some diagrams the is-a relationship between entities is too weak to capture the entire desired semantics of the diagram; a FactType Derivation Rule is placed on the predicates, namely the ORM roles.

The Derivation Rules we are focusing on are Subtype Derivation Rules. We want to formalise those rules in order to activate the reasoning on those diagrams with Subtype Derivation Rules. To achieve this goal it is important to take into account that the reasoning lays on logic, so we need to find a way to encode derivation rules into a logical language. First of all we need to understand how a derivation rule is made from a structural point of view, in other words we need to detect a clear syntax. Then, at the syntax is assigned a corresponding semantics and in the end an encoding into a logical language is performed in order to made the reasoning possible. In [16] is provided the full methodology used to formalise those rules and their mapping into OWL.

We provide an example with the graphical notation implemented by NORMA. For example, the diagram in Fig. 1 does not tell us which are *exactly* the people in the entity Citizen and *exactly* the people in the entity Visitor. We only know that a person can be a citizen or a visitor, but in the ORM standard notation there are no constraints able to capture these sets. If we want to use this knowledge we have to use the ORM Derivation Rules like in Fig. 3.

As we can see, a derivation rule is defined by an asterisk on entities and a text which defines the meaning of the rule. We state that all the people which are identified by an id card are citizens; all the people which are identified by a visa are visitors. It

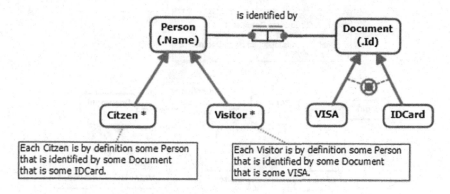

Fig. 3. ORM diagram example with derivation rules

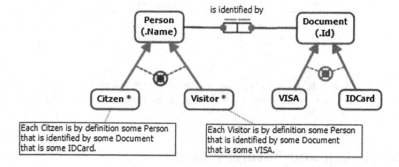

Fig. 4. Inferred disjoint and covering constraints

is important to observe that the text is not just a collections of words, instead it is in controlled-natural language format that is to say it is well defined by a precise syntax.

What can be the outcome of the diagram in Fig. 3? The answers is in Fig. 4. We obtained a disjunction and covering between the entities Citizen and Visitor. The disjunction is inferred because there is no chance to find a common element between the entity Visa and IDCard. Since the derivation rules capture separately the two sets, visitors and citizens, even the corresponding entities have no element in common. What about the covering? Since Visa and IDCard cover Document and since Person has the mandatory constraint on the relationship *is identified by*, each person must participate to this relation; in addition to this, the two derivation rules ensure that the sum of the instances in Citizen and Visitor are exactly those who are in Person. So it is not possible to find an instance which is not in Citizen of in Visitor. To prove this, now we add the entity Illegal: people without documents, neither a visa nor id card. The outcome of the reasoning is shown in Fig. 5. Illegal is red because it is an empty set. This means that the only consistent world is given by the entity Illegal with no instances, because there are no instances in Illegal which satisfy the rules of the diagram. Again, this is because Person has the mandatory constraint on the *is identified by* relationship and because the set of Person is already taken by the entities Citizen and Visitor. Therefore, there

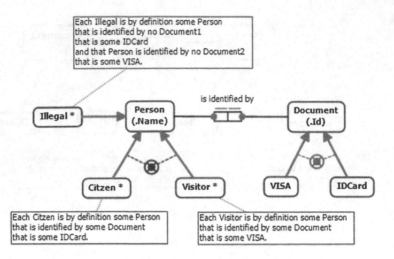

Each Illegal is by definition some Person
that is identified by no Document1
that is some IDCard
and that Person is identified by no Document2
that is some VISA.

is identified by

Illegal *

Person
(.Name)

Document
(.Id)

Citzen *

Visitor *

VISA

IDCard

Each Citzen is by definition some Person
that is identified by some Document
that is some IDCard.

Each Visitor is by definition some Person
that is identified by some Document
that is some VISA.

Fig. 5. Illegal is inconsistent

is no way an instance in Illegal could be in Person. The counter-example is trivial: if we remove the mandatory constraint on Person then Illegal would not be inconsistent anymore.

6   Conclusion and Future Works

We have seen in this paper an extension of the current ORM formalisation introducing ORM Derivation Rules. These rules express knowledge that is beyond ORM capabilities, but they are far expressive than standard ORM. Therefore, we have formalised a non-trivial decidable fragment in order to enable the reasoning over those ORM diagrams equipped with these rules. The reasoning procedure detects relevant formal properties as inconsistencies, redundancies or implicit constructs, helping the modeller to prevent unexpected software behaviours.

We presented UCM Framework, a tool which is specifically designed to enrich with reasoning any conceptual modelling software. It supports popular conceptual modelling languages like ER, UML and even ORM. A tool which makes use of this framework is ORMiE, a Microsoft Visual Studio plugin used to manage ORM diagrams.

The research and development of UCM Framework continues on two tracks: from one side we plan to add the explanation feature in order to enhance the understanding of the diagram by the user perspective, since this service explains why and how something went wrong during the modelling; while on the other hand we plan to extend the reasoning even on the instances of the conceptual schema.

Finally, an ongoing theoretical work concerns the formalisation of Fact Type Derivation Rules in order to capture a relevant decidable fragment that will be implemented in the UCM Framework.

References

1. Halpin, T.A.: Object-role modeling: Principles and benefits. IJISMD **1**(1), 33–57 (2010)
2. Halpin, T.: A Logical Analysis of Information Systems: static aspects of the data-oriented perspective. PhD thesis (July 1989)
3. Curland, M., Halpin, T.: The NORMA software tool for ORM 2. In: Soffer, P., Proper, E. (eds.) CAiSE Forum 2010. LNBIP, vol. 72, pp. 190–204. Springer, Heidelberg (2011). doi:10.1007/978-3-642-17722-4_14
4. Sportelli, F.: NORMA: A software for intelligent conceptual modeling. In: Proceedings of the Joint Ontology Workshops 2016 Episode 2: The French Summer of Ontology co-located with the 9th International Conference on Formal Ontology in Information Systems (FOIS 2016), Annecy, France, 6–9 July 2016 (2016)
5. Jarrar, M.: Towards automated reasoning on ORM schemes. In: 26th International Conference on Conceptual Modeling, ER 2007, pp. 181–197 (2007)
6. Calvanese, D., De Giacomo, G., Lenzerini, M.: Identification constraints and functional dependencies in description logics. In: Proceedings of the Seventeenth International Joint Conference on Artificial Intelligence, IJCAI 2001, Seattle, Washington, USA, 4–10 August 2001, pp. 155–160 (2001)
7. Jarrar, M.: Mapping ORM into the SHOIN/OWL description logic. In: On the Move to Meaningful Internet Systems 2007: OTM 2007 Workshops, OTM Confederated International Workshops and Posters, AWeSOMe, CAMS, OTM Academy Doctoral Consortium, MONET, OnToContent, ORM, PerSys, PPN, RDDS, SSWS, and SWWS 2007, Proceedings, Vilamoura, Portugal, 25–30 November 2007, Part I, pp. 729–741 (2007)
8. Keet, C.M.: Mapping the object-role modeling language ORM2 into description logic language dlrifd. CoRR, abs/cs/0702089 (2007)
9. Franconi, E., Mosca, A., Solomakhin, D.: The formalization of ORM2 and its encoding in OWL2. In: International Workshop on Fact-Oriented Modeling (ORM 2012) (2012)
10. Franconi, E., Mosca, A.: Towards a Core ORM2 language (Research Note). In: Demey, Y.T., Panetto, H. (eds.) OTM 2013. LNCS, vol. 8186, pp. 448–456. Springer, Heidelberg (2013). doi:10.1007/978-3-642-41033-8_58
11. Artale, A., Franconi, E.: Extending DLR with labelled tuples, projections, functional dependencies and objectification. In: Proceedings of the 29th International Workshop on Description Logics (2016)
12. Franconi, E., Mosca, A., Oriol, X., Rull, G., Teniente, E.: Logic foundations of the OCL modelling language. In: Fermé, E., Leite, J. (eds.) JELIA 2014. LNCS, vol. 8761, pp. 657–664. Springer, Cham (2014). doi:10.1007/978-3-319-11558-0_49
13. Halpin, T.A., Morgan, T.: Information Modeling and Relational Databases, 2nd edn. Morgan Kaufmann, San Francisco (2008)
14. Fact++ reasoner. http://owl.man.ac.uk/factplusplus/
15. Moreira, J.L.R., Sales, T.P., Guerson, J., Braga, B.F.B., Brasileiro, F., Sobral, V., Menthor editor: An ontology-driven conceptual modeling platform. In: Proceedings of the Joint Ontology Workshops 2016 Episode 2: The French Summer of Ontology co-located with the 9th International Conference on Formal Ontology in Information Systems (FOIS 2016), Annecy, France, 6–9 July 2016 (2016)
16. Sportelli, F., Franconi, E.: Formalisation of ORM derivation rules and their mapping into OWL," in On the Move to Meaningful Internet Systems: OTM 2016 Conferences - Confederated International Conferences: CoopIS, C&TC, and ODBASE 2016, Proceedings, Rhodes, Greece, 24–28 October 2016, pp. 827–843 (2016)

Author Index

Printed in the United States
By Bookmasters